# *Current Issues in*

# SELECTED READINGS

1995 EDITION



Edited by

*Michael L. McKinney*
University of Tennessee - Knoxville

*Robert L. Tolliver*
University of Tennessee - Knoxville

West Publishing Company
Minneapolis/St. Paul ❖ New York ❖ Los Angeles ❖ San Francisco

**WEST'S COMMITMENT TO THE ENVIRONMENT**

In 1906, West Publishing Company began recycling materials left over from the production of books. This began a tradition of efficient and responsible use of resources. Today, up to 95% of our legal books and 70% of our college texts and school texts are printed on recycled, acid-free stock. West also recycles nearly 22 million pounds of scrap paper annually—the equivalent of 181,717 trees. Since the 1960s, West has devised ways to capture and recycle waste inks, solvents, oils, and vapors created in the printing process. We also recycle plastics of all kinds, wood, glass, corrugated cardboard, and batteries, and have eliminated the use of Styrofoam book packaging. We at West are proud of the longevity and the scope of our commitment to the environment.

Production, Prepress, Printing and Binding by West Publishing Company.

Photograph courtesy of Bill Ingalls/NASA.

Printed in the United States of America
02 01 00 99 98 97 96 95 8 7 6 5 4 3 2 1 0

ISBN 0–314–06109–6

# CONTENTS

**Part Four: Resources and Pollution**

**Part Six: The History of Life—Origin, Evolution, and Ecology**

## Preface

The Earth's geologic processes are nearly timeless, but the impact geology has on humans is very much a current issue. To represent the wide variety of topics upon which the study of geology touches human activity and inquiry, the editors of this volume have collected seventy-one articles from a number of general interest magazines, science magazines, and *The Citizens' Guide to Geologic Hazards* published by The American Institute of Professional Geologists. This 1995 edition contains twenty-nine new articles covering new issues and discoveries from the previous year.

The editors have carefully chosen articles to supplement material a student might encounter when taking a course in physical geology, historical geology, environmental geology, dinosaurs, and earth science. Often, there is an overlap of subject areas taught among these courses. For example, a considerable portion of an environmental geology course usually includes material on physical geology. To help the reader identify these overlaps, the editors have divided the book into seven parts. Parts one through three cover physical geology topics, parts four and five cover environmentally-related articles, and the final two parts cover historical geology.

In this book, each article opens with a brief overview and discussion of the issues or concerns generated by the topic. Following the articles, the editors ask a few questions to help the reader focus on the issues and apply what they have learned from their geology classes.

## Acknowledgments

The editors and West Publishing wish to express their sincere thanks to the many magazines and the American Institute of Professional Geologists for allowing us to reprint their articles. We are indebted to Cendi and John Davis who typed and produced the camera-ready pages for this book. They deserve special thanks for their taking in stride last-minute replacements of recently published articles. Their round-the-clock work helped make this book as up-to-date as possible.

# Part One

## The Origin of the Earth and Its Internal Processes

1

*For the last two centuries the planets in our solar system were thought to have formed directly from condensation of a gaseous disk surrounding the young sun. Recently this view has begun to change. Now the accumulation of solid particles is seen as an important mechanism in planet formation. The inner, rocky planets would have begun their development with the accumulation of microscopic grains. Farther from the sun, where temperatures were lower, ice could accumulate to form the cores of the outer giant planets. As small particles continually collided and stuck together, they eventually would have reached the size of our present planets. This period of planet formation may have taken as little as 100 million years. Much of this early history of planet formation can be studied by examining meteorites, which were formed between 4.56 and 4.57 billion years ago.*

# A Rocky Start

## Pinning down the time of the solar system's hurly-burly birth

by Ivars Peterson

When Pierre-Simon de Laplace set out nearly two centuries ago to imagine how the solar system might have formed, he turned to the motions of the planets and their satellites for clues.

Laplace noted that all the planets move in the same direction around the sun, following nearly circular orbits that lie in roughly the same plane. He believed the planets spun in the same direction as their forward motion around the sun. Indeed, as far as he could tell, the sun, the planets, and their satellites all rotated in the same sense.

Such regularities inspired Laplace to suggest that the sun's atmosphere once extended beyond the orbits of all the planets. Its rotation produced a flat, gaseous disk, and subsequent cooling and contraction made the disk spin faster. This whirling, ebbing disk spun off one gas ring after another, and each ring later condensed into a hot, fluid ball. Finally, while the sun's atmosphere retreated to its present limit, these balls cooled and solidified into planets.

"One can thus conjecture that the planets were formed on the successive limits of the atmosphere, through condensation of zones of vapor, and on cooling they had to be released by this atmosphere in the plane of its equator," Laplace wrote in 1796.

The inventory of the solar system's contents has grown considerably since Laplace's era. It now includes two additional planets (Neptune and Pluto), a host of rocky asteroids between Mars and Jupiter, complete or partial rings of particles around the larger planets, and a cloud of comet nuclei at the solar system's outskirts.

Today's blueprint for the solar system's creation must take into account not only the solar system's assorted denizens but also a number of surprising quirks in the characteristics of these bodies. For example, Venus, Uranus, and Pluto actually spin in a direction opposite to their orbital motion. Neptune even has a large moon that orbits in the opposite direction of the planet's own rotation and of the motion of its other satellites.

Moreover, the composition of the planets seems to vary systematically. Small, rocky planets orbit near the sun, while gas giants—with the exception of tiny, icy Pluto—occupy the outer orbits. At the same time, Earth and Pluto have moons large enough to qualify these systems as double planets. Mercury has a surprisingly large core, and Uranus has an unusual tilt, tipped almost on its side.

Developing a coherent, consistent model of the solar system's birth that accounts for all these observations has proved remarkably difficult. In the last few years, the consensus among astronomers and planetary scientists has swung away from the notion that the planets formed directly from condensing gas. Instead, they now emphasize the importance of the accumulation of solid particles into planetesimals and the subsequent rapid aggregation of these objects into planets.

"The formation stage was very brief and turbulent," asserts Douglas N.C. Lin of Lick Observatory at the University of California, Santa Cruz.

What emerges is a portrait of an unsettled era—perhaps no longer than 100 million years—that saw the runaway growth of modest chunks of solid material into hefty bodies. It was a time of innumerable collisions and near misses, of drastic changes in orbit, of planet-size masses recklessly careening around the sun with devastating consequences.

Lin described recent observational evidence and theoretical work favoring this particular view of the onset and duration of planetary formation at the American Association for the Advancement of Science annual meeting, held last month in Boston.

Like Laplace, present-day astronomers generally start their creation scenarios with a gaseous disk surrounding a newborn star. Observations of young stars suggest that such disks form quite commonly in regions of our galaxy where star formation is taking place (SN: 1/16/93, p.36).

These young stars typically emit more infrared radiation than one would expect from their temperature and composition. Astronomers attribute this discrepancy to the presence of orbiting dust and gas that is much cooler than the star itself. In similar stars just a few million years older, most of this circumstellar material has apparently disappeared (SN: 10/29/88, p.280).

"The evolutionary time scale of many of these disks is of the order of a few million years," Lin notes. Thus, observations of young stars surrounded by disks furnish revealing snapshots of the various stages in which the solar system may have formed.

"It is this type of information that suggests that we may indeed be identifying the nursery out of which planets form," Lin says.

Meteorites provide the best evidence of when the first specks of solid material appeared in the gaseous disk, or solar nebula, out of which the solar system emerged. In particular, dark, carbon-bearing meteorites known as carbonaceous chondrites contain clumps of crystalline grains that apparently solidified after the disk material had cooled to temperatures less than 1,500°C.

Over the last few years, researchers have determined the ages of a variety of these mineral grains by measuring the proportions of different radioactive isotopes. The measurements consistently point to a time of formation between 4.56 billion and 4.57 billion years ago.

"They all agree fairly well," says Timothy D. Swindle of the University of Arizona in Tucson. "We think we know when this process [of planet formation] . . . started. What we don't know are the exact time scales [of the various stages]."

Nonetheless, a number of new dating techniques, based on relatively short-lived isotopes, suggest that these grains all formed within a few million years.

Because different types of grains have different freezing points, the chemical composition of the crystals in carbonaceous chondrites also provides an indication of what the temperature may have been in various parts of the solar nebula. Moreover, the hodgepodge of crystals typically found in a carbonaceous chondrite hints at the environment in which they formed.

"This suggests that the original solar nebula may have been very turbulent, and the grains were gathered together from regions that had somewhat different temperatures," Lin says.

Collisions and a natural stickiness helped these tiny grains grow into larger objects, including those eventually captured by Earth as meteorites. Oxides of the heavier elements in the solar nebula agglomerated into rocky bodies, and water froze into chunks of ice. At the same time, the sun's gravity gathered these particles into a thin, rotating "pancake" resembling an immense version of Saturn's rings.

Precisely how this clumping of grains into boulders a kilometer or more across occurred isn't completely settled yet. But several groups of researchers have shown in computer simulations that a slight tendency of particles to stick together after collisions is probably enough to form the loose aggregates of material known as planetesimals.

As the planetesimals collided and grew in size and mass, gravitational interactions began to play an increasingly important role. Even when a close encounter didn't result in a collision, the force of attraction between the two bodies could drastically change their speeds and orbits, boosting the likelihood of collisions.

Lin and his collaborators have investigated how the dynamical properties of planetesimals may have regulated their growth in the primordial solar nebula.

"Our numerical simulations show that, provided there is a sufficient supply of low-mass planetesimals, runaway coagulation can lead to the formation of protoplanetary cores with masses comparable to a significant fraction of an Earth mass," Lin and his co-workers report in the Jan. 20 Astrophysical Journal.

Lin estimates that it would take only about a million years for an Earth-size object to form from planetesimals.

Rings around planets also provide a useful picture of the kinds of gravitational interactions possible in a thin layer crowded with particles of various sizes. "Planetary rings are our best, closest examples of celestial disk systems," says Carolyn C. Porco of the University of Arizona. "Many of the processes going on in planetary rings are very similar—though different in detail—to the processes that likely occurred in the formation of the solar system."

In the turbulent environment of the solar nebula, collisions and close encounters probably occurred not only between small rocks and large planetesimals, but also between large bodies of comparable size. Such interactions would result in chaotic modifications of orbits and extensive mixing of materials.

George W. Wetherill of the Carnegie Institution of Washington (D.C.) has modeled the behavior of about 500 planetesimals, each one roughly the size of the moon, in the region now occupied by the terrestrial planets. His computer simulations show that these objects merge into planets generally resembling those now found in the inner solar system. Moreover, the resulting planets contain a mixture of materials gathered from widely distributed regions of the inner solar system.

Large-body collisions probably played an important role in establishing the final configuration of the inner solar system. Earth's moon, for example, may have been a by-product of a wayward, Mars-size body crashing into Earth. Mercury may have lost its rocky outer layers in a similar collision. Another encounter may have shifted Mercury to its present orbit close to the sun.

Among planetary scientists, there seems little doubt now that the smaller, solid worlds of the inner solar system built up from microscopic grains. Mainly rock, these planets formed in regions of the solar nebula hot enough to boil away ice. In contrast, the more distant, giant planets may have formed around cores of solid ice.

Some researchers have suggested that icy material in the outer reaches of the solar system built up into protoplanets about 10 times more massive than Earth. These icy giants were then big enough to start gravitationally picking up and retaining hydrogen and helium gas from the solar nebula.

This accumulation of gas eventually stopped when the protoplanet grew so large that its gravitational pull opened up a gap in the nebula similar to the gaps found in Saturn's rings. "Once [it] opens up a gap, the [protoplanet] can no longer accrete gas from the solar nebula," Lin says. "The object is on a kind of self-regulating diet."

The problem with this scenario is that young stars of roughly the sun's mass apparently lose most of the hydrogen and helium in their disks—perhaps blown away by an intense stellar wind—within 10 million or so years. That seems to allow too little time for the accumulation of sufficiently large ice cores, particularly for Uranus and Neptune.

But researchers continue to look for ways around this problem, and the notion of planetary formation by the runaway accumulation of material remains the most viable model.

Refined and modified, Laplace's original notion that the planets condensed out of a gaseous disk long dominated the thinking of astronomers. Now it is largely out of fashion. The present "standard" model of solar system formation, which emphasizes the rapid aggregation of dust and granular material into planets, does a better job of accounting for many features of the present solar system.

Nonetheless, the collision-accumulation scenario leaves a variety of details unexplained, especially in the construction of the gas giants. Whether researchers can resolve all of these problems in the context of runaway planet formation in a turbulent nebula remains an open issue. ❑

**Questions**:

1. The period of planet formation may have been very turbulent with grains accumulating from regions of different temperature. What feature of meteorites suggests this?

2. How long would it take for an Earth-size object to form?

3. How did the Earth's moon form?

Answers on page 301

# 2

*When the Earth first formed, the sun was about 30 percent dimmer than at present. This would mean that any surface water would have been frozen. But the oldest known sedimentary rocks, rocks deposited by water, were formed 4 billion years ago. The explanation for this paradox has been that the quantities of carbon dioxide in the atmosphere were much higher than at present. Recent work by Gregory Jenkins has provided an alternate explanation. Early in the Earth's history a day may have been only 14 hours long. The Earth also would have had a much smaller continental area. These two factors would have led to a climate warm enough to allow water to exist in a liquid form without extremely high carbon dioxide levels.*

## The Fast Young Earth

Four billion years ago, when the sun was young and dim, Earth may have stayed warm in part because it was spinning faster.

by Tim Folger

The sun, astronomers believe, wasn't always as bright as it is today. Four billion years ago it was probably about 30 percent dimmer. Although the sun had the same mass then as it does now, it was a little larger and less dense. Over the eons, as it contracted under its own gravity, it grew hotter and brighter. Astronomers have no problem with this picture—it seems to be the typical way stars evolve—but geologists do. If the young sun was so feeble, Earth should have been a frozen wasteland. And yet the oldest sedimentary rocks—which were deposited in water—are themselves 4 billion years old. Somehow Earth's surface was warm enough for liquid water despite a wan sun.

Geologists have weaseled out of this "faint young sun paradox" by assuming that hyperactive volcanoes erupted vast clouds of heat-trapping carbon dioxide into the atmosphere, warming the young Earth. Some models call for carbon dioxide levels 1,000 times higher than today's. But to Gregory Jenkins, an atmospheric scientist at the National Center for Atmospheric Research in Boulder, Colorado, that solution has always seemed ad hoc. "You have to put so much carbon dioxide into climate models in order to get an effect that it doesn't seem realistic," he says. "There is no direct evidence to show that carbon dioxide levels were ever a thousand times higher."

Jenkins thinks he has a better way out of the paradox. He and two colleagues at the University of Michigan, Hal Marshall and William Kuhn, propose that Earth's rapid rotation rate and the absence of large landmasses 4 billion years ago were the keys to the planet's warmth.

Four billion years ago Earth may have spun on its axis at least once every 14 hours. (Since then the moon's tidal drag has gradually slowed the planet's rotation and thereby lengthened its days.) Jenkins and his colleagues have constructed a sophisticated computer model to study the effects the revved-up rotation would have on Earth's climate. The model allowed them to follow the movements of air masses in three dimensions—something that had never been done before for early Earth. "None of the previous studies had looked at how the entire system would be changed under different rotation rates," says Jenkins. "There was a big gap."

In their model, Jenkins and his colleagues used a 14-hour day and also assumed that Earth had no land. That is not a bad approximation, says Jenkins: most landmasses formed less than 3.5 billion years ago. Before then, Earth's surface was probably one vast ocean dotted with volcanic islands.

This landlessness itself would have made the planet warmer; water reflects less sunlight than land and absorbs more, which means it has more energy to radiate to the air as heat. What surprised Jenkins and his colleagues, though, was the dramatic effect of the rapid rotation rate. On their fast-spinning model Earth, most storms and clouds were confined to the equatorial and subtropical regions, and global cloud cover was 20 percent less than today's.

There were fewer clouds in temperate latitudes, Jenkins explains, because the rapid rotation rate was making it difficult for warm, moisture-laden air to spread from the tropics—where evaporation is greatest—toward the poles. Through the so-called Coriolis effect, Earth's rotation deflects moving air to the right in the Northern Hemisphere and to the left in the Southern. The faster the rotation, the greater the deflection.

With Earth rotating once every 14 hours, Jenkins's model suggests, moist, cloud-forming air would get deflected back toward the equator even before it left the tropics.

And with clear skies over more of the planet, more sunlight would have reached the surface. Jenkins and his co-workers did include some carbon dioxide in their model, but the amount was only eight times higher than current levels, a tremendous reduction from the standard scenario. Even without a massive greenhouse effect, their model suggests, and even in the face of a faint young sun, the fast young Earth would have been warm enough to keep its ocean from freezing. In fact, it would have been about 10 degrees warmer than it is today. ❑

**Questions**:

1. Explain the "faint young sun paradox."

2. How would the lack of continents warm the Earth? How would the faster rotation of the Earth warm it?

3. Why has the Earth's rotation slowed over the last 4 billion years?

Answers on page 301

*The processes of plate tectonics are driven by heat within the Earth. Knowing the amount of heat stored in the Earth's core is important to fully understand Earth processes. A recent study suggests that the Earth's core may be much cooler than previously thought, with a temperature at the inner core-outer core boundary of 4,800 kelvins. Studies of the Earth's core involve the use of iron, which is presumed to make up the bulk of the core. This iron is heated and pressurized as high as possible. The behavior of iron at different pressures indicates how much heat it will give off. This information can also help us understand what may occur at the core-mantle boundary and within the lower mantle, factors that are important for understanding the processes driving plate tectonics.*

# Cooling the Vision of Earth's Hot Core

by B. Wuethrich

Heat trapped deep within Earth during its formation provides the energy that ultimately moves continents, powers volcanoes, and triggers earthquakes. So to understand better the planet's workings and its 4.6-billion-year geologic evolution, geophysicists want to know how much heat is stored inside Earth's iron-rich core.

New experiments now suggest a cooler core than previously thought.

To determine Earth's reserve of internal heat, scientists must know the melting temperature of iron at the boundary between the solid inner core and the molten outer core. Since that boundary lies 5,100 kilometers below Earth's surface at a pressure of 3.3 million atmospheres, it cannot be reached directly, nor can such high pressures be created in a laboratory.

Previously, the pressure limit at which scientists could hold iron was just 1 million atmospheres. Now, geophysicist Reinhard Boehler of the Max Planck Institute for Chemistry in Mainz, Germany, has pushed that to 2 million atmospheres. And he has extrapolated his data to 3.3 million atmospheres, calculating a temperature of 4,800 kelvins at the inner core-outer core boundary—a much lower figure than prior estimates of up to 8,000 kelvins.

Boehler reports his findings in the June 10 NATURE.

"We have reached the upper limit of pressure and temperature that can be achieved using this equipment," says Boehler, who used a diamond anvil that squeezes tiny iron samples between two diamond crystals.

While Boehler cranked up the pressure exerted by the diamonds, he heated the iron sample with a laser. Heating iron shifts its color from red to blue. Boehler determined the iron's melting temperature under different pressures by monitoring the sample's changing spectra.

He also estimated the heat loss at the boundary between the liquid outer core and the overlying solid mantle—an area of dramatic physical changes resulting in large part from the huge drop in temperature between the two regions, the geophysicist says. The temperature drops from 4,000 kelvins at the edge of the outer core to 2,700 kelvins at the bottom of the mantle, he calculates.

This transfer of heat can cause huge plumes, or currents, of solid rock to inch slowly upward, ultimately driving the motion of crustal plates. Yet as large as the temperature gap is between core and mantle, it is still much less than predicted by previous experiments.

This unexpectedly low temperature and Boehler's newer research on mantle rocks radically alter the picture of the lower mantle. In unpublished work on perovskite, the mineral that makes up most of the lower mantle, he finds evidence that perovskite's melting temperature is 7,000 to 8,000 kelvins—much higher than previously calculated.

Putting the two findings together, Boehler proposes that the lower mantle will prove to be a rigid area with limited flow and little chance of chemical reactions between the mantle and core. "All predictions of flow in the lower mantle have to be reworked," Boehler contends.

Geophysicist Raymond Jeanloz of the University of California, Berkeley, says that only four or five scientific groups worldwide are carrying out work similar to Boehler's and emphasizes the need for cross-checks. Jeanloz has carried out high-pressure diamond anvil experiments, with different results. He anticipates that further experiments will resolve those differences in numbers and illuminate the true dynamics of Earth's inaccessible core. ❑

**Questions**:

1. How is iron pressurized and heated?

2. What happens at the core-mantle boundary? What effect does this have?

3. Perovskite, a mineral common in the lower mantle, has a very high melting temperature. This, along with the large drop in temperature at the core-mantle boundary, suggests what?

Answers on page 307

4

*To most people, the more perfect a diamond is the more valued it is. But to geologists, the imperfections are often the most valued aspect of a diamond. Diamonds are formed more than 180 kilometers beneath the Earth's surface. Some of the tiny imperfections in diamonds are other minerals formed in the mantle. These other minerals may tell us something about the amount of mixing within the mantle. The mantle appears to be divided into an upper and lower part at about 660 kilometers deep, based on information from changes in the velocity of seismic waves. But whether or not the two layers mix is uncertain. The Brazilian diamonds contain minerals that should have formed under the temperatures and pressures found in the lower mantle. The mineral assemblages found in these diamonds suggest that mantle mixing does occur.*

# Bits of the Lower Mantle Found in Brazilian Diamonds

by Richard A. Kerr

Diamonds are esteemed for their rarity, hardness, and flash, but to some mineralogists their most valuable features are their imperfections. Formed at depths of more than 180 kilometers, then blasted to the surface by deep-seated volcanic eruptions, diamonds can capture traces of surrounding minerals and ferry them to within human grasp. And if mineralogist Ben Harte of the University of Edinburgh and diamond specialist Jeffrey Harris of the University of Glasgow are right, some Brazilian diamonds have just delivered a long-sought prize: microscopic bits of mineral from the lower mantle, the rocky region of the planet more than 660 kilometers down.

Harte and Harris aren't the first to claim that prize, but other researchers think they have the strongest case yet. "It's an outstanding discovery, and everything is compatible" with a lower mantle origin, says Stephen Haggerty of the University of Massachusetts, Amherst, who holds claim to a find from 300 to 400 kilometers down (*Science*, 10 May 1991, p. 783). The new discovery, announced at this year's meeting of the European Union of Geosciences in Strasbourg, is much more than a depth record. The lower mantle, the deepest rocky layer of the planet, is the focus of one of Earth science's longest running debates: Does the planet's internal heat churn the lower and upper parts of the 2900-kilometer-thick mantle into a uniform mass, or are the two layers segregated physically and chemically? The chemistry of mineral flecks like the ones found by Harte and Harris may hold clues to the answer.

The discovery is the fruit of university-industry cooperation. If academic researchers are interested in unusual mineral inclusions in diamond, it behooves them to be on good terms with the people who see a lot of diamonds—the diamond miners. As a consultant to the DeBeers Corp. on diamond mineralogy, Harris had the right connections. They paid off when DeBeers passed on to Harris some diamonds obtained by Sopemi (their associate organization in Brazil) from mines near Sao Luis that turned out to be chemically unusual.

Using a type of mass spectrometer called a microprobe, Harris and Harte were able to identify the 50- to 200-micrometer-sized bits of minerals trapped in the diamonds. Their composition suggested that the minerals and the diamonds containing them had originated hundreds of kilometers deeper than any accepted deep samples. One type of inclusion, a mineral type called ferropericlase-magnesiowüstite, had a magnesium-iron oxide composition that, according to theory and lab experiments, most likely formed in the lower mantle. Another inclusion, a magnesium silicate, can form in the upper mantle, where it is called enstatite. Together with the ferropericlase, however, it supports a lower mantle origin for the inclusions: At shallower depths the two minerals should com-bine rather than existing separately.

Still, the two weren't enough to make an iron-clad case for a lower mantle origin. In the 1980s ferropericlase inclusions had been found in seven other diamonds at three sites on three continents; one of these diamonds also had enstatite. Barbara Scott-Smith of the Anglo-American Research Laboratories and her colleagues, including Harris, suggested that at least a few of the diamonds came from the lower mantle, but, as Harris says, "not many people took up [the suggestion]." A

***Richard A. Kerr. "Bits of the Lower Mantle Found in Brazilian Diamonds." <u>Science</u>, September 10, 1993, v. 261, p. 1391.*** 

single lower mantle mineral was intriguing, it seemed, but hardly convincing.

What really strengthens Harte and Harris' case, says other researchers, is the presence in their diamonds of a second type of lower mantle mineral. In three diamonds from Sao Luis, enstatite and ferropericlase inclusions coexist with inclusions of calcium silicate, another mineral expected to form in the lower mantle. The combination of the three minerals in individual diamonds convinced Harte and Harris that their diamonds originated more than 660 kilometers down.

The very fact that minerals from such depths made it to the surface strengthens the case for mantle mixing, say some researchers. After all, the diamonds probably weren't blown straight to the surface by a volcanic eruption drawing on the lower mantle. Mineralogists debate the depth from which volcanism can raise diamond-bearing rock, but the conventional view puts the limit at about 200 kilometers. Rising plumes of hot rock must have swept the diamonds up to the depths that can be plumbed by volcanism, says Haggerty. And that in turn implies that the boundary between upper and lower mantle at 660 kilometers can't be impenetrable, he notes.

A few mineralogists, however, think that diamond-bearing eruptions can have deeper roots, all the way down to the 660-kilometer boundary between the upper and lower mantle. In that case, there would be no need to invoke mantle mixing to carry the minerals along the first leg of their journey. But the chemical makeup of the inclusions should help resolve the issue. If the mantle doesn't mix, the lower layer should be a reservoir of heavier rock. Lowermantle minerals should be denser than those from the upper mantle, most likely because of an extra measure of iron or silicon. In the earlier, less certain lower mantle inclusions, Susan Kesson and J. D. Fitz Gerald of the Australian National University in Canberra found iron and silicon abundances not much different from those seen in the samples of the upper mantle—evidence that the mantle is well mixed.

On the average, the iron content of the Sao Luis ferropericlases also fits the mixed mantle picture, but only on average. Iron in individual inclusions ranges to extremely high levels. Harte thinks melting and recrystallization of minerals could have skewed their iron concentrations and thus garbled any information on mixing. And Harte and Harris have found another mineral inclusion associated with ferropericlase—a garnet—whose existence in the lower mantle does not fit anyone's model of mantle chemistry, whether the mantle is mixed or layered.

Such nagging uncertainties leave mineralogists eager for more glimpses of the lower mantle. Sad to say, they won't be getting them from Sao Luis diamonds, at least not for a while. DeBeers has abandoned the site in the face of hordes of local diggers, an armed insurrection, and general lawlessness. For any bits of the lower mantle remaining at the site, the journey from the depths of the planet was the easy part. ❑

**Questions**:

1. What two lower mantle minerals are found in the Brazilian diamonds?
2. What is an alternative explanation for the presence of lower mantle minerals in diamonds?
3. How would the chemical makeup of the inclusions help resolve the question of origin?

Answers on page 301

# 5

*The last decade has seen significant strides in understanding the nature of Earth's mantle. Evidence from such diverse fields as experimental petrology, seismology, and computer modeling are helping to sort out a major debate in Earth sciences: does mantle convection create a completely mixed mantle or does mantle convection occur in distinct layers with little or no mixing. The most recent studies appear to support a stratified mantle with an identifiable boundary between 650 and 670 kilometers in depth. At this boundary, subducting slabs are slowed in their descent into the mantle. Some of these slabs get deflected and do not enter the lower mantle. Others, however, do eventually penetrate the lower mantle, allowing some mixing of material between the lower and upper portions of the mantle.*

## Scrambled Earth

### Researchers look deep to learn how the planet cools its heart

by Richard Monastersky

In the beginning, there was heaven and a sizzling chunk of rock called Earth. Asteroids bombarded the surface of the infant planet, while radioactive elements seethed below, building up so much heat that most of the globe eventually melted. From a distance, the world would have looked like a giant drop of liquid rock circling the sun.

Since that time, Earth has slowly cooled, releasing much of its pent-up original heat. Even today, the quenching continues beneath our feet. The escaping energy pushes continents around the planet's surface through a process called plate tectonics. It causes the ground to quake and volcanoes to blow, killing thousands of people every year.

Yet for all its planet-wrenching consequences, this global cooling remains a mystery. For more than 2 decades, researchers have tried, with little success, to decipher exactly how Earth lets off steam.

The effort is beginning to pay off, however. Recent discoveries from several fields of research are now converging, allowing investigators to zero in on key issues. Tools as disparate as computer models and diamond-studded vises are bringing the planet's hidden interior into much sharper focus than ever before.

When reduced to its simplest form, the cooling question hinges on the mechanics of mixing. Like a pot of steaming soup, Earth loses heat through the stirring action of convective currents that draw hot material toward the surface, where it cools and then sinks. Much of this convection takes place within the mantle—the great rocky layer that surrounds the core and makes up 83 percent of the planet. Although the mantle is solid, the pressures are so intense that the deep stone actually flows, albeit at speeds of only a few centimeters per year.

Geophysicists can agree on that general picture, but arguments break out when they discuss how deep the mixing goes. Since the 1970s, scientists have debated two competing theories of convection. One camp, call them the lumpers, believes that convection currents stir the entire mantle, mixing both the upper and lower parts of this layer. Another group, call them the splitters, argues that the upper and lower mantle remain separate, each convecting on its own like the stacked pots of a double boiler. In this case, the upper mantle would act as a thermal blanket, insulating the lower mantle and slowing the escape of heat from the core.

The lumpers and splitters may both have it wrong, however. The newest findings suggest that Earth does not follow either of these simple patterns but might combine elements of both.

The current debate over mantle mixing extends an intellectual revolution that started nearly 30 years ago, when the theory of plate tectonics swept the geosciences. This powerful concept revealed that Earth's outer shell—the lithosphere—is broken into separate blocks that migrate around the globe like bumper cars in extra slow motion. Where two plates crash together, they build giant mountain ranges such as the Himalayas. When one plate slips beneath another, it creates a deep ocean chasm like the Mariana Trench. If two plates slip-slide past each other, they form quake-making faults like the San Andreas.

The theory of plate tectonics succeeded because it provided the intellectual framework to explain the planet's surface, the part that geologists can feel directly under their boot heels. But the theory only goes skin deep; it addresses just the top 100 kilometers of a planet that spans 6,370 km from surface to center. Even under the penetrating light of plate tectonics, the inner Earth has remained a terra incognita.

Scientists are therefore striving for deeper knowledge. "The new revolution is to project the global view of plate tectonics downward into the third dimension," says Raymond Jeanloz of the University of California, Berkeley.

Because researchers cannot reach into the lower mantel to measure its properties, experimental geophysicists such as Jeanloz turn the problem around, bringing the inner Earth into the laboratory. Aiding them in this quest is the diamond anvil cell, a brick-size instrument that allows researchers to recreate the hellish pressures and temperatures of the planet's center.

To address the convection debate, Jeanloz's group put the squeeze on a rock called periditite, which comes from the upper mantle. Although these rocks normally lie under Earth's surface crust, geologists find periditite exposed in places where interplate collisions have thrust part of the upper mantle above the ground.

Jeanloz would like to work with actual material from the lower mantle, but no one knows what exists there. The rocks could contain the same ingredients as those of the upper mantle, or they could have a different composition.

As a test, Jeanloz used a diamond anvil to transport upper mantle periditite into a laboratory version of the lower mantle. The rationale vaguely resembles the Cinderella story. In that tale, if the prince can fit a foot into a glass slipper, he has found his princess. In Jeanloz's case, if the squeezed periditite fits the expected qualities of lower mantle rock, then he has found the long-sought stone that might fill this hidden layer of Earth.

To measure the fit, Jeanloz uses X-ray diffraction to gauge the density of the compressed periditite. He knows the actual density of the lower mantle because seismologists have previously measured that value.

Periditite, it turns out, is no Cinderella. Even intense squeezing and heating couldn't shoehorn periditite into a proper fit. It has a density 3 to 5 percent lower than what scientists have measured for the deep mantle, Jeanloz says.

The Berkeley group first found hints of that discrepancy several years ago while conducting separate diamond anvil experiments with each of the minerals in periditite. But they wondered whether the whole rock might behave differently than the sum of its parts. The team recently confirmed their original findings with actual periditite, Jeanloz reported last December at a meeting of the American Geophysical Union.

Teams at the Carnegie Institution of Washington (D.C.) and the University of Tokyo obtained similar results, adding weight to the idea that the lower mantel and upper mantle have slightly different compositions.

If true, then the two regions must remain essentially separate, maintaining two distinct convecting systems, says Jeanloz. Material from the upper mantle could not sink into the lower mantle, otherwise that mixing would homogenize the two regions within a fraction of Earth's history—thus wiping out the density difference, he says.

Chalk one up for the splitters.

Jeanloz acknowledges that these conclusions are controversial because the densities differ by only a few percent, keeping alive the debate over the lower mantle's composition. Yet he sees newer results tipping the balance in favor of the chemical layering theory. "Six years ago, we were sort of a lone voice in the wilderness," he says. "For better or worse, now a couple of other groups have joined in with us. It doesn't mean we're right. But at least that's the direction the pendulum is swinging."

Seismologists hear a different story when they eavesdrop on the murmurings of the inner Earth. These scientists study the mantle by analyzing the seismic waves that echo through the entire planet after earthquakes and nuclear explosions crack the crust.

The broad-brush picture from seismology reveals that the mantle has two parts: a thin upper shell surrounding a much thicker lower layer. The boundary between upper and lower mantle lies at a depth of between 650 and 670 kilometers.

Looking at an even finer level, seismologists can discern the edges of surface plates that are sliding into the mantle via ocean trenches. This process, called subduction, is Earth's own recycling system—a way to carry ancient lithosphere back into the mantle from which it formed. As the old lithosphere sinks, it makes room for new rock to grow at the youngest edge of the plate. Subduction also helps chill the inner Earth by carrying cool surface rocks into the torrid interior.

Because the mantle has a natural division, seismologists originally assumed that the two parts did not mix. Subducting plates could sink to the bottom of the upper mantle, but they would hit a barrier at a depth of 650 km that denies access to the lower mantle.

Thomas Jordan of the Massachusetts Institute of Technology

challenged that view in the 1970s with seismic evidence that subducting plates did indeed penetrate the lower mantle. Jordan and his colleagues suggested that the mantle mixed from top to bottom, thus launching the debate between lumpers and splitters. One of Jordan's chief rivals in the debate is Don Anderson of the California Institute of Technology, who served as Jordan's graduate school mentor in the late 1960s.

The arguments have bounced back and forth for 2 decades, with little resolution. But a new picture of subduction is emerging from tomographic studies—a seismological technique that explores the inner Earth much the same way CT scanning resolves structures within the human body.

By analyzing millions of earthquake recordings from around the world, Rob van der Hilst of the Australian National University of Canberra has tracked the path of subducting plates. Yoshio Fukao of Nagoya University in Japan conducted a similar investigation. These tomographic studies can pick out sinking slabs of lithosphere because such rocks are relatively cool and thus slow down seismic waves.

The work by van der Hilst and Fukao suggests that all plates were not created equal. Some slabs bend when they reach the 650 km boundary and remain in the upper mantle; others sink into the lower mantle. According to these studies, the boundary between the upper and lower mantle impedes the progress of sinking slabs, but it doesn't prevent all mixing.

Stephen P. Grand from the University of Texas at Austin has reached a somewhat similar conclusion through a different type of tomographic study. By concentrating on recordings from 180 earthquakes, he produced high-resolution maps of the mantle beneath the Americas.

Grand's maps support the lumpers' case by showing cool slabs reaching down into the lower mantle under South and North America. In fact, a patch of lithosphere under North America has apparently dropped all the way to the top of the core, he reports in a paper soon to appear in the JOURNAL OF GEOPHYSICAL RESEARCH.

But the slabs don't slide into the lower mantle easily. By reconstructing the position of the plates back in time, Grand reasons that the one beneath North America must be sinking at only one-fifth the speed that plates move when they first slip into the mantle. "Slabs encounter resistance at the 650 km depth but eventually they penetrate through that boundary and sink into the lower mantle at slow velocities," he says.

The idea that plates get held up before entering the lower mantle gains support from researchers who study artificial versions of Earth, created inside supercomputers. When used to simulate mantle convection, these three-dimensional numerical models show some surprising behavior.

Two modeling teams—one led by Paul J. Tackley of the California Institute of Technology in Pasadena and another led by Satoru Honda of the University of Hiroshima in Japan—recently examined what happens to subducting slabs as they sink. For these experiments, the researchers assumed that the lower mantle contains the same basic ingredients as the upper mantle but that the rocks assume a more compact crystalline structure in the lower mantle. Geologists know that this type of phase transformation does indeed occur at a depth of 670 km.

In the models, subducting rock sinks to the boundary but cannot penetrate at first. Only after enough cool material accumulates atop the boundary does it finally break through to the lower mantle, according to two simulation studies reported last year (SN:2/27/93, p.133).

Taken at face value, these results present a picture akin to that emerging from seismology. But many regard the models as too simplistic to offer much help in resolving the convection debate.

Modelers thus far have fudged their simulations by ignoring the stiffness of the lithospheric plates, a characteristic that, if included, would increase the cost of computing. According to some researchers, if the models represented the plates more realistically, these stiff slabs might break through the boundary instead of pooling above it. Tackley is currently trying to include the rigidity of plates in a three-dimensional simulation of mantle convection.

Despite the problems inherent in the models, Tackley believes the recent computer simulations, along with the seismic tomography results, have helped push the debate forward. He sees geophysicists now embracing the idea that the mantle does not mix as either one layer or two layers, but rather exhibits some intermediate type of flow.

"The debate is shifting to a question of the degree of the stratification in the mantle instead of whether the mantle convects as one layer or two," says Tackley, who sees some drift toward agreement between the opposing camps. "Even people who in the past have held extreme views are moving in favor of a compromise."

They may not admit it, however. Both Caltech's Anderson and MIT's Jordan regard their positions as unchanged.

Anderson thinks the debate has converged recently by shifting toward the idea that he has long championed—namely, that the mantle is stratified. The recent model results, he says, show at least partial layering within the mantle because sinking material gets delayed at the boundary. Anderson suggests that the models would produce even more complete stratification if they included differences in chemistry between the upper and lower mantle—a possibility supported by high-pressure experiments.

Jordan, not surprisingly, reads the situation differently, He contends that the new modeling studies don't contradict his original conclusion that plate material sinks all the way into the deep mantle. Whether the plates get delayed at the 650 km boundary is less important, he says. In fact, Jordan, Tackley, and others reported last year that seismic records do not show evidence of major pools of cold material sitting at the boundary. That study, he says, indicates that slabs cannot linger there for too long before breaking into the lower mantle.

Jordan agrees, however, that the layered mantle camp holds one wild card—the possibility of chemical differences between the upper and lower mantle. "Of all the arguments that were being erected [in the 1980s] in favor of mantle stratification, they've all fallen by the wayside except for the one that Jeanloz is discussing. The data that have come out recently support it, and it's a major fact that needs to be explained," says Jordan.

Don't expect an easy answer, though. With so many divergent ideas and observations seemingly contradicting each other, it may take a while to sort out the secrets of Earth's plumbing. "The pendulum seems to swing one way and then it swings the other," says Peter Shearer, a seismologist at the Scripps Institution of Oceanography in La Jolla, California. "It's not a problem that I expect to be resolved soon," he adds.

But rest assured that researchers will continue to battle it out. After all, the answer to the convection question ultimately determines how long Earth's inner fires will burn. Researchers have calculated that if the mantle mixes from top to bottom, it will be come a cold corpse with a few billion years. "But with multiple-layer convection, it takes 5 to 10 times as long to cool the planet down," says Jeanloz. Earth remains geologically vigorous, active, young, and healthy for a much longer period of time."

## A Tiny Window into Earth

Forget what the song says: Diamonds are actually a high-pressure physicist's best friend. Taking advantage of the jewel's unrivaled strength, researchers can mimic the intense stresses present inside Earth and other planets.

Charles E. Weir and his coworkers from the National Bureau of Standards first introduced these gems into high-pressure studies in the 1950s, using diamonds that the U.S. Customs Service had confiscated from a smuggler. Now in wide use, the diamond anvil cell works like a vise, squeezing a sample between the pointed ends of two brilliant-cut diamonds typically about one-third of a carat in weight.

Because the jewel points have facets that measure only a few hundredths of a millimeter wide, the diamond anvil can generate intense pressures with only the force of a hand-turned screw. Laboratories around the world routinely recreate pressures equal to those at the center of the planet, 3.6 million times atmospheric pressure. Conditions like this would shatter materials weaker than diamond.

Strength is but one of the attributes that diamonds offer. Because the gems are transparent, researchers can heat a squeezed sample thousands of degrees by shining a laser beam directly through it. By hooking a microscope up, they can look directly through the diamond to monitor the material under pressure. Other techniques involve bombarding the sample with X rays, neutrons, or laser beams, which can reveal properties of the object in question. ❑

**Questions**:

1. What percentage of the Earth's volume is the mantle?

2. Raymond Jeanloz determined that the lower mantle may not be composed of the mineral peridotite. On what did he base this conclusion?

3. Would the Earth cool faster if the mantle mixes from top to bottom or if it is stratified?

Answers on page 301

6

*During the history of the Earth, its magnetic field has periodically reversed poles. This has occurred every few hundred thousand years. Between reversals the magnetic field maintains a stable, dipolar shape. The transition between reversals takes between 2,000 and 20,000 years. Studying the record of these transition periods has proved difficult. The magnetic field during transitions appears to be exceedingly complex with no discernible pattern. New findings suggest that during reversals the magnetic field follows one of two paths: one through the Americas and another through Asia. If this is true it means that there must be some long-term structure to the Earth's mantle that affects the path of the magnetic field. This is still a controversial idea, and many scientists disagree with these results. If these findings are correct it may help us understand not only the magnetic reversal, but also the processes occurring within the core and mantle.*

# The Flap over Magnetic Flips

## What happens when Earth's magnetic field reverses itself?

by Richard Monastersky

The way Carlo Laj tells it, the 1980s were a dead time for scientists studying Earth's peculiar habit of flipping its magnetic field every few hundred thousand years. Laj and other paleomagnetists spent the decade roving the world, accumulating records of times when the magnetic field reversed itself, switching north pole for south and then back again. But even as their databases grew, the study of geomagnetic reversals foundered for lack of new theories to guide the research.

"The community was slowly going to sleep," says Laj, a geophysicist at the Centre des Faibles Radioactivités in Gifsur-Yvette, France.

Two years ago, however, Laj's group and Bradford M. Clement of Florida International University in Miami sent out a wake-up call that roused the paleomagnetic community and packed the halls of scientific conferences dealing with geomagnetic reversals. Working independently, Clement and Laj's team concluded that the magnetic field, quite unexpectedly, had displayed a persistent pattern of behavior during recent reversals—a finding that promised to reveal fundamental insights into the workings of the planet's deep interior.

Their far-reaching claims sparked a heated debate that simmers to this day. One after another, scientific teams across the world have joined the fray, with some groups supporting the 1991 observations while others attack Clement and Laj's proposal on a variety of fronts.

"This is one of the hottest things going on today in the earth sciences," offers Kenneth A. Hoffman, a geophysicist at California Polytechnic State University in San Luis Obispo.

Clement and Laj made their discoveries by taking a detailed look at what happened during the magnetic turnovers that have occurred during the last 12 million years. To trace the field's behavior backward in time, they measured the orientation of magnetic minerals within sedimentary rocks. When such grains dropped to the floor of an ancient ocean, they aligned themselves with the direction of the magnetic field existing at the time, eventually becoming locked into place as the sediment layers hardened.

During stable times, such as today, Earth's predominant field has a simple dipolar shape, as if the planet's core held a giant bar magnet. Field lines flow out of a magnetic pole near the coast of Wilkes Land, Antarctica, arc above the planet, and then dive back down into a second magnetic pole in arctic Canada, Geophysicists describe today's field as having normal polarity, while the opposite orientation has a reversed polarity.

In between these two stable states are the relatively short reversals, which can take anywhere from 2,000 to 20,000 years to complete. During a switch, the stable field weakens and then reestablishes itself with the opposite polarity, an act that most recently happened 730,000 years ago.

Geophysicists have long sought to understand what goes on during reversals, hoping this might give them some insight into how swirling currents of liquid iron within the core generate Earth's magnetic field.

While studies of reversals progressed rapidly in the 1960s and early 1970s, they hit a lull soon thereafter because accumulating evidence indicated that the magnetic field grew extremely complex during turnovers. This doused many researchers' hopes for deriving fundamental laws about the reversal process.

Unlike the simple field during stable times, the transitional field did not appear to retain a dipolar shape. Rather, it seemed to assume a form with a multitude of magnetic poles, suggesting a dizzying arrangement of field lines that loop in and out of the planet like a Spirograph pattern.

Evidence of such polar complexity appeared when scientists examined rocks at different sites around the world and tried to trace the shifting locations of the south magnetic pole back in time. If rocks at one site showed that the south pole passed through the Atlantic as the field reversed, rocks at another site suggested that the south pole traveled through Asia, and samples from a third site revealed yet another path, through Africa. In other words, the rocks seemed to indicate that a welter of south poles appeared during reversals, following no particular track when the field switched.

But Clement challenged the prevailing view in 1991 when he reported that pole paths from several locations showed a strikingly simple pattern that persisted for different reversals. Laj and his colleagues announced finding the same pattern in their own data. Rather than showing a random arrangement of numerous pole paths, both studies revealed that during reversals, the south pole tended to follow only two specific tracks: one through the Americas and another through Asia, on the opposite side of the globe.

Most important, these so-called preferred paths showed up again and again in reversals widely separated in time. To Clement and Laj, it seemed that some enduring factor inside Earth steered the pole along a particular pair of routes time and again over a period of 12 million years.

Looking to explain such persistence, both researchers knew they weren't likely to find an answer in the outer core, the region that generates the magnetic field. The problem is one of timing: The outer core consists of liquid iron, which swirls far too quickly to preserve flow patterns for millions of years.

Just above the fast-moving outer core, however, lies the sluggish mantle, a region of solid rock squeezed and heated so much that it slowly flows, moving at about the speed of a growing fingernail. At that clip, the mantle remains consistent over millions of years, making it a natural spot to look for a pattern that persisted through dozens of reversals.

In an influential map that appeared on the cover of the June 6, 1991 Nature, Laj fingered one suspect that might have guided the polar path during reversals. The well-traveled tracks through the Americas and in Asia happen to lie above regions in the deep mantle that have unusually low temperatures, Laj pointed out. Such cool spots at the bottom of the mantle could stimulate flow patterns in the liquid core that would recur during most reversals, creating the paths seen in the sedimentary records.

Researchers had long surmised that the base of the mantle influences the magnetic field, but they had lacked any direct evidence to support that idea. Clement and Laj felt they had finally stumbled across a sign that the mantle does indeed play some role in the reversal process.

The connections may not stop there, though; they could extend all the way to the planet's surface. Many earth scientists believe the cool spots in the base of the mantle hold sunken pieces of Earth's outer shell that have been pushed deep into the planet through the process of plate tectonics. That would imply a previously unappreciated link between the planet's skin and its core, 3,000 kilometers underfoot.

During the last year, scientific journals have published a volley of papers alternately attacking and supporting the observations made by Clement and Laj, making the debate appear something like a tag-team tennis match with continuously rotating players.

Geophysicists C.G. Langereis, A.A.M. van Hoof, and P. Rochette hit a hard return shot when they argued that the evidence for preferred paths may simply reflect an artifact in the way sedimentary rocks record the geomagnetic field. Langereis and van Hoof hail from the Paleomagnetic Laboratory at Fort Hoofddijk in Utrecht, the Netherlands, while Rochette works at the Faculté des Sciences Saint-Jérôme in Marseille, France.

In their argument, published in the July 16, 1992 Nature, the three researchers focused on the way sedimentary rocks form. Because it takes thousands of years for sediments to accumulate on the ocean floor and get locked into place, rocks created from such layers do not present a crisp, detailed portrait of how the magnetic field behaves during a reversal. Instead, sedimentary rocks smear together different stages, producing a blurred image of events that occurred over many millennia. Langereis and his co-workers contended that this tendency could present a problem for people trying to study a relatively short event such as a reversal.

The task of finding an accurate reversal record seems all the more difficult because the magnetic field weakens considerably when it switches direction. Langereis' group argued that the sediments deposited during a reversal might therefore bear a blurry overprint of the much stronger, stable field immediately preceding or following the reversal. Examining a sedimentary record from Crete, the team showed that the reversal recording at this site apparently reveals more about the field before and after the reversal than about the field that existed during the event.

A group led by Jean-Pierre Valet of the Institut de Physique du Globe in Paris is also exploring ways in which sedimentary records distort the picture of a reversal. A former student of Laj's, Valet notes that investigators have not collected sediment cores in a uniform pattern around the world. Rather, he says, the sampling sites tend to group into particular bands of longitude that could introduce some bias in the paleomagnetic data, producing the false impression that the magnetic field behaves similarly during different reversals.

On the flip side, several researchers have weighed in with evidence supporting the idea that the mantle exercises control from one reversal to another. Hoffman, for instance, has reviewed data on the magnetic orientations stored within lava, a recording medium that doesn't pose the same problems encountered in sediments. Because lava hardens quickly after an eruption, it provides an instant snapshot of the field's direction.

Volcanic records do pose a disadvantage, however: They provide information only for times when a volcano erupts. If a mountain blows just once during a particular reversal, a researcher studying that volcano will find a solitary snapshot recording the event.

Examining the lava data, Hoffman found that the magnetic field does behave similarly in several different reversals. In that way, the lava flows support the sedimentary record. But the two records also harbor important differences, he reported in the Oct. 29, 1992 NATURE.

Unlike the sedimentary record, the volcanic data do not show the pole repeatedly following a line of longitude up through the Americas or Asia. Instead, they indicate that the pole tends to return time and again to two discrete places in the southern hemisphere: one off the east coast of southern South America and another off the west coast of Australia. What makes things interesting is that these two clusters fall within or close to the preferred paths discerned in the sedimentary data.

The coincidences don't end there. The clusters detected by Hoffman also correspond with the locations of irregularities in the current magnetic field. These spots represent weak "nondipole" elements of the field—in other words, the part of the magnetic field left over when researchers theoretically subtract the dominating dipole.

Hoffman suggested that such nondipole elements are persistent parts of the magnetic field. To explain the clustering, he proposed that when the dipole weakens at the start of a reversal, it tends to reestablish itself temporarily in a tilted position, oriented in a way that corresponds with the field's nondipole component. That would put the south pole at one of the clustering sites in the South Atlantic or near Australia.

As Hoffman envisions it, the south pole hops to different spots around the globe in a seemingly random way throughout the reversal, but it lingers for the longest spells in one of these special patches. Although each reversal looks different in the lava data, Hoffman found that the pole tended to seek the same two patches in many reversals over the past 10 million years. That evidence pushed him toward the conclusion reached by Clement and Laj two years earlier—that long-lived features inside Earth direct the behavior of the flipping magnetic field.

Other geophysicists, especially in Europe, say the evidence of a mantle influence remains far too flimsy at this point. For every study that suggests some type of recurring reversal pattern, another study appears that bashes the idea.

Pierre Camps and Michel Prévot of the University of Montpellier in France have reviewed the existing lava data for all reversals during the last 16 million years. Using this more complete record, Camps and Prévot could not confirm the clustering that Hoffman found in a limited number of reversals. They reported their findings at a meeting of the European Union of Geosciences, held in Strasbourg in April.

"We don't observe any longitudinal confinement of the magnetic poles," says Prévot. "They are not preferentially located over the Americas. They correspond to a uniform distribution."

While the band of skeptics remains strong, one of their ranks has recently undergone a reversal of his own. In the May 5 GEOPHYSICAL RESEARCH LETTERS, van Hoof presents a detailed study of sedimentary rocks from Southern Sicily that apparently supports Hoffman's idea of the magnetic pole lingering in select sites during a reversal.

So the debate will continue, as geophysicists circle the globe looking for new data to test the theory that the mantle helps steer the course of geomagnetic reversals. For his part, Laj grants that the additional sedimentary and volcanic records could well disprove the idea, but he would prefer to see that simple theory win out in the end.

"It gives some sort of unity to the Earth, from the upper part to the deep interior," he says. "I hope it is true, because it is so very elegant." ❑

**Questions**:

1. Where does the Earth's magnetic field originate?

2. How could the mantle retain a signal of internal processes for millions of years?

3. Explain Kenneth Hoffman's findings.

Answers on page 301

7

*Rifting is one of the fundamental surface expressions of plate tectonic processes. Rifts are important because they are places where new crust is formed. However, they are hard to study because the entire ridge system, with the exception of Iceland, is beneath the ocean. Recent work by British and French scientists used seismic profiling to study the structure of rift margin rocks off the coast of northwestern Europe. This research has identified two distinct types of rifting. The younger, more northern rift margin contains extensive lava flows above continental crust and massive igneous intrusion beneath. Rifts of this type are called hot rifts. The older, more southern rift margin contains very little igneous rock but has many faults and tilted blocks in the upper continental crust. Rifts of this type are called cold rifts. The different response to rifting of these two margins appears to be related to differences in the mantle. Hot rift margins are generated where the underlying mantle is unusually hot. The hot mantle plume that formed the northern rift margin is still active today, creating that one exception: Iceland.*

# Breaking Up Is Hard to Understand

When continents split, a new ocean is born. Until recently, however, the deep waters guarded the secrets of what happened during the breaking-up of a continent
by Bob White

When continents collide, they make mountains; when they divide, a new ocean forms. Geologists can study the history of collisions, because erosion gradually reveals the structure of the rocks inside mountain ranges. But when a continent tears into two, the broken edges vanish beneath the waves. Now geophysicists have found a way to examine the remnants of these continental edges or margins. Seismic surveys of the rocks beneath the sea floor show that the continental break-up which generated the North Atlantic Ocean was accompanied by extensive volcanism—but only in certain parts. From these surveys, geophysicists have begun to piece together the history of the splitting of one continent which took place tens of millions of years ago. Last week, an international team of geologists, working on board the research ship Joides Resolution, started to collect samples of rocks from a continental margin off the coast of Portugal. These samples, geophysicists hope, will help them to confirm their theories, and to understand why the continents split apart in the first place.

When a continent breaks, the process is known as rifting. It is happening today in the Red Sea, where the Arabian Peninsula is moving slowly away from Africa. The continent stretched and thinned until the continental crust finally broke into a series of faults and then subsided. New ocean crust is now forming under the Red Sea as molten rock wells up from below to fill the gap. Millions of years ago, Greenland and North America were near neighbours of Britain. But here, too, the thick continental crust stretched like toffee until it finally broke, allowing the North Atlantic to form. A few years ago, geophysicists from the Universities of Cambridge and Durham put to sea on the Natural Environment Research Council's research ships, the Charles Darwin and the Discovery, to carry out seismic surveys of Hatton Bank, a continental margin formed when Greenland broke away from northwestern Europe 55 million years ago.

Seismic profiling has become a powerful tool for earth scientists. Seismic waves generated underwater by explosive charges or bursts of compressed air travel into the sea floor where they reflect and refract as they pass through rocks with differing physical properties. Researchers can deduce the structure of the rocks below from the signals that return to the surface. Seismic waves have low frequencies (between 5 and 50 hertz) and propagate well through rock, giving researchers a chance to "see" a long way into the Earth. For the Hatton Bank project, researchers used seismometers placed on the sea floor, a refinement of the technique enabling them to pick up signals that had travelled up to 100 kilometres through the Earth's crust.

Computer simulations of the seismic profiles at Hatton Bank led the team to a model of the structure of the continental edge that included far more volcanic rock than expected. Continental crust was stretched and thinned towards the new ocean, as expected, but the seismic reflections indicated that there were many thin sheets of rock near the surface, gently sloping out to sea. These have since been drilled and identified as lava flows, but at the time no one suspected that there would be so much igneous rock, solidified from lava, at the edge of the continent.

**Startling Lava Flows**

The association of continental break-up and volcanism is not a new observation. Signs of extensive volcanic activity have been found across western Britain and Ireland dating from the time when the North Atlantic opened—examples include the volcanoes and basaltic lavas of Skye and the Hebrides, and rock formations such as the Giant's Causeway in Antrim. It was the scale of the lava flows that was startling: in places, the layers of rock were 5 kilometres thick, making up much of the thickness of the crust at the margin. Mapping the sea floor by seismic reflection methods shows that there are similar lava flows along the continental edge on both sides of the North Atlantic. They stretch from Edoras Bank to the northern tip of Norway on the east side of the Atlantic and along the entire Greenland coast to the west. When the Atlantic first opened, perhaps as much as 2 million cubic kilometres of lava poured out in only one or two million years—almost overnight in geological terms.

The seismic sections across Hatton Bank held another surprise: there is between three and four times as much igneous rock trapped near the base of the continental crust. This means that some 10 million cubic kilometres of igneous rock were generated when the North Atlantic first broke open. This is an enormous amount of rock: if spread across the continental landmass of the US, it would make a layer 700 metres thick.

In 1987, a joint French and British team, working aboard the French research ship Le Surôit, recorded data across another part of the Atlantic margin, near the Goban Spur at the mouth of the English Channel. Ocean started to form here as North America broke away from Europe, about 120 million years ago and some 65 million years before the Hatton Bank margin. The Le Surôit results showed that there was very little igneous rock, either as lava flows near the surface or as intrusions lower down in the crust. The structure in the upper levels of the crust was different, too. Faults, and the tilting of blocks of rock associated with fault movements, played a much bigger part in the break-up of the continent than they did in the Hatton Bank section.

Why did these two parts of the continental margin behave so differently? The answer lies in the interior of the Earth. The crust is part of the lithosphere, the rigid outer layer of the Earth which is about 100 kilometres thick. Beneath the lithosphere is the mantle, which continues 3000 kilometres down to the core. Although mantle rocks are hot enough to flow—typically reaching 1350 °C at 100 kilometres down—the pressure is so high at these depths that they remain solid. But if the pressure drops for some reason, the mantle will begin to melt. This is what happens as the lithosphere stretches when continents break apart: the mantle beneath wells up to fill the gap, decompresses and partially melts. The melt is less dense than the mantle around it, so it trickles upward to the surface. Eventually it solidifies to make new ocean crust.

At the Goban Spur, the crust of the continent was stretched and grew thinner and thinner until the upper part finally fractured and broke into series of faulted blocks. The same stretching took place lower in the crust, but there the rocks were hotter and under higher pressure, so they deformed without breaking up into faults. The crust continued to stretch until it was only 5 kilometres thick, about one-sixth of its original thickness, before molten rock welled up to form new ocean crust.

The amount of molten rock generated when the mantle decompresses is highly dependent on the temperature of the mantle. If the mantle heats up by 140 °C, a rise of just 10 per cent, it produces three times the volume of molten rock. Normally the ductile part of the mantle is at a uniform temperature and only small volumes of molten rock form under rifts. But from time to time, enormous plumes of unusually hot mantle rise towards the surface. This is what happened beneath the region of the present North Atlantic that lies between Ireland and Greenland. Shortly before the continent split, a mantle plume had begun to form, giving rise to volcanism over an area 2000 kilometres across. This plume is still going strong; it is responsible for the active volcanoes of Iceland, having already built the whole island from volcanic rocks.

Why, though, did the bulk of the molten rock stay within the crust at Hatton Bank? Why did it not all erupt as lava? The answer is a matter of density. The molten rock is less dense than the mantle that melted to form it, so it rises. But magma is more dense than continental crust. Most of it is trapped beneath the edges of the rift, where it solidifies to form intrusions of igneous rock injected into the continental crust.

Hatton Bank represents one extreme type of split, called a hot rift; Goban Spur the other, known as a cold rift. The mantle plume beneath Hatton Bank generated large-scale volcanism and the igneous rocks thickened the crust. The rocks were warmer, so faulting was a less important mechanism in the overall thinning of the crust. At Goban Spur, there was no plume and little volcanism, so faulting was a significant mechanism in the stretching of the colder and more brittle rocks.

The temperature differences in the underlying mantle also affect how quickly the rifted continental margins sink to form the new sea. The continental crust is thicker than oceanic crust, so it "floats" higher in the mantle. This means that the formation of a new ocean always involves subsidence; the stretched and thinned crust on the continental margin usually sinks 2–3 kilometres, while the even thinner ocean crust eventually sinks more than 5 kilometres. At the Goban Spur margin, for example, the thinned continental crust margin subsided by several kilometres as it stretched and continued to sink over the following 50 million years or so. But at Hatton Bank, the underlying plumes added so much new igneous rock to the crust that, initially, the continental margins were elevated above sea level. Hot rifts therefore tend to rise as the continent splits. They do subside later as the ocean widens, but never by as much as a cold rift.

The differences between hot and cold rifts have major implications for researchers investigating how continental margins form and develop. In the North Atlantic, the contrast between two areas, just 1000 kilometres apart on a continental margin 10, 000 kilometres long, represents a challenge to the theorists. How did the Atlantic margin change as it grew northwards? Do these two types of margin represent very different types of rift, or are they just variations on a larger overall theme?

Geophysicists are still conducting experiments to try to shed light on these and other problems. On the Charles Darwin last summer, a team of scientists from the Universities of Cambridge and Edinburgh took seismic soundings at Edoras Bank to probe the layers of lava and the underlying crust. Edoras Bank, once joined to the southern tip of Greenland, was at the edge of the Iceland plume when the North Atlantic was opening, so the results should help them find out how the presence of the plume affected the development of the continental margin.

Last week, an international team of scientists, led by Bob Whitmarsh of the Institute of Oceanographic Sciences in Surrey and Dale Sawyer of Rice University, Houston, started to gather physical evidence of what happened at a cold rift. They are aboard Joides Resolution, a ship especially adapted for drilling boreholes and collecting samples of rock from beneath the sea floor as part of the international Ocean Drilling Program. Equipped with 9150 metres of drill pipes that can be suspended through a hole in the hull, the ship will spend eight weeks drilling into the thinned continental crust at the Iberian margin, an area off the west coast of Portugal that was once joined to Newfoundland in North America.

During October and November, the Joides Resolution will attempt to drill into a hot rift off the east coast of Greenland, an area that was joined to Hatton Bank before the North Atlantic was formed. Geophysicists hope that the samples from the two sets of drilling will provide the physical evidence on the subsidence history of the continental margins and composition of lava flows to confirm their theories based on the seismic soundings. In addition, analysis of lava samples should also tell them more about the mantle plume that was responsible for the formation of 10 million cubic kilometres of igneous rock that poured out when Greenland broke away from northwest Europe.

Combining data from these different investigations into the geology of rifted margins should soon help geologists to reconstruct the history of the break-up which generated the North Atlantic. It may even illuminate a different, and important aspect of the geology of Britain's western seaboard: possible oil and gas deposits far out on the continental edge. Understanding how the continents rifted in the first place, and whether they were hot or cold rifts, will give oil geologists a head start when their search takes them to these remote and dangerous waters. ❑

## Questions:

1. Where is continental rifting occurring today?
2. Why doesn't most of the molten rock in a hot rift erupt to the surface?
3. What happens to the elevation of hot rifts as the continent splits apart, and why?

Answers on page 301

8

*Until recently it was thought that little water remained in subducting rocks below 60 miles in depth. Any water carried down with a subducting slab would get squeezed out by increasing pressure. Recent research has indicated that water may in fact be present in significant quantities as deep as 250 miles. In the 1980s it was discovered that less water was being expelled from volcanoes than was being subducted and that phases of rock existed that could contain water at greater depths of up to a few hundred miles. Recently, evidence from the velocity of seismic waves and the presence of anomalously high levels of helium indicated the presence of large quantities of water beneath areas of the Earth's crust where ancient subduction zones existed millions of years ago. The presence of such large quantities of water deep within the Earth may have an important effect on tectonic processes and on the distant future of the Earth.*

# The Ocean Within

## Old oceans never die; they are preserved, in scattered fossil form, in the rock of Earth's mantle, 250 miles down

by Carl Zimmer

Jules Verne had an impressive track record of predicting technological progress, having foretold in his books the invention of rocket flight, the submarine, and television. At first glance he would seem to have had less luck with geology; in *Journey to the Center of the Earth* Professor Otto Lidenbrock and his nephew Axel travel underground and discover that our planet is hollow. But the Lidenbrocks also sail across a vast underground ocean, and geologists now think Verne was onto something there. Where researchers once envisioned nothing but dry rock, they are now discovering signs of huge reservoirs of water deep underground—250 miles deep—that may play a key role in the evolution of Earth.

When the theory of plate tectonics was being formulated in the 1960s and 1970s, most researchers would have told you that water could go underground—but not very far. The plates that make up Earth's crust form from up-welling hot magma at mid-ocean ridges and move like a conveyor belt until they encounter another plate; at that point one of the two colliding plates may dive into Earth's mantle, in a process called subduction. Seawater jumps onto this conveyor belt at several stages. It works its way into the crystal structure of the hot rock at the ridge, and it squeezes itself into the pores that open as the rock cools. In the ensuing millions of years, as the plate moves away from the ridge, soggy organic matter rains onto it incessantly, coating it with a waterlogged veneer of sedimentary rock. When the plate finally dives into Earth's interior, most of this water goes down with it.

Until recently it was thought that the water then quickly came back up. By the time the slab dived 60 miles, geologists figured, the increasing pressure would squeeze the rock like a sponge, sealing its pores and forcing it into a new and denser phase (an arrangement of the rock's crystal lattice) that had no space for water. The liberated water would rise into the rock above. Like salt spread on ice, the water would dramatically lower the rock's melting point. Heat from the rubbing of the plates would then melt the rock, and the magma would rise to the surface to form volcanoes, which would spew water vapor into the atmosphere. Later the water would rain to the ground, flow to the ocean, and eventually be sucked back into the mantle.

This neat circuit of plumbing sprang a leak in the mid-1980s. As geologists improved their bookkeeping, they discovered that much more water was being subducted into Earth than was coming out of subduction-zone volcanoes. At the same time, research with diamond-anvil cells—devices that squeeze slivers of rock between diamond tips to mantle pressures while heating them with lasers to thousands of degrees—was revealing new phases of rock that could hold on to water as a plate plunged into the mantle. Only when the descending slab reached a depth of hundreds of miles would its water be forced into the surrounding rock. It became conceivable that there were reservoirs of water down there.

Now it is becoming more than conceivable: there is evidence that the reservoirs really exist. Some of the evidence comes from seismic tomography, a technique that essentially allows researchers to CT-scan Earth. Earthquake waves travel fastest through rock that is cold and dense, slower through rock that is hot. By analyzing seismic waves from different earthquakes received at many different seismometers, a computer can reconstruct a three-dimensional picture of Earth's interior. Princeton seismologist Guust Nolet and one of his graduate students, Alet Zielhuis, recently made tomograms of the mantle beneath eastern Europe. There they discovered what they think is a buried ocean—not one you could sail on, to be sure, since the water is in the form of scattered molecules and droplets trapped in the rock, but still a huge volume of $H_20$.

Between 500 and 400 million years ago there was an ocean at the *surface* of this part of the world. The ocean was rapidly disappearing, though, as two continental blocks were slamming together to create Europe. In the process an oceanic slab was thrust into the mantle below what is now Poland and Ukraine. Near this slab, at a depth of 250 miles, Nolet and Zielhuis found two regions, each hundreds of miles wide, where seismic waves traveled 4 percent more slowly than average.

That is a much greater slowdown than normal variations in rock temperature or composition could produce. It is the sort of slowdown seismic waves undergo when they travel through rock that is actually molten. "Rocks are like particles connected by springs," Nolet explains, "and when you pull one, a wave travels through the springs. You can think of melting rock as breaking some of them, which slows down the seismic wave."

There is only one problem with this explanation: at a depth of 250 miles, mantle rock doesn't normally melt. The pressure is too great. Nolet realized that something abnormal was going on—or rather, two things. The first is water. As the laboratory research had suggested, Nolet thinks the oceanic slab that was buried under eastern Europe carried water down 250 miles and then injected it into the surrounding rock, lowering its melting point.

Under normal conditions, the water would have gone unnoticed. But eastern Europe is abnormal in a second way. Over the past hundreds of millions of years, as Europe has shifted back and forth, it has twice crossed over a hot spot: a narrow plume of hot rock rising from deep in the mantle. Hot spots sometimes poke through the crust and make volcanoes; the one that eastern Europe passed over is now under Mount Etna in Sicily. Under eastern Europe, though, all the plume did was heat the mantle—and melt those regions that were rich in water and thus easily meltable. The two seismically slow regions under eastern Europe, says Nolet, are places where the hot-spot plume hit a trapped ocean of water released millions of years earlier by the ancient subduction zone. The plume acted like a flashlight, exposing parts of the hidden ocean to the scrutiny of seismologists.

There may be other such oceans. The White Mountains of New Hampshire are geologically similar to eastern Europe: they were the site 300 million years ago of a subduction zone, and 100 million years ago they passed over a hot spot. No one has done a seismic tomogram of the White Mountains, but geochemist Tom Torgerson of the University of Connecticut has found what he says is another sign of an underground water reservoir: high levels of the light form of helium, helium 3.

Once helium gets into the atmosphere, it quickly floats out to space, so any helium you find near the surface is gas that has recently seeped up from the mantle. In White Mountains groundwater, Torgerson measured helium 3 levels 1,000 times higher than are typically found elsewhere on Earth—which is evidence, he thinks, that a mantle plume and an underground water reservoir have been conspiring to bring lots of helium up from the depths. When a plume heats up water-rich rock, Torgerson explains, the rock melts in bubbles. "Those bubbles suck up lots of gases like helium, just as when you boil water the bubbles suck in oxygen and carry it away," he says. "The helium is the smoking gun for Nolet's hypothesis."

Nolet envisions hidden oceans all over the world, wherever plates once plunged into the mantle. Rough calculations suggest that a volume of water equal to that in the oceans on Earth's surface may be trapped in these reservoirs. This water might be crucial to many geologic processes. If the mantle plume below New Hampshire hadn't hit water, for example, creating bubbles of buoyant magma, the White Mountains might be much smaller today. And when water squirts out of the subducted slabs, it may make the dehydrating rock shudder, creating the enigmatic earthquakes that occasionally rumble deep inside the planet—like the one under Bolivia last June that was felt as far north as Toronto.

The buried oceans may also be important to the surface ones we're familiar with. The old water cycle that researchers once envisioned is being replaced by a new one: some subducted water quickly goes back to the surface through volcanoes, but some flows down to reservoirs 250 miles underground. Mantle plumes that pass through them force the water out and gradually back up to the surface. But Nolet points out that in coming eons, as the interior of Earth cools, mantle plumes may occur less frequently. Then more water will go down than comes up. The surface oceans will gradually drain into

Earth's interior, making our planet a dry world like Venus or Mars. In the end Jules Verne may have the last laugh: of all Earth's oceans, only his may remain. ❑

**Questions**:

1. What internal geologic feature helped scientists to identify the presence of large quantities of water within the Earth?

2. Why was the presence of high levels of helium 3 considered to be an anomaly?

3. What may happen in the distant future as the Earth cools?

Answers on page 301

9

*Scientists have always wanted to sample the depths of the Earth's crust and mantle. Geophysical methods, such as seismic profiling, can supply much needed information on the underlying structure of the Earth, but most geologists are never really satisfied until they can get their hands on actual rocks. Until recently, reaching down to depths of 10 to 12 kilometers has been beyond our technology. Now a drilling project in Europe has reached a depth of 7.5 kilometers and is projected to reach a depth of 10 kilometers. Already the results of this drilling project are causing scientists to reevaluate some of their ideas. Some unexpected results included warmer crust, rocks with more of the mineral ilmenite and less graphite, and much larger volumes of fluid. Eventually, this may well help explain the role of the crust in plate movements.*

# Looking Deeply into the Earth's Crust in Europe

by Richard A. Kerr

Call it a wishing well for Earth scientists. If you cleared all the equipment from the drill hole near the Bavarian village of Windischeschenbach in southeastern Germany and tossed in a coin, you would have several minutes before a faint tinkle of coin hitting bottom reached you from 7.5 kilometers down. German geoscientists, though, are not stopping with a pfennig or two. During the next year and a half they will throw the last of a third of a billion dollars into their Kontinentales Tiefbohrprogramm der Bundesrepublik Deutschland (KTB) hole, all the while wishing for geoscience riches: access to a region of Earth's crust whose nature has so far only been guessed at.

The hole, the second deepest ever drilled into hard basement rock after a Soviet project, still has another 2.5 kilometers to go to reach its target depth of 10 kilometers. After a kilometer of the most difficult drilling yet, money is short, and project managers have had to request an additional $20 million from the Federal Ministry of Research and Technology to fund the remaining drilling. Yet the hole has already yielded unexpected payoffs. Geologists had previously inferred a picture of the crust under the drill site by examining rocks brought to the surface by erosion and making electrical and seismic measurements that probe deep beneath the surface. Since drilling began in September 1987, however, they have had to redraw large parts of that picture. The subsurface faults and folds look quite different than predicted. An unexpected set of mineral "circuits" seems to open conductive pathways in the deep rock. And at depths once thought to be bone-dry, the drill is penetrating abundant reservoirs of hot brine.

These and other KTB findings are helping convince the geophysical community of the value of deep drilling. Many had wondered whether deep holes were worth the trouble and expense, especially after watching the Soviet Union's 15-year effort to drill the world's deepest hard-rock well, 12 kilometers deep on the Kola Peninsula near Murmansk. Its cost, though never disclosed, was thought to be astronomical, and western scientists were skeptical about some of the scientific results.

"I never was a big enthusiast for drilling," says Alfred Duba of Lawrence Livermore National Laboratory, who has studied samples from the KTB hole. "I thought the engineering costs were too large for the science you got. But the more we drill, the more we find out how little we know. There is a place for drilling." Yet Duba still adds, "I just wish it didn't cost so much." In spite of this caveat, the KTB hole has converted him and many of his colleagues to the idea of exploiting technology developed in the project for a broader international program of drilling.

In keeping with the well's tendency to deliver surprises, both of the features that initially drew the 300 KTB researchers to the Windischeschenbach site have turned out to be mirages. One was the expectation of moderate temperatures in the deep rock, which would have allowed drilling to a depth of 12 kilometers before heat overwhelmed the drilling equipment. But the first surprise from KTB—temperatures far higher than predicted—forced a retrenching to a target depth of 10 kilometers. The other attraction seemed to be an opportunity to drill through the buried boundary between two tectonic plates that collided 320 million years ago to help form the present Eurasian plate. But the

***Richard A. Kerr. "Looking Deeply Into the Earth's Crust in Europe." Science, July 16, 1993, v. 261, pp. 295–297.*** 

suture, first predicted to slant under the KTB site at a depth of about 3 kilometers on the basis of surface geology, failed to show up at 3 kilometers, or at 5 kilometers as later hoped. And at 7.5 kilometers, researchers still "haven't seen any sign of a dramatic change" that would mark the boundary between the two plates, according to Jörg Lauterjung of the KTB project.

**Ground Truth**

Other predicted boundaries in the rock have proven equally elusive. One object of the hole was to provide a kind of ground truth for seismic reflection profiling, the radar-like technique that creates images of subsurface structures from the manmade seismic waves that they reflect. "You see all kinds of [seismic] reflectors around the world," says KTB operations leader Peter Kehrer, "but we don't know what they are." Distinguishing among faults, changes in rock type, fluid-filled cracks, or other possibilities has been largely guesswork—and the KTB hole suggests an extra measure of caution. Seismic profiles predicted the position of some—but not all—of the structures encountered so far, and even those that showed up where expected were often different from what researchers had assumed they were.

Many surprises, however, have been more positive. Duba, for example, prizes the evidence that "a rock we get at the surface is not a very good indicator of what goes on at depth," at least in his specialty of subsurface electrical conductivity. Like seismic profiling, conductivity surveys can reveal aspects of the deep crust without the need for expensive drilling. When researchers inject electric current into the ground at the KTB site, the Earth's electrical response points to regions of high conductivity deep beneath the surface. From experience in other regions, Duba had expected the conductivity to reflect a high proportion of conductive graphite in the deep rocks. Rocks exposed at the surface by erosion didn't conduct well enough to account for the observations, but Duba guessed that chemical alteration and weathering of deep rocks as they gradually worked their way to the surface must have disrupted their continuous graphite "circuits."

If so, rocks extracted from the depths of the well should have intact graphite films that, at least under deep-Earth pressures, should readily conduct electricity. When Duba compressed KTB samples in a press and passed an electric current through them, they conducted as well as he'd hoped. So far, so good. But he couldn't find the expected amounts of graphite in the deep rocks. The samples' conductivity instead seemed to come from an unexpected source: the mineral ilmenite, an iron titanium oxide. To Duba, that implies that variations in the subsurface conductivity, at the KTB site and elsewhere, may reflect a far more complex web of circuits than had been assumed: not just graphite but also ilmenite and other oxide minerals, and even the brines that the drilling has been turning up.

**Water from the Stone**

The brines are another surprise that is opening researchers' eyes to the merit of deep drilling. "When I started 25 years ago, the idea was that the deeper you go into the crust, the drier it gets," says Kehrer. Conventional wisdom had it that kilometers of overlying rock squeeze shut any cracks, cutting off the fluid flows that deposit ores and chemically alter the rock at shallower depths. But after the drill bit had penetrated more than 3 kilometers of dry rock, it broke into water aplenty. Core samples retrieved from 3.4 kilometers were veined with open cracks more than a centimeter wide that had presumably carried fluids. That was only a hint of what was to come at 4 kilometers, where more than half a million liters of a gas-rich, calcium-sodium-chloride brine twice as concentrated as seawater poured into the well. Abundant fluids gushed from depths as great as 6 kilometers. "This has been a real sensation," says Kehrer. "The surprise is that there are fluids of *that* amount."

Geophysicists had had some earlier hints from the Kola hole, where drilling "mud" pumped into the hole to lubricate and cool the bit returned to the surface with its chemistry subtly altered. The Soviets took the changes as evidence that brines were flowing into the hole from surrounding rock, but western scientists remained unconvinced. Now the obvious inflows in the KTB hole give more weight to the Kola claims. "The Kola well and our own have shown that a deep crust of dense, hot rock is definitely not the case," says Kehrer. "There are large amounts of highly saline brine in the crust that migrate, carrying metals around and depositing them as minerals."

A minority of geophysicists, including Lawrence Cathles of Cornell University, had suspected that at least some permeability would remain at great depths, enough to allow fluids to circulate. But the large volumes flowing into the KTB hole are welcome confirmation. "It's beginning to look like the lower parts of the crust can be fairly permeable," Cathles says, making room at relatively great depths for ore formation.

The hole shows the potential for settling another debate about the lower crust—specifically, about the role it plays in transmitting the forces that shift tectonic plates. Theory and experiments on rocks in the lab suggested that down to about 10 kilometers, increasing pressures strengthen rock. At greater depths, the weakening effects of increasing temperatures should overwhelm the effect of pressure. Thus, the thinking went, the strong top 10 kilometers of crust should carry most of the stress that moves the entire 100-kilometer thickness of a plate. In recent years, however, some investigators have argued that the upper crust weakens so rapidly with depth that the lower crust and the mantle beneath drive plate motions. If so, surface motions, including earthquakes, would simply reflect whatever the underlying rock was doing. As Mark Zoback of Stanford University puts it: "Does the top drive the bottom or the bottom the top?"

Stress measurements in shallower wells down to 3 kilometers had showed that, at least to that depth, the strength of the upper crust increases with depth, supporting Zoback's expectations, but he saw an opportunity for a more convincing test at KTB. With colleagues at the University of Karlsruhe and the KTB project, Zoback measured the stress at the bottom of the hole when it was 6 kilometers deep—halfway through the upper crust—and found that the strength of the rock was still increasing, just as he predicted.

The KTB drillers are hoping that the hole starts defying predictions again as it approaches 8 kilometers. If not, the drilling may end prematurely. The reason: A seismic reflector at about that depth, which Ewald Lüschen and his colleagues at the University of Karlsruhe suspect could be a reservoir of fluids, might spell trouble. If the fluids are voluminous and highly pressurized, the drill rig could find them hard to contain—and the resulting delays might be enough to end the money-strapped project. KTB officials find all this highly speculative, but any sort of problem could [prove] fatal. "If all goes well, we will reach 10 kilometers in October of next year," says Kehrer. "If any larger unknown things happen, this could be the end of it." Making one more wish at his deep, deep well, he adds, "We have to have luck from now to the end." ❑

## Questions:

1. What is the deepest well drilled to date, and where was it drilled?
2. Why was graphite expected to be present?
3. The strength of the crust was increasing down to 6 kilometers. What does this suggest about plate tectonic processes?

Answers on page 301

# Part Two

# Earthquakes and Volcanoes

*On June 28, 1992, there were two earthquakes in southern California. Two months earlier there had been another earthquake in the same general area. None of these earthquakes occurred along the San Andreas Fault. They, along with some other earthquakes earlier in this century, form a line running north of the San Andreas Fault. These quakes may indicate the development of a new fault system that may eventually replace the San Andreas as the dominant fault along which the North American plate and the Pacific plate slide past each other. However, not everyone agrees with this proposal.*

# Finding Fault

by Rosie Mestel

If the two earthquakes on June 28 didn't frighten southern Californians out of their wits, they at least frightened people into supermarkets to finally get those emergency kits together. Bottled water, canned food, and crowbar sales boomed; radios and TVs boomed, too, with dire warnings that the San Andreas Fault, just to the south of the quake areas, could shortly unleash its long-awaited "big one." Ironically, this recent spate of quakes could be telling Earth scientists something else: that the mighty San Andreas may one day be an abandoned relic, replaced by a new, more active fault.

The two quakes—a magnitude 7.3 at Landers and a smaller one at Big Bear, to the west—got their energy from an ongoing scuffle between two gigantic plates scraping against each other. The San Andreas Fault marks the main border between these plates. It runs roughly north-south. To its east the North American plate carries most of the United States; to its west the Pacific plate is chugging up toward Alaska at a speed of two inches a year, dragging a skinny strip of coastal California along with it.

Stanford geophysicist Amos Nur points out that the predawn Landers temblor was directly in line with the epicenter of a magnitude 6.1 quake that went off two months earlier at nearby Joshua Tree, and with other quakes that struck earlier this century. This line of quakes runs due north from an area near Los Angeles where the San Andreas begins to kink to the west for 100 miles or so. Nur says this line may be the sign of a gradually emerging fault system that will eventually replace the San Andreas as the plate border. He argues that the sliding plates, like truck drivers, may be having a tricky time navigating the L.A. area and may have opted to avoid it altogether (as truck drivers only wish they could) by detouring north before they hit it. "Faults rotate with time," says Nur. "Eventually they get locked up because they're unfavorably oriented for the plate movements, and then new ones have to form."

In 1989, Nur first claimed something like this was happening, based on several 1970s quakes that fell on the line. "Then Joshua Tree fell on the line, and the rupture direction did, too," he says. "So suddenly this became very important. Then, while I was writing it up and trying to figure out what it all means, we had the Landers earthquake—and it fell on the line, too."

But there are problems with Nur's proposal, says Lucile Jones, a seismologist at the U.S. Geological Survey's Pasadena office. The western edge of the North American plate is being stretched out along the San Andreas, and this new crack could simply be a way of relieving some of that strain. Other cracks are being formed in the same area. Instead of continuing farther and farther north through mountains and thick continental crust, a course requiring a lot of energy, Nur's crack could just become part of a network of fissures along the edge, giving the plate a little flexibility. So instead of supplanting the San Andreas, Jones says, the new fault may more likely be an adjunct to it.

Even if the San Andreas's days are numbered, the number of days involved in creating Nur's fault is very large, and Californians should put away the champagne for a few million more years. Right now the fault is still dangerous. The Landers and Big Bear tremors ran along small faults forming two sides of a triangle resting on the San Andreas, and the triangular block of land they enclose has shifted northward, away from the fault. That effectively released a geologic brake pad, making it easier for the San Andreas to slip. What's worse, a chunk of land on a different fault section, but on the opposite side, moved 15 feet toward the San Andreas. "So we have a situation where on one part of the fault we've let off the brake," says Jones, "and on the next section we're pedaling harder."

Still, seismologists aren't certain where or when the San Andreas will next shake. "We've got a camel," says Jones, turning from bikes to vehicles better suited to the San Andreas's desert route. "And we know we've put a big straw on its back. But we don't know how strong the camel is. And we don't know how much straw it was carrying to begin with." ❑

**Questions**:

1. How large was the Landers earthquake?
2. How might these earthquakes make a large earthquake along the San Andreas more likely?
3. What geologic features in California might make the development of this new fault particularly difficult?

Answers on page 301

# 11

*Three of the largest earthquakes in U.S. history occurred in the center of the continent, far from any plate boundaries. The New Madrid, Missouri earthquakes of 1811–1812 have always been puzzling. All three quakes were estimated to have had magnitudes of 8.0 or greater. One explanation for these earthquakes is the presence of a weakness within the crust. New Madrid is located above an ancient failed rift. This feature may provide a preexisting weakness where the crust could break. A more recent explanation is that the mantle beneath New Madrid is hotter and more malleable and does not provide a strong base for the overlying crust. This would allow the crust in this region to be more easily compressed than surrounding areas.*

## Midcontinent Heat May Explain Great Quakes

by Richard Monastersky

While most North Americans think of California as "The Earthquake State," three of the largest tremors in U.S. history struck the nation's heartland near New Madrid, Mo., during the winter of 1811–1812. Why such massive jolts should rock the continent's otherwise stable center has long puzzled geologists. This week, two researchers proposed that excess heat under the New Madrid region may explain its seismic unrest, which is expected to continue in the future.

At a meeting of the American Geophysical Union in Baltimore, Lanbo Liu and Mark D. Zoback of Stanford University suggested that heat in the mantle underneath the New Madrid area has weakened this portion of the North American plate, making it more susceptible to earthquakes.

"On the basis of our calculations, the strength of the New Madrid seismic zone is much, much less than in the surrounding region," Liu says.

Most earthquakes occur along the edges of the dozen large tectonic plates that cover Earth's surface like a cracked egg shell. When two plates crash together as in the Himalayas, or when they grind past each other as in California, their margins absorb the brunt of impact, leaving the stronger interior land undeformed.

In U.S. history, large jolts have rattled only two sites within the stable eastern half of the North American plate: the New Madrid area and Charleston, S.C., which although on the coast lies several thousand kilometers from the plate boundary in the mid-Atlantic. The three New Madrid quakes had estimated strengths of magnitude 8.0 or greater; an 1886 Charleston quake had an estimated strength of 7.8. For every one-point increase in magnitude, the power of an earthquake increases 30-fold.

Geologists traditionally seek to explain such intraplate earthquakes by focusing on weaknesses within the crust. By this thinking, previous tectonic injury in a particular location would fracture the upper crust there, predisposing the plate to break again in the same spot. New Madrid, for instance, sits atop a scar formed 600 million years ago after a great rent started, but failed, to rip the North American plate in two.

Liu and Zoback took a different approach by considering the upper mantle, which forms the underside of the plate. In most of eastern North America, the upper mantle is relatively cool and stiff; it provides a strong layer that keeps the plate from breaking under the tectonic force pushing North America away from Europe. But in the New Madrid area, they suggest, the mantle is too hot and malleable. Without the support of a strong mantle, the crust in this region cannot stand up to the force, so it breaks and causes earthquakes.

According to Liu and Zoback, several lines of evidence suggest the mantle underneath New Madrid is warmer than surrounding areas. Heat coming out of the crust averages 58 milliwatts per square meter near New Madrid; heat flow values in other areas of the eastern United States average 20 percent lower.

Seismic waves passing through the lower crust underneath New Madrid move slower than elsewhere, providing another indication that the upper mantle has excess heat, Liu says.

Lastly, he cites evidence that molten rock rose up into the crust underneath the New Madrid region 40 million years ago, relatively recent by geologic standards. Heat from that volcanic episode would have lingered in the mantle even until today, he says.

The new theory would explain measurements reported last year by Liu and others, showing that the New Madrid region is being squeezed at an extremely high rate. While the plate tectonic forces push on the entire eastern United States, New Madrid compresses more than others because the lower part of the plate can't support the force, Zoback says. ❑

**Questions**:

1. Why does relatively a cool mantle provide support for the crust?
2. What evidence suggests that the mantle beneath New Madrid is hotter than the surrounding areas?
3. How much more powerful is a magnitude 8.0 earthquake than a 7.0 earthquake?

Answers on page 301

12

*In the last 5 years two earthquakes in California—Loma Prieta and Northridge—caused close to $30 billion in damage. Both of these earthquakes occurred on previously unrecognized faults. Both, however, occurred in areas known to be seismically active. This may require that we refocus our energies in regards to earthquake hazard mitigation, which consists of two parts: characterizing earthquake source areas and predicting the effects of earthquakes. We may benefit from a greater emphasis on predicting the effects of earthquakes in areas with known seismicity than in putting most of our energy into detailed characterization of the faults themselves. This change in emphasis may be required to reduce the costs in property and lives in future earthquakes; however, it must be emphasized that both parts of earthquake hazard mitigation are important.*

# Predicting Earthquake Effects—Learning from Northridge and Loma Prieta

by Thomas L. Holzer

The continental United States has been rocked by two particularly damaging earthquakes in the last 4.5 years, Loma Prieta in northern California in 1989 and Northridge in southern California in 1994. Combined losses from these two earthquakes approached $30 billion. Approximately half these losses were reimbursed by the federal government. Because large earthquakes typically overwhelm state resources and place unplanned burdens on the federal government, it is important to learn from these earthquakes how to reduce future losses. My purpose here is to explore a potential implication of the Northridge and Loma Prieta earthquakes for hazard-mitigation strategies: earth scientists should increase their efforts to map hazardous areas within urban regions.

Earth science contributes to earthquake hazard mitigation primarily by conducting two activities: (i) assessing or characterizing earthquake sources and (ii) predicting earthquake effects. The Northridge and Loma Prieta earthquakes raise a question about how completely earthquake source zones must be characterized for earthquake hazard mitigation. Could adequate mitigation be undertaken if source zones were only generally known but hazardous areas were comprehensively mapped? In economics terms, does the marginal rate of return on the investment in source-zone characterization diminish as identified seismic sources increase?

Although the specific locations of both the Northridge and Loma Prieta earthquakes were surprises in that they initiated on unidentified faults, their general locations were not. Both earthquakes occurred in structurally complex and seismically active areas. The 1987 Whittier Narrows earthquake already had alerted scientists to the earthquake potential of reverse faults that do not offset the land surface in the Los Angeles basin. Thus, only the specific location of the Northridge earthquake came as a surprise. The Loma Prieta earthquake was similar in this respect. It too occurred on a deep, unknown, reverse fault but within a region that was known to have many faults and high earthquake potential. The adjacent segment of the San Andreas fault had been recognized in an earlier government study in 1988 as having a 30% probability of producing a significant earthquake in the next 30 years (1).

Perhaps the most significant finding of either earthquake was the high level of ground shaking that was recorded in the Northridge epicentral area. Peak horizontal ground accelerations were approximatley 1.7 times values predicted by data from earlier earthquakes. Once the new data are incorporated into the ground motion database, near-source ground motion predicted by empirical methods will increase.

In contrast to the Northridge observations, ground motion during the Loma Prieta earthquake was more distinctive for its high value at unusually long distances from the epicenter. Approximately 70% of the property losses caused by the earthquake occurred in localized areas 100 km away from the epicenter (2). This is twice the distance at which significant damage normally is observed. The localization of damage was not a surprise, however. Much of the area with damage is underlain by soft clayey soils that amplify ground shaking. On maps published in 1975, these areas were identified as areas subject to elevated levels of shaking (3). Forebodingly, some of these areas suffered damage in 1865 from an earthquake that may also have been on the Loma Prieta fault plane. In addition, these areas suffered heavy damage in 1868 and 1906 from nearby earthquakes.

A mitigation strategy that emphasizes delineating hazardous areas in urban regions would encourage society to focus more directly on the effects of earthquakes and the severity of the hazard. For example, mapping soils with potential to amplify seismic waves identifies areas that will be more frequently exposed to damaging ground motions. This approach would have priority because the region is large that contains seismic sources with potential to cause damage. This point is illustrated by a fault map of the greater San Francisco Bay area. Based on the Loma Prieta earthquake, any magnitude >7 earthquake within approximately 100 km of San Francisco may cause damage in the areas underlain by soft soils. An important aspect is that the earthquake potential of many mapped faults is poorly known. In addition, some seismogenic faults may be either buried or not yet recognized at the surface.

By emphasizing hazardous areas in a mitigation strategy, uncertainties in locating specific seismic sources become less critical. San Francisco has implicitly adopted such a strategy in its retrofit program for unreinforced masonry buildings. Priority for low-interest loans in this program is given to nonconforming structures underlain by soft soils. The state of California adopted a similar philosophy following the Loma Prieta earthquake when it enacted the Seismic Hazards Mapping Act of 1990. By this legislation, cities and counties require land developers to address earthquake hazards in areas delineated by the state as subject to high levels of ground shaking, liquefaction, or seismic landslides.

This emphasis might be particularly applicable to the eastern United States where seismic source zones are poorly characterized. Despite intensive investigations in the epicentral regions of the two largest historic earthquakes in the east, the earthquake potential of these regions as well as the remainder of the eastern United States remains unknown (4). This uncertainty deters earthquake hazard mitigation in this region.

In conclusion, the Northridge and Loma Prieta earthquakes suggest the merit of increasing efforts to identify and map those parts of urban areas where shaking hazards are greatest. By delineating and evaluating these hazardous areas, we could at least focus on areas where the impact from earthquakes is likely to be greatest. These efforts would include mapping areas susceptible to near-source ground motion where source information is adequate. In addition, liquefaction or landsliding potential could be mapped where appropriate. Clearly, source characterization remains an essential part of earthquake hazard mitigation strategies. In addition, it is necessary to identify the parts of urban areas with potential for near-source ground motion. While we continue to pursue specific seismogenic faults, however, we could begin a systematic national effort to identify the parts of urban areas with the greatest potential hazard.

The Federal Flood Insurance Program provides a precedent for mapping hazardous areas. This program delineates areas subject to flooding with an average recurrence interval of 100 years. It requires parties in those areas who seek mortgages from federaliy insured institutions to purchase flood insurance. This is not to say that parties outside the mapped flood zone will not be flooded. As midwesterners painfully learned in the summer of 1993, flooding occurs outside the 100-year flood boundary. Government, however, has elected to concentrate its mitigation efforts, here an insurance program, on areas where the hazard is greatest. Similarly, maps of hazardous areas within seismically active regions would allow society to concentrate its mitigation efforts on areas at greatest peril. ❑

References and Notes


1. The working group on California earthquake probabilities, *U.S. Geol. Surv. Open-File Rep. 88-398* (1988).
2. T.L. Holzer, *Eos* **75**, 299 (1994).
3. R.D. Borcherdt, Ed., *U.S. Geol. Surv. Prof. Pap. 941-A* (1975).
4. Evidence regarding recurrence intervals is conflicting for large earthquakes in the New Madrid, Missouri, seismic zone of the Central United States, the location of three large earthquakes in 1811-1812. Some studies suggest that there were multiple large earthquakes [E.S. Schweig and M.A. Ellis, *Science* **264**, 1308 (1994)], whereas others suggest that nothing comparable to the 1811-1812 earthquakes has occurred in the last 5000 to 10,000 years [S.G. Wesnousky and L.M. Leffler, *Seismol. Res. Lett.* **63**, 343 (1992)]. Field studies near Charleston, South Carolina, site of a large earthquake in 1886, have not been able to identify the geologic structure responsible for the earthquake. Although relics of sand boils that document prehistoric strong ground shaking have been found [S. Obermeier, R. Weems, R. Jacobson, G. Gohn, *Ann. N.Y. Acad. Sci.* **558**, 183 (1989)], the likelihood of recurrence of a Charleston-type earthquake is unknown.


7. I thank R.D. Brown Jr. and R.E. Wallace for critical reviews of early drafts.

**Questions**:

1. How do high temperatures prevent lockup of two crustal plates?

2. Why is the locked-up zone much narrower in the Pacific Northwest than other subduction zones?

3. If there was an earthquake in the Pacific Northwest, where would it most likely occur? Would it be likely to cause much loss of life?

Answers on page 301

13

*Tsunamis are one of nature's most serious hazards in many parts of the world, particularly in and around the Pacific Ocean. Two features of tsunamis make them particularly hazardous: the earthquakes which cause them are often not felt by humans and they can travel at such great speed that even if the tsunami is discovered there may not be time to warn communities within its path. Recent technological advancements may help to address both of these problems. Modern seismometers are more sensitive to the lower-frequency seismic waves that are generated by tsunami-causing earthquakes, and new sensors placed on the ocean floor can distinguish tsunami waves from other waves in the ocean. These advancements may increase the likelihood of identifying tsunami-causing earthquakes and allow warning agencies sufficient time to evacuate people from potentially hazardous areas.*

# Waves of Destruction

Tsunamis have always been mysterious monsters—mountain-size waves that race invisibly across the ocean at 500 mph, drain harbors at a single gulp, and destroy coastal communities without warning. But now some researchers are trying to take the mystery away.
by Tim Folger

Like most people in Nicaragua, Chris Terry didn't feel the mild earthquake that shook the country at about 8 p.m. on September 1, 1992. He didn't notice anything out of the ordinary until some minutes later. Terry and his friend Scott Willson, both expatriate Americans, run a charger fishing business in San Juan del Sur, a sleepy village on Nicaragua's Pacific coast. On the evening of the earthquake they were aboard their boat in San Juan de Sur's harbor. "We were down below," says Terry. "We heard a slam." The sound came from the keel of their boat, which had just scraped bottom in a harbor normally more than 20 feet deep. Somehow the harbor had drained as abruptly as if someone had pulled a giant plug.

Terry and Willson didn't have much time to contemplate the novelty of a waterless harbor. Within seconds they were lifted back up by a powerful wave. "Suddenly the boat whipped around very, very fast," says Terry. "It was dark. We had no idea what happened."

The confusion was just beginning. As Willson and Terry struggled to their feet, the boat began dropping once again, this time into the trough of a large wave. Willson was the first to get out to the deck. There he found himself staring into the back side of a hill of water rushing toward the shore. "He was seeing the lights of the city through the water," says Terry. "And then the swell hit, and the lights went out, and we could hear people screaming."

One of those on the shore was Inez Ortega, the owner of a small beachfront restaurant. She hadn't noticed the earthquake either. While preparing dinner she glanced out at the harbor and noticed that the water seemed unusually low. "I didn't pay much attention at the time," she says. But when she looked up again a swell of water at least five feet high was racing up the beach toward her restaurant.

"I started running, but I didn't even get out of the restaurant when the wave hit," she says. Ortega and several of her customers spent about half an hour swimming in the debris-filled stew before they managed to drag themselves out of the water.

Ortega and everyone else in San Juan de Sur looked about themselves in stunned silence. The waves had swept away restaurants and bars lining the beach, as well as homes and cars—and people—hundreds of yards inland. Terry and Willson managed to ride out the disaster on their boat. Still reeling, they witnessed the receding wake of the last wave.

"When the wave came back out, it was like being in a blender," says Terry. Collapsed homes bobbed in the water around their boat.

Terry, Willson, and Ortega had survived a tsunami, a devastating wave triggered by an undersea earthquake. Although the waves that hit San Juan del Sur were extremely powerful, they rose only 5 to 6 feet high. Other parts of Nicaragua weren't so lucky. All told, the offshore earthquake sent tsunamis crashing along a 200-mile stretch of the coast and newspapers reported 65-foot waves in some places (though seismologists consider that figure unlikely; a more realistic wave height might be about 30 feet). The waves killed about 170 people, mostly children who were sleeping when the waves came. More than 13,000 Nicaraguans were left homeless.

Destructive tsunamis strike somewhere in the world an average of once a year. But the period from September 1992, the time of the Nicaraguan tsunami, through last July was unusually grim, with three major tsunamis. In December 1992 an earthquake off Flores Island in Indonesia hurled deadly waves against the shore, killing more than 1,000 people. Entire villages washed out to sea. And in July 1993 an earthquake in the Sea of Japan generated one of the largest tsunamis ever to hit Japan, with waves washing over areas 97 feet above sea level; 120 people drowned or were crushed to death.

In Japanese tsunami literally means "harbor wave." In English the phenomenon is often called a tidal wave, but in truth tsunamis have nothing to do with the tame cycle of tides. While volcanic eruptions and undersea landslides can launch tsunamis, earthquakes are responsible for most of them. And most tsunami-spawning earthquakes occur around the Pacific rim in areas geologists call subduction zones, where the dense crust of the ocean floor dives beneath the edge of the lighter continental crust and sinks down into Earth's mantle. The west coasts of North and South America and the coasts of Japan, East Asia, and many Pacific island chains border subduction zones. There is also a subduction zone in the Caribbean, and tsunamis have occurred there, but the Atlantic is seismically quiet compared with the restless Pacific.

More often than not, the ocean crust does not go gentle into that good mantle. As it descends, typically at a rate of a few inches a year, an oceanic plate can snag like a Velcro strip against the overlying continent. Strain builds, sometimes for centuries, until finally the plates spasmodically jerk free in an earthquake. As the two crustal plates lumber past each other into a new locked embrace, they sometimes permanently raise or lower parts of the seafloor above. A 1960 earthquake off Chile, for example, took only minutes to elevate a California-size chunk of real estate by about 30 feet. In some earthquakes, one stretch of the sea bottom may rise while an adjoining piece drops. Generally, only earthquakes that directly raise or lower the seafloor cause tsunamis. Along other types of faults—for example, the San Andreas, which runs under California and into the ocean—crustal plates don't move up and down but instead scrape horizontally past each other, usually without ruffling the ocean.

Seismologists believe the sudden change in the seafloor terrain is what triggers a tsunami. When the seafloor rapidly sinks—or jumps—during an earthquake, it lowers (or raises) an enormous mountain of water, stretching from the seafloor all the way to the surface. "Whatever happens on the seafloor is reflected on the surface," says Eddie Bernard, an oceanographer with the National Oceanic and Atmospheric Administration (NOAA). "So if you imagine the kind of deformation where a portion of the ocean floor is uplifted and a portion subsides, then you'd have—on the ocean surface—a hump and a valley of water simultaneously, because the water follows the seafloor changes."

One major difference between the seafloor and the ocean surface, however, is that when the seafloor shifts, it stays put, at least until the next earthquake. But the mound of water thrust above normal sea level quickly succumbs to the downward pull of gravity. The vast swell, which may cover up to 10,000 square miles depending on the area uplifted on the ocean floor, collapses. Then the water all around the sinking mound gets pushed up, just as a balloon bulges out around a point where it's pressed. This alternating swell and collapse spreads out in concentric rings, like the ripples in a pond disturbed by a tossed stone.

Although you might think a tsunami spreading across the ocean would be about as inconspicuous as a tarantula walking on your pillow, the wave is, in fact, essentially invisible in deep ocean water. On the open sea, a tsunami might be only ten feet high, while its wavelength—the distance from one tsunami crest to another—can be up to 600 miles. The tsunami slopes very gently, becoming steeper only by an inch or so every mile. The waves so feared on land are at sea much flatter than the most innocuous bunny-run ski slope; they wouldn't disturb a cruise ship's shuffleboard game. Normal surface waves hide tsunamis. But that placid surface belies the power surging through the water. Unlike wind-driven waves, which wrinkle only the upper few feet of the ocean, a tsunami extends for thousands of fathoms, all the way to the ocean bottom.

Tsunamis and surface waves differ in another crucial respect: tsunamis can cross oceans, traveling for thousands of miles without dissipating, whereas normal waves run out of steam after a few miles at most. Tsunamis are so persistent that they can reverberate through an ocean for days, bouncing back and forth between continents. The 1960 Chilean earthquake created tsunamis that registered on tide gauges around the Pacific for more than a week.

"You've got to remember how much energy is involved here," says Bernard. "Look at the size of these earthquakes. The generating mechanism is like a huge number of atomic bombs going off simultaneously, and a good portion of that energy is transferred into the water column."

The reason for tsunamis' remarkable endurance lies in their unusually long wavelengths—a reflection of the vast quantity of water set in motion. Normal surface waves typically crest every few feet and move up and down every few seconds. Spanning an ocean thus involves millions of wavelengths. In a tsunami, on the other hand, each watery surge and collapse occurs over perhaps 100 miles in a matter of minutes. For a large subduction-zone earthquake—magnitude 8 or more—the earthquake's impulse can be powerful enough to send tsunamis traveling across the Pacific—from the Chilean coast to Japan, Australia, Alaska, and all the islands en route as well.

For much the same reason, tsunamis can race through the ocean at jetliner speeds—typically 500 miles an hour. To span a sea, they need to travel a distance equal to just a few dozen of their own wavelengths, a few swells and collapses. That means the wave only has to rise and fall a handful of times before the surge reaches its destination. The outsize scale of a tsunami makes an ocean seem like a pond.

As a tsunami speeds on its covert way, undersea mountains and valleys may alter its course. During the 1992 Indonesian earthquake, villages on the south side of Babi Island were the hardest hit, even though the source of the tsunami was to the north of the island. Seismologists believe that the underwater terrain sluiced the tsunami around and back toward the island's south coast.

Only when a tsunami nears land does it reveal its true, terrible nature. When the wave reaches the shallow water above a continental shelf, friction with the shelf slows the front of the wave. As the tsunami approaches shore, the trailing waves in the train pile onto the waves in front of them, like a rug crumpled against a wall. The resulting wave may rear up to 30 feet before hitting the shore. Although greatly slowed, a tsunami still burst onto land at freeway speeds, with enough momentum to flatten buildings and trees and to carry ships miles inland. For every five-foot stretch of coastline, a large tsunami can deliver more than 100,000 tons of water. Chances are if you are close enough to see a tsunami, you won't be able to outrun it.

As Inez Ortega and Chris Terry witnessed in San Juan del Sur, the first sign of a tsunami's approach is often not an immense wave but the sudden emptying of a harbor. This strange phenomenon results from the tremendous magnification of normal wave motion. In most waves, the water within the crest is actually moving in a circular path; a wave is like a wheel rolling toward the shore, with only the top half of the wheel visible. When that wave is 100 miles long, the water in the crest moves in long, squashed ellipses rather than in circles. Near the front and bottom of the wave, water is actually on the part of the elliptical "wheel" moving backward—toward the wave and out to sea. If you've ever floated in front of a wave, you've probably felt the pull of the wave as water sloshes back toward the crest. With a tsunami, that seaward pull reaches out over tens of miles, sometimes with tragic results: when an earthquake and tsunami struck Lisbon in 1755, exposing the bottom of the city's harbor, the bizarre sight drew curious crowds who drowned when the tsunami rushed in a few minutes later. Many people died in the same way when a tsunami hit Hawaii in 1946.

Although seismologists and oceanographers understand in broad terms how tsunamis form and speed across oceans, they are still grappling with some nagging fundamental questions. One of the major mysteries is why sometimes relatively small earthquakes generate outlandishly large waves. Such deceptive earthquakes can be particularly devastating because they may be ignored by civil agencies that are charged with issuing tsunami warnings.

The Nicaraguan earthquake is a case in point. By conventional measures, it shouldn't have produced a tsunami at all. The earthquake registered magnitude 7.0 on the Richter scale, not puny by any means, but not large enough, seismologists believed, to pose much of a tsunami risk. The quake's epicenter was 60 miles offshore, distant enough to dampen the tremors on land. Yet people who had not even felt the quake found themselves swept out to sea minutes later.

Hiroo Kanamori, a seismologist at Caltech, has made a point of studying the earthquakes that spring these unexpected tsunamis. Such earthquakes, he says, are responsible for some of the most damaging tsunamis on record. In 1896, for example, an earthquake in Japan was followed by a tsunami that drowned 22,000 people, even though survivors reported only mild shaking before the wave. And a relatively moderate 1946 quake in the Aleutian Islands sent a huge tsunami tearing across the north Pacific and into Hawaii. where it inundated much of the city of Hilo.

Twenty years ago Kanamori proposed an explanation for these surprise tsunamis. But to test his ideas he needed seismometers sensitive to a broad spectrum of ground movement, and the instruments of the 1970s just weren't up to the job. Only in the past few years, in fact, have seismometers become sophisticated enough for his purposes. And the 1992 Nicaraguan tsunami proved an ideal test case.

Kanamori thinks some earthquakes may release their energy very slowly, over a minute or more, rather than in a brief, spastic lurch. This could happen, he says, if soft ocean sediments were sandwiched between two interlocked crustal plates. The lubricated plates would slide past each other smoothly, without sharp, building-shaking convulsions. "If you have two blocks of hard rock," says Kanamori, "usually the friction between them is very high, so you can accumulate large amounts of stress. And when it slips, it slips very fast. But if you have lubricating materials in between, it can slip at relatively low stress, and when it slips, it goes slowly."

The seismic energy from such a quake moves Earth's surface in long, slow undulations. Humans don't feel them. We notice only the shorter, sharper shivers from earthquakes, the type that rattle foundations. Nevertheless, the slow "tsunami earthquakes," as Kanamori calls them, can be as quietly dangerous as a pristine hillside of snow on the verge of an avalanche. The Nicaraguan quake, he estimates, moved a 120-mile-long section of the seafloor more than three feet in about two minutes when part of the Cocos plate, a wedge-shaped slice of the Pacific ocean floor, slid under North America. This slow-motion shift, imperceptible to humans, sent the tsunami on its destructive way.

When Kanamori first suggested this model, seismometers weren't able to reveal the true nature of these long-period vibrations. Part of the problem was that most seismometers depended ultimately on a pendulum to trace earthquake-generated movements onto a graph. The pendulum typically hangs by a spring from a supporting arm. Nestling right up against it is a recording drum—essentially a revolving scroll of paper—anchored firmly to the ground. During an earthquake, the recording drum bumps up and down. But because the pendulum is suspended in midair by a spring, its inertia makes it lag behind the shaking drum. In a sense, it floats freely, while the earth (and drum) jiggles underneath it. A pen attached to the pendulum scribbles a jagged portrait of the ground motion on the drum's paper.

In these older seismometers, the long-period waves that Kanamori was looking for tended to get buried under the mountain of squiggles left by the far more numerous short-period waves. Newer instruments also rely on pendulums, but their coils are suspended in magnetic fields and attached to electronic recording systems rather than rolls of paper. When the seismometer shakes during an earthquake, the motion of the coil generates a current that can be analyzed digitally by a computer, revealing even subtle patterns such as the long-wavelength signatures typical of slow tsunami earthquakes.

The newer technology allowed Kanamori to study the deceptive Nicaraguan quake in all its complexity. Any earthquake, large or small, releases a symphony of shock waves. Some of these waves, particularly the jarring, high-frequency tremors that cause most of the damage near an earthquake's epicenter, die out quickly and don't travel far through the earth. In the Nicaraguan quake, these tremors never made it to the mainland with enough force to be felt on shore. Lower-frequency waves, however, resonate strongly in the planet's interior, lingering like the deep boom from a bass drum in a concert hall. These tremors cyclically raise and lower the ground about every 20 seconds or more. Seismometers in Nicaragua and elsewhere in the world were able to pick up some of these signals, but not all of them.

As it turns out, the conventional measure of earthquake magnitude—the Richter scale—isn't an accurate gauge of all an earthquake's overtones. "In the 1930s," explains Northwestern University seismologist Emile Okal, "when Charles Richter designed the scale, he had instruments that recorded 20-second-long seismic waves very well but that didn't record waves with much longer periods." As a consequence, standard seismic measuring devices may underestimate the size of an earthquake. The values on the Richter scale are derived primarily from a ratio of two numbers: the amount of ground-surface motion during an earthquake—as recorded on a seismometer, either electronically or on paper—and the number of vibrations per second. Simply put, lots of movement in a short time span means a big earthquake. Some earthquakes, however, like the one in Nicaragua, may release their energy very slowly, generating signals with periods longer than 20 seconds.

Last year, in the British scientific journal *Nature*, Kanamori argued that the magnitude 7.0 assigned to the Nicaraguan quake was far too low. By including the longer, slower seismic signals in his calculations, Kanamori upped the quake's magnitude to 7.6. Since an increase of one on the scale corresponds to a tenfold increase in the size of an earthquake, the Nicaraguan quake, by Kanamori's calculations, was five times bigger than originally estimated.

Emile Okal often does research at a French seismological station in Tahiti, one of a handful of stations in the world equipped with the technology to detect the longer signals of slow earthquakes. The Tahiti station, says Okal, received signals from the Nicaraguan quake even though it was more than 4,000 miles to the northeast, and immediately recognized the potential for disaster.

"The only problem," says Okal, "was that by the time we picked up the signal, the Nicaraguan coast had been totally ravaged." It took about 25 minutes for the signal from the quake to travel through Earth's crust to Tahiti, Okal explains. And although seismic waves travel 15 to 20 times faster than tsunamis, this tsunami had to cover only 40 miles before hitting the Nicaraguan coast.

Unfortunately, even the network of seismological stations Nicaragua did have had greatly deteriorated during the country's civil war. Had more modern seismometers been available, and had researchers been aware of the importance of the long-period seismic waves, says Kanamori, there might have been enough warning time to tell people to run for high ground. Kanamori hopes that more countries—especially those on the earthquake-prone Pacific rim—will invest in technology capable of detecting a wider range of earthquake tremors. "No matter what we do, tsunamis are going to happen," he says. "The question is whether we can have a very effective tsunami warning system."

The United States has two tsunami warning centers, one near Honolulu and another in Palmer, Alaska, just north of Anchorage. Both were built in the 1960s after tsunamis from two large earthquakes, one in Alaska in 1964 and the other in Chile in 1960, caused millions of dollars' worth of damage in the two states. The tsunami from the 1964 earthquake also hit Crescent City in northern California, killing 11. The warning centers, staffed around the clock, collect data via satellite from dozens of seismological stations in more than 20 countries bordering the Pacific.

For earthquakes above about magnitude 6.5, says Michael Blackford, a seismologist at the Hawaii warning center, the center alerts the warning systems in other countries around the Pacific. If the earthquake is far away—say in Chile, Japan, or Alaska—the seismic signal, pulsing rapidly through Earth's crust, will arrive in Hawaii hours ahead of any potential tsunami, giving the center time to alert the state civil defense via a hot line.

In addition to seismographic data, the warning systems also receive readings from tide gauges scattered in harbors in Alaska, Hawaii, and Pacific-rim countries: if, accompanying an earthquake, a harbor's water level drops a few feet in a few minutes, the quake may have triggered a tsunami. While that information would be too late to help local residents, more distant communities could be forewarned.

Even with seismographic and tide-gauge data, tsunami prediction is a hit-or-miss proposition. False alarms outnumber real tsunamis by more than two to one, and this can cause problems besides simply jading the public. In 1986, after a magnitude 7.7 earthquake in Alaska, the Hawaii center issued a tsunami warning. Television, radio, and even air-raid sirens were used to get people to evacuate coastal areas. The exercise cost some $20 million. Yet the tsunami that arrived was barely three feet high.

"I wasn't here at the time," says Blackford, "but I got a lot of feedback from people who were. It's well remembered. The tsunami arrival time was to be in the late afternoon, so the civil defense just blew the sirens. The word was, everybody should evacuate from the coastal areas. They closed offices and turned everybody loose. This resulted in virtual gridlock on the roads in Oahu. Some of the roads are right next to the sea. People have told me about this chaotic situation: you're sitting in your car with the ocean lapping alongside you there, and you're wondering, 'Why am I here? I'm supposed to be away from the ocean and I'm stuck in this traffic jam.'"

Part of the reason it's so hard to issue reliable warnings lies in the inherent difficulty of detecting a tsunami barreling invisibly across the ocean. That's a problem now being addressed by Eddie Bernard and Frank González, oceanographers with NOAA's Pacific Marine Environmental Laboratory in Seattle. For the past few years González and Bernard have been working on developing a deep-sea tsunami sensor. Six of their devices now rest on the ocean floor, one 300 miles directly west of the Oregon-Washington border, one 30 miles southwest of Hawaii, and four more spread out a few hundred miles south of the Aleutians. The instruments are remarkably sensitive. "In 12,000 or 15,000 feet of water they are capable of sensing a change in sea level of less than a millimeter," González says.

These sensors measure the weight of the water column above them. When a wave passes across the surface, it increases the height and therefore the weight of the water column above the sensor. As a trough comes by, the height and weight of the water column decrease. The sensor consists of a small metal tube about four inches long, which floats just above the ocean bottom; it is held in place by an anchor (usually an old iron railroad wheel). Partially enclosed within the metal tube is a small device—called a Bourdon tube—shaped like a comma with a very long tail. The end of the tail sticks out of the bottom of the metal tube, exposed to the ocean, and is open like a straw. The other end of the Bourdon tube is closed. When a wave passes overhead, increasing the weight of the water column, the pressure slightly straightens the Bourdon tube in the same way a paper noisemaker unfurls when you blow into it. When a trough comes by, the tube curls up again. As the Bourdon tube alternately straightens and curls, it pushes and pulls on a sensitive quartz crystal, which in turn produces an electric signal that varies along with the changes in pressure of the water above.

The pressure sensors remain on the bottom for 12 months at a time, storing data electronically every 15 seconds. To recover the sensors—and their data—González and Bernard need to send a ship to the ocean site above the sensor. When the ship reaches the site, it broadcasts a signal telling the sensor to release its anchor. The entire package (which includes an orange marker buoy and a signal transmitter) then floats to the surface.

Tsunamis are easy to spot in the sensors' records, says González. "Tsunamis typically have periods of anywhere from 3 to 30 minutes or so," he says. "On the other hand, tides and signals generated by storms have periods on the order of hours and tens of hours." (Passing waves or ships generate pressure changes too small to reach the ocean floor.)

For now the sensors aren't tied into any warning system. But with very minor changes, says González, they could be. Instead of being picked up periodically by a ship, they would have to transmit data every few seconds to a receiver on a surface buoy. If a tsunami rolled by, the buoy would relay the steady, maybe half-hour-long increase in pressure to a satellite or directly to a land station, where researchers could quickly work out the wave's size and heading. Sensors between Hawaii and Alaska could warn Hawaiians several hours in advance of an Alaska-born tsunami heading their way.

"We have experience in all of these components," says González. "It's just a matter of putting them together to get them to do what we want in this specific application."

NOAA has proposed setting up seven such stations, four off Alaska and three off the northwest coast. Altogether the stations would cost about $700,000 to install and another $250,000 a year to maintain, according to Bernard.

One of the reasons Bernard is eager to push ahead with a warning system is that he worries a tsunami could catch the West Coast by surprise. Hidden beneath the waves about 50 miles off the northwest coast is the Cascadia subduction zone, where the Pacific floor plunges under North America. The zone stretches from Vancouver Island to northern California, just a few hundred miles northwest of San Francisco. Many seismologists fear that it's just a matter of time before an earthquake convulses the area.

Although no major earthquake or tsunami has battered the Northwest in historical times, some geologists believe they have found evidence of past tsunamis in the region. Sand deposits resembling the debris left by modern tsunamis have been found in a number of places along the northwest coast and appear to have been laid down about 300 years ago, before Europeans settled there. Native Americans in the area, moreover, have legends of great floods from the sea that sound eerily like tsunamis.

At a workshop in Sacramento last spring, Bernard met with a number of seismologists to discuss the risk of a Cascadia zone earthquake and tsunami. "We all talked for an entire day about what is the most probable earthquake, not the worst case, but what is the most probable earthquake you could expect from this area. From those discussions came a scenario earthquake of a magnitude of about 8.4, and its fault dimensions are about 140 miles long by about 50 miles wide. Using this as a basis, we set out to model what tsunami could result from that size earthquake."

Based on case studies of tsunamis generated by earthquakes of a similar magnitude, Bernard and his colleagues estimate that the earthquake would spawn a tsunami that might be 30 feet high when it hit the coast. A tsunami that size would threaten coastal cities in the Northwest, northern California, and Hawaii. But Bernard cautions that there is no way of predicting whether such a disaster will happen next year or in 300 years.

Does this mean that a tsunami could come rolling through the Golden Gate, drowning Alcatraz, Sausalito, and Fisherman's Wharf? Probably not. Most seismologists speculate that the headlands outside the Golden Gate would take the brunt of a tsunami's wrath. Others admit the very remote possibility that a tsunami could funnel right into San Francisco Bay. No one really knows.

"I want you to understand that this is a fairly speculative business. But we do have a public safety issue at hand. We have to play the 'what if' game," says Bernard. "Unlike Alaska and Hawaii, where we have warning systems in place to respond in five minutes, we have no such facility on the West Coast."

It's been 30 years since the United States last suffered through a major tsunami, and some seismologists think the lull has fostered a dangerous, false sense of security. Another tsunami assault is inevitable, seismologists say. And vulnerable coastal areas, like much of Hawaii—where development has skyrocketed since the last tsunami—may be especially hard hit.

Like others, Bernard hopes that NOAA's proposed warning system will be in place before the next big tsunami breaks on Pacific shores, but he's not oeptimistic. "You see, there's a trend here," he says. "We always build a warning center after a big tsunami. We built one in Hawaii after the 1960 event. We built one in Alaska after the 1964 event. *After,* I want to underline the word *after*. The question I have is, Are we going to wait until after the Cascadia subduction zone earthquake to build one in California and Oregon, or are we going to do it in advance?" ❑

**Questions**:

1. How does movement occur along faults during tsunami-causing earthquakes?

2. Are tsunami waves noticeable in the open ocean?

3. Wind-driven waves dissipate their energy fairly quickly, whereas tsunami-generated waves dissipate their energy very slowly. For how long can a tsunami generated-wave continue to travel through the ocean?

Answers on page 301

# 14

*The "Richter scale" is a household phrase that is often used by the news media to convey information on the size of earthquakes. Prior to this scale, developed in the 1930s, there was no way to accurately compare the absolute size of earthquakes. Richter devised his scale based on a recording device, located about 100 kilometers from a quake's epicenter. However, seismologists now realize that there are many serious problems with the Richter scale. These include: modern instruments that do not respond to seismic waves in the same way as Richter's original device, Richter's scale was calibrated for the California crustal rocks, the logarithmic scale is hard for many people to understand. As a result, the Richter scale is being modified, and some scientists want to abandon it.*

## Abandoning Richter

How a white lie finally caught up with seismologists

by Richard Monastersky

Seismologists, as a rule, tend to keep their cool, even when the ground heaves beneath their feet and buildings collapse around them. But these days, earthquake experts are scurrying for cover at the mere mention of two words. Such is the fallout over use of the term "Richter scale"—a household phrase that lies at the heart of a brewing controversy about conveying earthquake information to the public.

The rhetoric has reached such a pitch that one newspaper columnist pilloried the agency in charge of disseminating earthquake information. "A kick in the butt is what someone ought to give the U.S. Geological Survey for its dithering about how to define the magnitude of earthquakes," wrote Keay Davidson in the San Francisco Examiner following a large Bolivian tremor on June 8.

Davidson is by no means alone. Reporters, editors, and many seismically sensitive members of the public are having trouble sorting out how scientists measure earthquakes. Much to their dismay, people are learning that seismologists typically do not use the Richter scale to judge quake size.

It's almost like hearing that Santa Claus doesn't exist.

"What's going on is that we're just recovering from decades of telling a white lie, that's all," says seismologist Thomas H. Heaton half in jest. Heaton is president of the Seismological Society of America and a USGS researcher in Pasadena, Calif.

In one sense, the flap boils down to semantics. While seismologists generally do not use the original Richter magnitude scale, the measuring systems currently in vogue represent extensions of the type that Charles Richter developed nearly 60 years ago. That explains why some seismologists continue to use the term when addressing the press.

But the recent brouhaha goes beyond the question of the name itself. According to seismologists who frequently get up in front of the television cameras, the problems now surfacing reflect a deep-seated misunderstanding about earthquakes—one that has important consequences for how the public and even engineers respond to seismic hazards.

"The public gets extremely confused after they've been through a heavy shake, and they're frightened," says Heaton. "Then you say, 'Oh, by the way, we're expecting an 8 and that is 50 times bigger.' What they now imagine is 50 times the intensity of the ground motion they just felt, and they realize that nothing can survive it. At that point, they just stop talking about it."

It's only fitting that Davidson and other reporters feel such a strong connection to the Richter scale, because journalists played an important role in its origin. "[Richter] introduced it because he was tired of the newsman asking him about the relative size of earthquakes," recalls veteran seismologist Bruce A. Bolt from the University of California, Berkeley.

Prior to Richter's work, researchers in the United States had no way of judging an earthquake's absolute size, which remains the same no matter where it is measured. Instead, they dealt with a concept called intensity, which describes the strength of shaking at a particular location. Because tremors fade with distance from the epicenter, the intensity of a single

***Reprinted with permission from SCIENCE NEWS, the weekly newsmagazine of sicence, October 15, 1994, v. 146, no 16, pp. 250-252, copyright 1994 sy Science Service, Inc.***

quake varies considerably from point to point.

In the early 1930s, Japanese seismologist Kiyoo Wadati devised a method of comparing the sizes of quakes. He would take seismic recordings of various shocks and set them on an equal footing by factoring in the distance between the recording station and the earthquake.

But this method was not easily grasped by lay people, especially the reporters of quake-plagued southern California. In 1935, Richter dressed up the Japanese method to create an earthquake index—a simple numerical scale much like the stellar magnitudes used by his astronomical colleagues at the California Institute of Technology in Pasadena. Richter defined seismic magnitude in terms of a particular type of recording device, called a Wood-Anderson seismograph, situated at a standard distance of 100 kilometers from an earthquake's epicenter.

Richter also appropriated from astronomy the idea of a logarithmic scale—based on powers of 10—to accommodate the incredible range of earthquake sizes. (The smallest detectable tremors equal the energy of a brick dropped off a table, while monster quakes surpass the largest nuclear explosions.) By Richter's original definition, a shake of magnitude 1.0 would cause the arm of the Wood-Anderson machine to swing one-thousandth of a millimeter. A magnitude 2.0 temblor would make the arm swing 10 times as much, or one-hundredth of a millimeter.

In theory, the scale had no upper limit. But in practice, magnitudes could not top 7.0. "You would never see an earthquake bigger than magnitude 7 [on the original magnitude scale], or at least we hope you never would because everything would be dead," Heaton says.

Of course, scientists rarely had a Wood-Anderson seismograph stationed exactly 100 kilometers from an earthquake. But by comparing the arrival of slow versus fast seismic waves at a recording station, they could calculate what one of the devices would have detected at the standard distance.

The magnitude index, as originally defined, could only measure southern California earthquakes because Richter calibrated the scale for the crust there. What's more, it only worked for jolts within a few hundred kilometers of a Wood-Anderson seismometer.

Recognizing these limitations, Caltech's Beno Gutenberg and Richter devised a more general magnitude measurement to handle distant earthquakes. To avoid confusion, they denoted the new magnitude $M_s$, because it depended on measurements of surface waves rippling through Earth's crust with a period of about 20 seconds. The original magnitude scale—based on waves with periods of 0.1 to 3.0 seconds—became knows as $M_L$, or local magnitude.

Even the new and improved magnitude formula had problems, however, because deep earthquakes do not produce many surface waves. So Gutenberg and Richter invented $m_b$, measured from body waves, which travel through the planet's interior. This yardstick proved helpful in distinguishing nuclear explosions from actual earthquakes.

In the 1970s, seismologists realized that all existing magnitude methods underestimated the energy of truly large earthquakes. To circumvent this limitation, Hiroo Kanamori, a successor of Richter and Gutenberg at Caltech, created a magnitude scale, $M_w$, that quantifies the total amount of seismic wave energy released in an earthquake.

But because such calculations are difficult, scientists usually approximate the energy by computing a quantity called "seismic moment," determined from long-period vibrations. In the case of great earthquakes, these vibrations have cycles longer than 200 seconds. Seismologists therefore refer to $M_w$ as the moment magnitude.

$M_w$ differs from all other types of magnitude in that it measures the earthquake source, Kanamori says. The Richter magnitude and most others gauge only the strength of vibrations sensed at Earth's surface. But to calculate moment magnitude, seismologists use the long-period waves to decipher the dimensions of the fault rupture that produced the quake.

In other words, moment magnitude measures the cause rather than the effect.

Although researchers have developed more than a dozen other ways of calculating earthquake magnitude, moment magnitude remains the figure of choice among seismologists, especially for earthquakes larger than magnitude 6.5.

Confused?

With $M_L$, $M_S$, $m_b$, $M_w$ and a litany of other Ms floating around, it's no wonder that many seismologists took the easy way out over the years by giving reporters what they thought the media wanted. When pressed for details, researchers typically simplified the issue by calling any magnitude a Richter magnitude, even though this term applies only to the local magnitudes determined by Richter's original formulation.

"The problem is that seismologists have used the term 'Richter scale' in a very loose way, and now it's catching up with them. We didn't use it among ourselves because it doesn't mean anything," Heaton says.

These days, seismologists hope to clean up the magnitude morass in their dealings with the public. The USGS put out a statement in July explaining how the newer measurements do not renounce the Richter scale but rather extend the original magnitude both to greater distances and to larger earthquakes.

At the USGS' National Earthquake Information Center in Golden, Colo., director Waverly J. Person says his staff balances the need for timelines with the desire to report moment magnitudes, which take an hour or two to compute.

Immediately after an earthquake, the center releases a preliminary measurement, which could be a surface wave magnitude, a body wave magnitude, or even a local magnitude (similar to Richter's original formulation except that modern seismographs have replaced Wood-Anderson ones.) After determining the moment magnitude, they release this number, which may fall above or below the preliminary one.

As for the use of the term "Richter scale," the USGS has dodged any decision. "The question of labeling these magnitudes as 'Richter scale' is a matter of tradition, semantics, and personal perspective. The USGS has no official scientific position on the use of the term," declares the July statement.

The USGS' Heaton, who works across the street from Richter's old Pasadena office, says he wants to avoid the term entirely. "You probably wouldn't catch us using the term 'Richter magnitude' around here, even though this was the home of Richter."

Other seismologists note that while the public feels comfortable with the term, they often lack even a basic understanding of what it means. Several scientists tell tales of people asking to see the Richter scale.

"It seems to be a popular misconception that it's actually a piece of equipment, like a bathroom scale," says Roger Musson of the British Geological Survey in Edinburgh. "Things have come to such a pass in today's press that I had an inquiry recently from the Sunday Times, no less, asking for a picture of the Richter scale. I said this was a bit like asking for a picture of kilometers."

Others describe the wild rumors that circulate after an earthquake. In the case of the Jan. 17 Northridge, Calif., jolt, the reports of different earthquake magnitudes—$M_s$ 6.6 versus $M_w$ 6.7—confused many Angelenos, prompting speculation that the USGS was underestimating the magnitude to save the federal government from spending disaster relief money.

"People thought we were lying on the magnitude," says an incredulous Heaton.

As journalists get more seismically sophisticated, they may head off some of the confusion. The Associated Press recently retired the term "Richter scale" in favor of the phrases "preliminary magnitude" and "moment magnitude." Unless further elaboration is required SCIENCE NEWS will continue its tradition of using the generic word "magnitude," which in the case of recent earthquakes refers to a determination of moment magnitude.

But simply tidying the terminology will not, on its own, help people better understand the size of an earthquake. After all, how can one number convey the power of something equivalent to a colossal nuclear explosion?

Even moment magnitude does not suffice, says its inventor. "The problem is everyone thinks that a single number determines everything. It's almost like asking how big you are," says Kanamori. "The question is whether you are asking height, weight, or width. Depending on how you measure a person, the answer can be very different. In the case of earthquakes, it's even more complex."

# Sizing up seismicity

When Charles Richter invented the concept of seismic magnitude, he made it easy to compare earthquakes. Anyone who can count to 10 will recognize that a magnitude 7.0 shock packs a bigger punch than a 6.0 quake. But the question "How much bigger?" is not so easily answered.

In the original definition of magnitude, a 1-point increase meant that peak waves recorded by a Wood-Anderson seismometer jumped by a factor of 10. So far, so good. But not all seismometers respond to seismic waves equally. Some measure different frequencies, and some are more sensitive than others. So newer instruments do not respond the way Wood-Andersons did in their era.

Delving even deeper, what does the seismometer measure anyway? It doesn't translate directly into the strength of the shaking felt by humans or buildings, because seismometers measure one band of frequencies whereas we feel a different range of waves.

An increase of one unit in magnitude therefore does not translate cleanly to 10 times more shaking. In fact, the force of the ground motion close to a tremor's epicenter rises much less than a factor of 10.

Going from a magnitude 6.5 to 7.5 jolt, the jerky shaking close to the quake may increase in strength only by a factor of 1.5 (equal to a 50 percent boost). On the other hand, a seismograph stationed halfway around the globe may measure a 10-fold difference in the surface waves that have managed to travel that far.

In terms of energy, magnitude units rise even faster. A step of one full unit increases the energy by roughly 33 times, so a magnitude 7.0 quake unleashes approximately 1,000 times the energy released by a magnitude 5.0 temblor.

Is there an easier way? Some feel that the logarithmic magnitude scale is just too difficult for the public to comprehend. "It's made a lot of confusion," says Thomas H. Heaton, president of the Seismological Society of America. "To be honest, I think Richter did us a disservice. We spend as much time explaining to the public what a logarithm is as anything else. We could have just given them a number in the first place and not bothered with a logarithm. Why not just say 1, 10, and 10 million."

Seismologists themselves compare earthquakes using seismic moments, which represent the length of the fault rupture multiplied by the amount of rock movement and then again by the stiffness of the rock. But moments are expressed in unwieldy numbers, such as $2 \times 10^{27}$ newton meters—clearly not an appealing figure for the public.

Pat Jorgenson, a USGS spokeswoman in Menlo Park, California, says she would prefer to discuss quakes in terms of something people can comprehend. "When the comet hit Jupiter this summer, it was reported that this was equivalent to so many atomic bombs. Why can't we report earthquakes like that?"

In that vein, a magnitude 1.0 earthquake would equal roughly 6 ounces of TNT. For a magnitude 5.0, think of 1,000 tons of TNT. A quake of magnitude 7.2 corresponds to a million tons of explosive—which is a little less than the energy locked in the swirling winds of a typical hurricane. The largest recorded earthquake, of moment magnitude 9.5 in Chile in 1960, equaled about 3 billion tons of TNT. ❑

**Questions**:

1. How does $M_w$ differ from all other types of magnitude?

2. Does the USGS have an official position on the use of the term "Richter scale"?

3. Moment magnitude measures the ____ rather than the ______.

Answers on page 301

15

*Our record of volcanic eruptions is fairly complete only for the last 100 to 200 years. However, even this record is often biased by the distribution of human populations and world events. Earlier records of volcanism are even less complete. Recent studies of volcanism have included many sources of information from archeological digs to ice cores. Ice cores have provided some particularly useful data including sulfuric acid concentrations and volcanic glass fragments. The information from these ice cores can be correlated with other information on past climates to help us interpret the relationship between volcanic eruptions and climates in the past.*

# Distant Effects of Volcanism—How Big and How Often?

by Tom Simkin

What do an archaeological dig, a leaf imprint, a stratospheric circulation model, and a Greenland ice core have to do with a large volcanic eruption in ancient Japan? Volcanologists studying the eruption's deposits can measure its size, but its date may be known only to the nearest century. Archaeologists digging at the base of the ash, however, find leaves that accurately identify the eruption's season, and that is just what the meteorologist needs to assess the likelihood that its fine aerosol could have been transported to Greenland and preserved there as an acidic annual layer in an ice core. Glaciologists studying a new 3053-m core (1) have used such varied information from disparate fields not only to date the eruption more precisely but also to use that layer to calibrate other ores and build a better chronology that can, in turn, link with tree-ring, oxygen isotope, pollen, varve, and other chronologies to illuminate climatic fluctuations of the past. This scenario illustrates the excitement of interdisciplinary convergence in contemporary volcanology.

The record of past eruptions is important because large eruptions are sufficiently varied and sufficiently uncommon that to understand them (and prepare for them), we must make the most of the historic and recent prehistoric record.

Major explosive eruptions cause short-term climate change when their gases, if sufficiently voluminous and blasted high enough into the stratosphere, form a widely dispersed aerosol of sulfuric acid ($H_2SO_4$) droplets. This fine mist acts as a filter, partially blocking solar radiation from reaching Earth's surface. The 1815 eruption of Indonesia's Tambora, history's largest, is widely believed to have caused June snowstorms in New England and severe crop failures at high latitudes during the following "year without summer." The substantially larger eruptions of prehistory call attention to volcanism as an important natural agent of climate change (2).

But volcanism has other significant impacts as well. In 1783-1884, fluorine-rich gas, ash, and rain destroyed crops and poisoned livestock throughout Iceland, killing 24% of the human population by starvation. One hundred years later, the Krakatau tsunamis (up to 40 m in height) killed over 34,000 people, including one as far away as Ceylon. Pumice from Krakatau floated in the Indian Ocean for 2 years, disrupting shipping at times and transporting organisms to distant shores. More recently, air transport has been affected when, on at least seven occasions, jumbo jet engines failed upon entering ash clouds. Two million years ago (just yesterday to geologists), a Yellowstone eruption produced 2500 $km^3$ of tephra in a matter of hours. Its effect on climate must have been substantial, but the sheer thickness of distant ashfall—covering 16 states and exceeding 20 cm at distances of 1500 km—would be considerably more than a nuisance if it were to be repeated tomorrow. Yes, we need to know more about these significant perturbers of our world.

Our record of historical volcanism shows a dramatic increase over the last two centuries that is apparent rather than real (3). It is closely proportional to the increase in global population during that same period, and it spans huge advances in communication, transportation, and record keeping. The two largest drops in apparent volcanism coincided with the two World Wars, when observers (and editors) were preoccupied with other things. And the larger eruptions—less likely to be missed—have been relatively constant through the last two centuries. Volcano watchers like to think that we have been capturing most eruptions since regular reporting started in the 1960s (yearly by the Volcanological Society of Japan and monthly by the Smithsonian), but we know we are still missing some. Even the record of large eruptions decays rapidly from the average of more than five per decade in recent centuries to 0.7 per decade before the global exploration and printing advances of the 15th century. Deep submarine eruptions are far more voluminous but are rarely documented even today. The hard fact is that our historical record of volcanism, beyond the last 100 to 200 years, is poor for most parts of the world, and in only a few of these has careful fieldwork, with lab dating of young products, filled in much of the record (4)

In recent decades, though, eruptive products reaching the stratosphere have been sampled by aircraft and balloon, analyzed for total $SO_2$ content by satellite, and measured for altitude and thickness by laser imaging. These results, on the moderate as well as large eruptive clouds of recent decades, have greatly advanced understanding of volcanism's far-researching effects. It is now clear that total $SO_2$ is at least as important to climatic impact as eruptive volume, the traditional volcanological measure of eruption size and the only one that can be conveniently estimated for ancient events. Furthermore, these two measures are not always correlated, and several high-$SO_2$ eruptions of modest size have put up significant stratospheric aerosols (5). These findings complicate attempts to evaluate climatic and other impacts of ancient eruptions.

But all these complications and weaknesses in the record help to underscore the importance of alternative approaches from other disciplines to the building of a reliable chronology of global volcanism. The $H_2SO_4$ aerosols eventually settle to Earth, and pioneering work by Danish glaciologists in the 1970s (6) showed that the resulting acidity layers in deep ice cores from Greenland provide a volcanic chronology. American and French groups found the same evidence in Antarctica and, by correlating several layers and confirming the common composition of their (rare and tiny) volcanic glass fragments, showed that some eruptions have a truly global distribution of products (7). These results have continued, through painstaking work on ever more cores, and the report by Zielinski *et al.* discusses results from the newest and deepest Greenland core (1). The authors suggest more accurate dates for several large eruptions and provide many new dates (particularly before 0 B.C.) from unknown sources. The largest signal in the last 7000 years, also detected in Antarctic cores (7), was from an unknown source around 1258 A.D. However, four larger signals were found in the 7th millennium B.C., marking this as easily the most volcanically active part of postglacial time.

The new results are exciting to all scientists interested in the volcanological record. The principal problem of this approach, however, is that aerosols move swiftly around the globe but their latitudinal spread is relatively slow. This means that an eruption from high north latitudes (Iceland, Alaska and Kamchatka) leaves a relatively large volcanic deposit on Greenland, whereas a comparable one from low latitudes leaves a much smaller record, and one from the Southern Hemisphere may leave none at all. Until more cores are obtained from mid- and low-latitude sites (not famous for their stable glaciers), substantial uncertainty will surround the identification and calibration of eruptive sources. Added to this problem is the danger of misinterpreting the completeness of volcanism's recent historical record. Very large eruptions may well have been missed only a few hundred years ago in some parts of the world, so the matching of sulfate spikes with poorly constrained dates from the volcanic record needs caution. Nobody should be surprised to learn tomorrow of a previously unreported larger eruption around that same time from another part of the world.

The ice core approach is enormously exciting, however, and holds great promise. The pieces of a large puzzle seem to be falling into place. The linking of these results to proxy records of past climate, such as tree-ring chronologies (8), offers the opportunity to refine both volcanic and climatic chronologies while gaining a more profound understanding of the relation between volcanism and climate change. ❑

References and Notes

1. G.A. Zielinski *et al.*, *Science* **264**, 948 (1994).
2. M.R. Rampino, S. Self, R.B. Stothers. *Annu. Rev. Earth Planet. Sci.* **16**, 73 (1988); H. Sigurdsson and P. Laj, in *Encyclopedia of Earth Systems Science*. W.A. Nierenberg. Ed. (Academic Press, San Diego, CA, 1992), vol, 1, p. 183; A. Robock, in *Greenhouse-Gas-Induced Climatic Change*. M.E. Schlesinger, Ed. (Elsevier. Amsterdam, 1991), p. 429.
3. T. Simkin, *Annu. Rev. Earth Planet. Sci.* **21**, 427 (1993); see also T. Simkin and L. Siebert. *1994 Volcanoes of the World* (Geoscience Press, Tuscon, AZ, ed. 2, 349 pp., 1994).
4. H. Machida and F. Arai, *Atlas of Tephra in and Around Japan* (Univ. of Tokyo Press. Tokyo, 1992); P.C. Froggatt and D.J. Lowe, *New Zealand J. Geol.* Geophys. **33**, 89 (1990).
5. G.J.S. Bluth, C.C. Schnetzler, A.J. Krueger, L.S. Walter, *Nature* **366**, 327 (1993).
6. C.H. Hammer, H.B. Clausen, W. Dansgaard. *ibid*. **288**, 230 (1980).
7. J.M. Palais, S. Kirchner, R.J. Deimas. *Ann. Glaciol.* **14**, 216 (1990).
8. V.C. LaMarche and K.K. Hirschboeck, *Nature* **307**, 121 (1984).
9. Comments by J. Luhr, S.Sorensen, and R. Fiske are appreciated.

**Questions**:

1. What was the largest historic eruption and when did it occur?

2. What world events led to a decrease in the apparent number of volcanic eruptions?

3. What two aspects of a volcanic eruption are most important in affecting climate?

Answers on page 301

16

*Although the power of volcanic eruptions is easily seen in the landscape afterward, the actual amount of energy released in an eruption had never been measured until 1991. This measurement took place at Japan's Unzen volcano on June 8, 1991. Shock waves were measured using three pressure-sensitive meters. The velocity of the shock wave was determined to be 75 meters per second. Based on this figure, the total energy released was estimated to have been equivalent to 12,000 tons of TNT.*

# First Direct Measure of Volcano's Blast

by D. Pendick

Since Japan's Unzen volcano awoke in 1990 from a 200-year repose, lava has oozed from a vent on its eastern slope, forming an unstable dome that looms menacingly over towns below. Periodically, part of the dome shears off or collapses, releasing a cascade of debris with explosive force.

Volcanologists yearn to measure directly the energy released during such volcanic events. But such close-up, detailed observations pose extreme danger, and the fury unleashed by a dome collapse can turn expensive instruments into scorched, shattered hulks.

Now, using a simple, rugged device designed to gauge military munitions and other explosives, Japanese scientists have achieved the first direct measurement of the energy released during a volcanic blowout. Volcanologist Hiromitsu Taniguchi of the Science Education Institute in Osaka and geologist Keiko Suzuki-Kamata of Kobe University report their findings in the Jan. 22 GEOPHYSICAL RESEARCH LETTERS.

The researchers measured the shock wave created by a dome collapse on June 8, 1991, and then calculated the pent-up energy required to generate it—the equivalent of about 12,000 tons of TNT. Previously, volcanologists relied on more approximate measurements. In one widely used method, researchers locate a chunk of debris and calculate the energy required to hurl it from the volcano to its landing place.

To make their measurements, Taniguchi and Suzuki-Kamata set up three meters within the volcano's destructive range. The pressure-sensitive part of the meter consists of a hollow chamber about two inches wide, covered with a thin lead plate. The sensor is mounted on a sturdy pole driven into the ground.

The researchers calculated the June 8 shock wave at 75 meters per second at the source on the basis of how severely the wave deformed the lead plate as it passed the meter at Taruki-daichi, a town 2,700 meters northeast of the lava dome. That's powerful enough to knock over a person standing in Taruki-daichi, they report.

The new method of measuring may reduce inaccuracies, says Richard B. Waitt, a volcanologist at the U.S. Geological Survey's Cascades Volcano Observatory in Vancouver, Wash. "It's a far more direct means of [making measurements]," he says. "This allows some calculations as to what the volcano is capable of."

Unzen has proved capable of quite a lot. The blast that the Japanese researchers measured came just five days after a massive flow of hot ash and debris from a dome collapse killed 43 people in Kitakamikoba, a town directly in the firing line of the volcano's east-facing vent.

Direct measurements of volcanic blasts may provide a means of checking the theoretical models some volcanologists have created to explore the physics of crumbling lava domes, says volcanologist Jonathan H. Fink of Arizona State University in Tempe, who helped develop such a mathematical tool.

"It's interesting that the number these [researchers] came up with—75 meters per second as the maximum velocity—is well within the range that we would calculate based on the model," Fink comments. ❑

**Questions**:

1. What was the possible range of latitudes within which the volcano could have occurred? How did researchers constrain the location of the eruptions from the ice core data?

2. What product of volcanic eruption found in the ice cores can be used to find the source volcano?

3. How might the eruption have contributed to the downfall of the Maya?

Answers on page 301

17

*Accurately predicting volcanic eruptions, as well as earthquakes, has been a challenge for scientists for many years. All eruptions display variations which hinder applying precursor behaviors of one volcano to the prediction of eruptions at another volcano. Studies of eruptions at Colombia's Galeras Volcano suggest that the pattern of sulfur dioxide release and long-period tremors may be a good predictive tool. The behavior of these two phenomena delineate three pre-eruption phases. These phases may relate to the closing of fractures which trap gases and allow pressure to build up, ultimately leading to an eruption. Applying these patterns to other active volcanoes has found some similarities, suggesting that this may be a useful predictive tool. However, not all volcanoes show this pattern, suggesting that some volcanic eruptions may still escape prediction.*

# Deadly Eruption Yields Prediction Clues

by Richard Monastersky

When volcanologist Stanley N. Williams visited Colombia's Galeras Volcano last year, he had no idea that burning chunks of rock would soon shoot from the crater, severely injuring him and killing nine people standing nearby. But from this tragedy, Williams and several colleagues have discovered geological clues that can help scientists foretell when pressures inside a volcano are reaching a critical level, according to research results they report this week.

"If we could have recognized this pattern, there might have been a chance [last January] to say it was too dangerous to go into the volcano. But we hadn't seen this pattern before," says Tobias P. Fischer, a graduate student studying with Williams at Arizona State University in Tempe. Fischer, Williams, and their colleagues discuss their work in the March 10 NATURE.

The volcanologists made their discovery by comparing earthquake behavior with the amount of sulfur dioxide gas that Galeras emitted prior to last year's Jan. 14 and March 23 eruptions. The researchers noticed a relationship between the gas releases and a particular type of tremor that has long-period seismic waves.

According to theory, volcanoes produce these long-period quakes when pressurized gases flow through underground fractures, setting up vibrations much like the clanging in water pipes. Over the last decade, researchers have gradually come to realize that swarms of long-period tremors often herald an eruption.

Indeed, such jolts shook Galeras before the fatal field trip last year, but the number of quakes was much lower than it had been, providing a false sense of security. "Basically, we were not worried at all about the volcano," Williams told SCIENCE NEWS.

When Fischer examined data for last year, however, he observed that three distinct phases preceded the eruptions. In the first period, roughly 6 weeks before the blasts, Galeras began releasing more sulfur dioxide than usual, indicating that gases were flowing unimpeded to the surface. Researchers can measure such gases remotely with a spectrometer because sulfur dioxide blocks ultraviolet rays from the sun.

In the second stage, a few weeks after the initial gas surge, emissions fell, while the energy of long-period quakes increased. These two trends suggest that fractures had started to close, reducing the gas leakage and increasing the pressure within the volcano.

During the final phase, the energy of long-period tremors dropped, signaling that the fractures had sealed completely and that gas could no longer escape. The internal pressure increased until it blew out the rocks that plugged the inner crater.

Volcanologists have monitored earthquakes and gas emissions at dangerous volcanoes for over a decade, but this study was the first to compare the two and decipher a distinct pattern, says Fischer. He suggests that researchers can better monitor the pressure within a volcano by tracking gases and seismicity.

Scientists studying other eruptions are finding similar, although not identical, patterns. At Mt. Pinatubo in the Philippines, for example, sulfur dioxide emissions did drop before the June 1991 eruption, even as the energy of long-period quakes increased, according to Christopher Newhall of the U.S. Geological Survey (USGS) in Seattle.

But Alaska's Mt. Redoubt showed no gas changes prior to erupting in 1989, says Bernard A. Chouet of the USGS in Menlo Park, California. Although volcanologists may identify precursors of some eruptions, Chouet warns that they cannot bank on seeing the same changes before all blasts. "Every volcano is going to have its own personal behavior," he says.

At Galeras, Williams and others intend to watch closely for any repeats of the past pattern. Although the recent eruptions there were minor, Williams worries that a much larger blast could occur. Almost 400,000 people live within 7 kilometers of the active crater, and their houses sit atop old volcanic flows. It was this danger that spurred researchers to gather there last year to discuss Galeras. ❑

**Questions**:

1. What produces long-period tremors?

2. What happens to sulfur dioxide emissions and long-period tremors during the second phase prior to eruption?

3. How did Alaska's Mt. Redoubt differ from the predictive pattern?

Answers on page 301

*In 1992 a small robot descended into Antarctica's Mt. Erebus to collect scientific information in an environment considered too dangerous for human investigation. It only made it 20 feet. Two years later a second robot, Dante 2, descended into Alaska's Mt. Spurr. This time the robot successfully descended into the crater, collected its data, and almost made it back out. The progress that is being made in robotics is allowing us to investigate areas that until now have been too dangerous or too inaccessible. These types of robots may soon be used to help us explore the planets and moons of our solar system.*

# Dante Conquers the Crater, Then Stumbles

by Richard Monastersky

The robot explorer Dante 2 discovered last week that the road home is often the roughest. After spending 8 days successfully navigating the steaming crater of Alaska's Mt. Spurr and enduring attacks by falling boulders, the spiderlike machine flipped over while climbing out. The Dante crew will attempt to airlift the helpless robot this week.

Despite the late setback, engineers connected with the mission say Dante performed beyond expectations in a harsh environment. "From the science and technology side, it's a clear victory and a success," says William"Red" Whittaker, a principal investigator on the project and a robotics researcher in Carnegie Mellon University in Pittsburgh.

The eight-legged robot descended 660 feet into the crater of the active volcano, analyzed gases escaping from several fumeroles, and collected terrain information for mapping the interior. For parts of the trip, the robot controlled its own movement by sensing the environment, picking a route, and planning where to place its feet. At other times, the Dante team controlled aspects of the robot's walking from a command center 80 miles away in Anchorage and from NASA's Ames Research Center in Mountain View, Calif.

Dante was connected by a tether to an electric generator and a satellite dish on the rim of the crater. During the first part of its trek, the robot descended snow-covered slopes of 20° to 30°; it later encountered bare rock, mud, and fields of large boulders. At certain points, the 10-foot-long, 10-foot-high robot climbed over rocks larger than itself.

Mt. Spurr erupted in 1992 and let loose smaller blasts in 1993. But even more than the threat of eruption, the primary danger inside the crater came from falling rocks. As the walker neared the crater floor, its video cameras showed several refrigerator-size boulders bounce close by. A 1-foot-diameter rock knocked one of the legs out from under Dante but did not cause major damage.

While in the crater, the explorer used onboard analyzers to test for sulfur compounds in the volcanic gases. Those tests did not detect hydrogen sulfide or sulfur dioxide, which suggests that no fresh magma lies at a shallow depth beneath the volcano waiting to erupt.

The robot had climbed about one-third of the way out of the crater when it rolled over while crossing a 35° slope and could not right itself. "We were just on terrain beyond our capability," explains Project Manager John E. Bares.

Dante 2 evolved out of a walking robot that Whittaker's team took to Antarctica's Mount Erebus in late 1992. This earlier version descended only 20 feet into the crater before broken connections in its tether ended the mission (SN:1/9/93, p.22).

The Dante 2 project pushed the bounds of robotics by exposing a walking machine to a range of demanding conditions never before experienced. "It sets the state of the art in walking on rough terrain. It's a pretty impressive accomplishment," says Roger D. Quinn, who designs walking robots at Case Western Reserve University in Cleveland.

NASA, which funded the mission, is encouraging the development of autonomous robots for planetary exploration. "The moon is the place for the fulfillment of this work," says Whittaker. ❑

**Questions**:

1. What produces long-period tremors?

2. What happens to sulfur dioxide emissions and long-period tremors during the second phase prior to eruption?

3. How did Alaska's Mt. Redoubt differ from the predictive pattern?

Answers on page 301

# Part Three

# External Earth Processes and Extraterrestrial Geology

19

*Life has played a dominant role in the formation of carbonate rocks since the radiation of multicellular organisms. So how did carbonate rocks form before the rise of multicellular life? The mineralogy and chemistry of carbonates are good indicators of the processes forming the carbonates and the ocean chemistry at the time of precipitation. So what do the mineralogy and chemistry of Precambrian rocks tell us about their formation? The role of microorganisms in the formation of early carbonates is of particular importance. Some stromatolites may have been precipitated without the aid of microorganisms, whereas others may have been due to the presence of microbes. Based on detailed studies of Precambrian carbonate rocks there appears to be a change in the dominant processes of carbonate precipitation: the Archean may have been dominated by inorganic processes and the Neoproterozoic by microbial processes, with a transitional period in between.*

# New Views of Old Carbonate Sediments

by John P. Grotzinger

Earth's environment has been modified on many time scales. Global changes that occur on time scales approaching Earth's age contrast with those that are anthropogenically induced. The Precambrian rock record is essential for investigating long-term change because it represents about 85 percent of Earth history. Within the last decade, research has shown that carbonate sediments are sensitive indicators of changes in the composition of the ocean and atmosphere, as well as biological evolution.

With the exception of its final 50 million years, the Precambrian fossil record provides no evidence of calcified animals and plants in the construction of carbonate platforms. But deposits as extensive as the Australian Great Barrier Reef are known to be as old as 2.5 billion years, and less substantial accumulations are observed in strata extending back about 3.5 billion years. Our paradigm for Phanerozoic carbonate production, which is solidly founded on the remains of calcified animals and plants, is challenged to explain carbonate sedimentation in a world inhabited only by microorganisms.

The issue is whether microorganisms had an active role in Precambrian carbonate production, or if sediments were mainly precipitated through abiotic processes. Breakthroughs in understanding Precambrian platform evolution have resulted from integrating regional stratigraphic and sedimentologic studies with detailed petrographic and geochemical studies of textures and diagenetic histories.

Recent attempts to synthesize the history of Precambrian carbonate sedimentation reveal substantial long-term changes in depositional facies that would constrain the modes of carbonate precipitation. Earlier studies viewed the entire Precambrian sedimentary record as a single time-rock unit, analogous to a Phanerozoic system, and masked important transitions that are the result of long-term environmental change.

Well-preserved early to middle Archean carbonates are rare. But late Archean carbonates, which were deposited in cratonic and non-cratonic settings, are locally very well developed.

Late Archean examples contain precipitated carbonates that occur in a variety of facies that represent all preserved depositional environments. Calcitic marine cements formed widespread beds of pure precipitated carbonate, which suggests that late Archean seawater was possibly substantially oversaturated with calcium carbonate. The cement also occurs as the internal microtexture of many stromatolites, sedimentary structures that may represent the sediment-accreting activities of vanished microbial mats. Much of this cement consists of an undescribed herringbone texture and may represent calcitization of primary high-magnesium calcite. Other marine cement formed botryoidal fans that grew on a variety of depositional surfaces up to 50 centimeters high. The crystal fans may represent calcitization of primary botryoidal aragonite.

It seems likely that at least some of the stromatolites from the late Archean may have been formed as inorganic precipitates. Layered, domal structures are also produced by inorganic processes and mimic features that are often described as stromatolites and interpreted to be the result of benthic microbial activity. Recent studies of late Archean carbonates imply caution in extracting biological information from Archean stromatolitic fabrics.

***Reprinted with permission from Geotimes, September 1993, pp. 12-15***

The general, first-order facies relationships are similar between Paleoproterozoic and platforms that are somewhat younger or older. Shelf-to-basin transitions are well represented. The spectrum of facies that document this transition is surprisingly independent of time. Subtidal, bedded marine cements, however, are far less common in Paleoproterozoic carbonates than in late Archean platforms. Locally, subtidal marine cements occur even in siliciclastic-dominated shelf sequences as condensed facies following maximum flooding of the shelf. In addition, Paleoproterozoic carbonates often include precipitated tidal-flat tufas, a unique facies that distinguishes them from most younger carbonates.

Tidal-flat tufas are expressed as millimeter-thick stratiform crusts, colloform crusts, or discrete microdigitate stromatolites. Given the association between tufas and stromatolites, tufas may have been associated with microbial activities that may have aided in the precipitation process. Previously, the radial-fibrous fabric was thought to represent replaced, calcified sheaths. In rare examples, the primary fabric is well preserved in early chert that predates inversion of the aragonite fabric and provides compelling evidence in favor of inorganically precipitated aragonite. Current interpretations of biological influences all feature shifts in local pH as driven by photosynthesis in live microbial mats or bacterial degradation of decomposing mats.

Precipitated marine cements in Paleoproterozoic tidal-flat facies represent an intermediate state between the cement-dominated late Archean carbonate record and the cement-poor Neoproterozoic and younger record. Mesoproterozoic carbonates show transitional textures. Consequently, the shift from Archean to Mesoproterozoic carbonates may document a decrease of calcium carbonate in the saturation state of the ocean. In the late Archean, precipitation of bedded marine cements was possible in unrestricted, subtidal, open marine environments. As more carbonate was deposited on newly formed cratonic platforms, the degree of oversaturation decreased and bedded marine cements could only form on relatively restricted tidal flats as tufas. As saturation further decreased, precipitation of marine cements occurred in pores between previously deposited sediments. At some point beginning in the Mesoproterozoic and onward, differences in the history of marine cementation are difficult to establish. The record of evaporites provides one of the most important constraints on the composition of the Precambrian ocean. The most important point is that bedded or massive calcium sulfate deposits formed in primary evaporative environments are absent in the Archean and Paleoproterozoic record. The record is limited to rare sediment- or barite-filled gypsum casts. Significant sulfate evaporites first appear in the Mesoproterozoic. Deposits are reported from the approximately 1.5 billion-year-old MacArthur basin (northern Australia) and younger basins of the Canadian Arctic Archipelago. Several Paleoproterozoic carbonate platforms provide evidence for halite precipitation directly following carbonate precipitation, with no evidence for an intermediate calcium sulfate phase. The appearance of calcium sulfate evaporites in the Mesoproterozoic would therefore represent a major transition in the evolution of Precambrian carbonate platforms and the composition of seawater. This transition might be due either to an increase in atmospheric oxygen or a change in the calcium-to-bicarbonate ratio of seawater.

Many interesting changes in the character of Precambrian carbonate sediments occur during the Neoproterozoic. One of the most important changes involves the prominent decline in stromatolite taxa. Each taxon is represented by a distinct carbonate microfabric, as seen in thin section. The success of using this approach to demonstrate trends in microbial evolution depends on the assumption that the carbonate textures observed in the rock today are the result of biological influences dominating over sedimentological processes, and on the absence of diagenetic overprints of the primary textures. However, as carbonate petrologists gain experience with stromatolite textures, it is increasingly clear that separating the abiotic components of a stromatolite enough to make the biological controls clear may be difficult. The good news is that whatever may be lost in biological information is gained in a wealth of new data that can be used to document environmental change.

Compilations of Proterozoic stromatolite taxa suggest that the diversity of stromatolites reached a maximum near the Mesoproterozoic/ Neoproterozoic boundary and declined during the late Neoproterozoic. This decline could be due to several factors, such as the advent of Ediacaran soft-bodied metazoans, competitive exclusion of microbial mats by higher algae, and changes in the composition of seawater. Since the number of stromatolite taxa is essentially a record of carbonate textures, the decline of Proterozoic stromatolites could be as much related to a global reduction in carbonate saturation and/or change in mineralogy through time, as to the competitive/ predatory pressures created by higher algae and animals.

A clear distinction cannot yet be made between the relative roles of biologic and abiotic processes in forming Precambrian stromatolites and other carbonate sediments. Sorting out where the role of biology ends in the accumulation of a carbonate deposit is difficult. For example, if microfossils are found in an ancient stromatolite, it is often uncertain whether the living microorganisms directly participated in forming the stromatolite by binding sediment or calcifying sheaths. The microorganisms may have helped only indirectly by inducing local changes in the pH of the sediment surface or interstitial pore water. Or, like many microbes today, they may have had no effect at all. Similarly, an environment where marine cements and organic films are interlayered and biologic processes have no effect on the kinetics of precipitation is hard to imagine. The biological and environmental factors are unlikely to be mutually exclusive; their effects should be seen in degrees. In this context, some distinctions can be made, particularly in establishing the origin of stromatolitic and other carbonate fabrics.

The record of Precambrian carbonate sedimentation should be viewed in discrete intervals marked by important yet subtle differences in the mode of carbonate production. The Archean record of carbonate precipitation appears to have been dominated by the inorganic process of sea-floor precipitation, and the Neoproterozoic characterized by microbial trapping and binding of sediment to form micritic stromatolites. Intervening periods were transitional and show features of both. Accordingly, the Archean record has examples of micritic stromatolites (Neoproterozoic preludes), and the Neoproterozoic record includes sea-floor cement fans (vestiges of the Archean). Combined with the possibly misinterpreted history of Precambrian evaporite sedimentation, these changes in the record of carbonate sedimentation provide a warrant for non-uniformitarian models of Earth evolution and environmental long-term change. ❑

**Questions**:

1. What features suggest that seawater during the Archean may have been substantially oversaturated with calcium carbonate?

2. What does the appearance of calcium sulfate evaporites in the Mesoproterozoic suggest about changes in the environment?

3. What are two possible causes of the decline in stromatolite diversity at the Mesoproterozoic/Neoproterozoic boundary?

Answers on page 301

*Water is perhaps the most important substance on Earth. It is necessary for life, it is an important component of climatic processes, and it is a required element in many, if not most, geologic processes. It is also important in the formation of coal and oil. It has generally been thought that organic reactions do not work well in water. But at high temperatures, water may be important to organic reactions. At elevated temperatures, both water and organic molecules begin to break apart into charged ions. These ions are much more reactive. Under these conditions, water can act as both an acid and a base, influencing different types of organic reactions. Additional benefits of this new knowledge are in the use of hot water to reduce the cost of processing fossil fuels and in reducing the waste by-products of petroleum-based materials.*

# Water, Water Everywhere

## Surreptitiously converting dead matter into oil and coal

by Elizabeth Pennisi

Just as everyone knows that oil and water don't mix, scientists know that organic reactions don't work well in aqueous solutions.

But to a small group of scientists studying how oil and coal form from carbon-rich decayed plants and algae, aqueous organic chemistry—reactions of carbon-based compounds in hot water—represents a better way of thinking about how the Earth created those vast underground energy reserves.

"We're promoting the idea that water is important in natural organic reactions," says Michael D. Lewan, a geochemist with the U.S. Geological Survey (USGS) in Denver.

In addition, sustained investigation into how hot water affects organic materials may lead to more efficient and environmentally friendly processes. Someday, water may aid in making—and recycling or cleaning up—plastics and other petroleum-based products.

Once again, this commonplace substance turns out to have some uncommon attributes. "We just take water for granted," says Theodore P. Goldstein, an organic chemist at Mobile Research & Development Corp. in Princeton, N.J. "We don't think its properties can change."

Little did he and others realize how mutable water can be.

Until now, scientists thought that coal forms when dying plants in soggy marshes get buried, creating a peat that simmers in this soupy environment. If no oxygen is present, chemical events slowly change peat, first into lignite and then, millions of years later, into bituminous coal. If temperatures climb high enough, anthracite coal forms.

Oil formation was viewed similarly. Dead marine microorganisms sink to the seafloor, then become buried by silt washing out of a river. If enough silt piles up, it creates a geologic Dutch oven, in which high temperature and pressure cause the organic debris to condense. A source rock—oil shale—forms. In its pores, chemical processes continue until oil oozes forth. The key requirement is getting this "oven" hot enough for a long enough time—or so everyone thought.

These explanations did not satisfy Andrew Kaldor, a researcher at Exxon Research and Engineering Co. in Annandale, N.J. He realized that ideas about oil and coal formation had evolved many years ago and had not really been updated to include new chemical and biological knowledge.

So Kaldor and Exxon organic chemist Michael Siskin decided to reexamine these ideas by first determining the chemical composition of organic materials in source-rock shales—an awesome job given the complex and highly variable nature of this starting material and the cascade of molecular transformations that occurs in forming oil and coal.

In both, plant matter decays into a potpourri of molecules that, depending on the conditions at hand, break up and clump in any number of ways. Carbon atoms get rearranged into assorted rings and chains to create a complex, interlocking network. Hydrogen atoms join, leave, and sometimes rejoin this network, as do other elements such as oxygen, nitrogen, or sulfur, eventually forming giant, insoluble macromolecules. "It's everything that winds up in the sediments," notes Goldstein.

Few chemists would even know how to begin teasing out the right combination of hydrocarbons to create oil or coal, but somehow nature manages to break these giant molecules in just the right places.

To understand this process better, the Exxon group collected samples of oil shale from different parts of the world. The samples included a series from oil shale under the North Sea, where rocks in different locations exhibit different degrees of transformation. Siskin then placed the samples into a pressurized reaction vessel and heated them individually to temperatures ranging from 570°C to 750°C. These hotter-than-natural conditions sped up the transformation from a geologic time frame of millions of years to one measured in days and hours.

Over the course of about two years, these and other tests helped the scientists piece together the locations of various atoms and side groups in representative molecular structures and in the intermediate products created in the transformation from molecular glob to oil.

"Knowing the structures at different stages of maturation can help us see the pathways," notes Siskin. He and Kaldor began to realize that heat alone could not break some of the bonds between these atoms.

To examine this inconsistency further, the scientists decided to look at "model" molecules—simple, commercially available hydrocarbon compounds that stood in for the organic matter in source rock or coal. Each represents a class of organic material—esters, amides, alkanes, for example—found in nature. The researchers wanted to see what conditions dissolved the linkages in these molecules. The experiments confirmed their suspicions.

"A lot of the molecules present in the structures we had developed were thermally not reactive," says Siskin. Certain groups of atoms key to holding the macromolecules together just would not fall apart, no matter how hot the chamber got.

The results perplexed the Exxon scientists, because they knew these groups of atoms disintegrated naturally—and at lower temperatures than those used in the experiments. Also, coal formed under similar conditions. They began to eye water for answers.

"We knew that oil forms in an aqueous environment," recalls Kaldor. "One natural question is, 'Is the water really benign, or does it play another role?'"

As early as 1979, Lewan, then working for an oil company, and his colleagues had demonstrated that they could simulate oil formation in the laboratory, but only if they added water to their system. Without water, "the products are seldom like that in natural crude oil," recalls Lewan. He and his colleagues then tried adding water to their source-rock samples. When they opened up their reactor at the end of the experiment, "we found a beautiful layer of oil on top of the water," he adds. He called this conversion hydrous pyrolysis.

Work at Exxon had also suggested aqueous influences on organic reactions. One group had discovered that water and carbon monoxide can enhance the liquefaction of certain types of coal. A different research team in Canada began using a mix of hot water and steam to increase the amount of heavy oil recovered from reservoirs buried under Cold Lake in Alberta. Also, Alan R. Katritzky, an organic chemist from the University of Florida in Gainesville, had decided to wet down some of his reactions to test whether water could help remove sulfur, oxygen, and nitrogen contaminants from ringed hydrocarbon compounds.

In 1985, Katritzky teamed up with Siskin to investigate water's role more intensively. "The folklore would have it that organic molecules will not react with water," says Siskin. But the two researchers realized they could make water more amenable to organic materials by putting it under pressure and heating it up. As water molecules get hotter, they become less polar and so are more likely to interact with nonpolar organic molecules. At 300°C, water acts like the organic solvent acetone at room temperature, Siskin adds.

The hot water molecules also tend to break apart, splitting into positive hydrogen and negative hydroxyl ($OH^-$) components. They become acidic and basic—and therefore much more reactive. Also, keeping the water under pressure ensures that it remains liquid; as such, "it can act as a solvent, it can act as a catalyst, and it can act as a reagent," Siskin says.

Siskin and Katritzky started adding water to their reaction vessels and experimenting with different combinations of heat, water, brine (salt water), minerals, and clay—conditions that might exist deep within Earth's crust. They discovered that high temperatures cause an organic molecule to break into fragments—and so does water and brine, sometimes more effectively.

"In these model systems, the results are that water is not always benign," Kaldor says. Some classes of organic molecules proved very susceptible to water's influence. In fact, water sometimes causes organic material to disintegrate into fragments that then transform into oil's hydrocarbons more readily than heat-induced fragments do.

The results indicate that hot water becomes a catalyst for a series of ionic reactions—creating a second pathway for the cascade of molecular transformations that leads to oil. The acidic and basic nature of hot water—rather than heat—drives this cascade.

For example, water may function first as a base, nibbling away at certain linkages in the organic material. As new molecular fragments build up and modify the reaction environment, water can change its catalytic nature. It can then act as an acid, accelerating different reactions. The resulting products attack parts of the remaining molecules, further speeding the breakdown. Siskin and Katritzky described these processes in the Oct. 11, 1991 SCIENCE.

"What we have learned is that ionic chemistry predominates in most cases and opens up pathways not accessible by the thermal route," Kaldor says.

These results bolster findings obtained by Lewan and complement research in hydrous pyrolysis by providing some details about the chemistry that could be occurring during these transformations. "It opens up the idea that you can't focus on one aspect, the organic aspect, of oil formation," says Goldstein. "You've got to focus on the chemistry of the whole system."

Lewan, too, has continued his investigations into the role of water. From his perspective, hydrous pyrolysis takes place because small amounts of water trapped in pores in source rock become awash in organic molecules, not because organic material dissolves in aqueous solution. Like Siskin, he runs his experiments at higher temperatures than exist naturally in order to speed up the process. Some of Lewan's ideas about the details of water's role differ from those of the Exxon group. Nevertheless, these and other results are building a convincing body of evidence.

"Over the past 10 years within the organic geochemistry community, it's been very controversial whether water is important," says Lewan. "I think what we'll argue over the next 10 years is *how* it's important."

Whatever the actual mechanism, the fact that water plays a role could wreak havoc on established ideas about oil formation. The results suggest that oil can mature faster than previously thought, says Kaldor. Consequently, not only do the ideas buck tradition, but, if right, they will require the revision of time parameters in computer programs now used to predict locations of new reserves. For that reason alone he expects the oil-exploration community to accept these ideas slowly.

"We have not convinced Exxon geoscientists that the model is correct," Kaldor says.

"But these experiments are making believers out of them," Lewan adds, citing his work as well as that of Exxon.

To help convince their colleagues that such revision is warranted, Kaldor and Siskin are seeking ways to verify their ideas. "You can't do real-time experiments," Kaldor says, "so you have to begin to look for clues in the natural environment." Those clues could be molecules that form only through water-initiated ionic reactions, for example. "But we don't have that yet," he says.

Meanwhile, the concepts that have come out of this work may have broad impact. "We started this whole quest with just an interest in oil," Kaldor says. "But it really has paid off in a generic way in natural processing. In enables people to begin to do things they wouldn't be able to do before."

Already, Exxon foresees the possibility of using hot water to introduce more hydrogen into coal—to make it more amenable to liquefaction and to reduce the cost of this process. Also, hot water and steam might help add hydrogen to low-quality oil deposits, improving and loosening the oil from pores in source rock so it will move easily to the surface.

Hot-water chemistry promises to aid other chemical processes as well. For example, it could increase the efficiency of the production of isopropyl alcohol by providing a way for a waste product, an ether, to be converted into more alcohol. Hot-water processing also offers ways to break down petroleum-based materials that might otherwise contaminate the environment. Even the U.S. Army has expressed interest in aqueous organic chemistry—as a way to destroy chemical warfare agents. In other instances, the use of hot water may eliminate the need for other catalysts that prove difficult to dispose of safely. "It opens up an entire area of synthetic chemistry," Goldstein says.

"We're at the stage where it's becoming more and more routine," adds Kaldor. "My guess is that in the next few years, we'll see commercial applications." ❑

**Questions**:

1. What two things happen to water molecules as they heat up?
2. As an acid, how does water benefit organic reactions?
3. How might water improve the recovery of low-quality oil deposits?

Answers on page 301

# 21

*The process of coal formation is still not well understood. Part of this may be due to the structure of one of the components of plants: lignin. Lignin gives plants their structural rigidity, allowing plants to rise up off the ground. Lignin is also one of the last molecules to remain as a plant makes the transformation to coal. As the plant decays, the lignin molecules change structure and composition. In trying to study this transformation one scientist has suggested that lignin molecules may be organized differently than previously thought.*

## From Brown to Black

How does a tree trunk become a coal lump? A chemist's answer suggests a new view of how a tree is built in the first place.

by Kathy Svitil

You probably think you know how coal formed: from trees and ferns and others plants that died and were buried in the muck of primeval swamps—buried deeper and deeper, until the heat and pressure reduced them to a black, energy-rich essence. This picture is true enough as far as it goes. But it doesn't go very far for Patrick Hatcher, a fuel scientist at Penn State. Fuel scientists want the details—what chemical reactions happened when during the hundreds of millions of years it took for living plants to become black rock. "It's almost an intractable problem," says Hatcher, "unless you can simplify it by narrowing in on a specific part of a specific type of plant, and then follow the evolution of that part."

For the past decade, Hatcher has focused on lignin, the molecule that strengthens the soft, sugary cellulose walls of wood cells the way reinforcing rods strengthen concrete. Lignin keeps trees from falling over while they're alive, and when they die it becomes the primary component of coal. In a series of laboratory experiments, Hatcher has analyzed samples of conifer lignin and coal—and of all the stages in between—to see how that transformation takes place. In the process he has turned up some controversial evidence that lignin, and thus trees, may be put together differently from the way researchers had imagined.

The first thing that happens when a tree trunk crashes into a swamp is that microbes eat away the cellulose. Within a thousand years or so, only the lignin skeleton of the wood cells is left. (The log remains virtually the same size, however, because lignin is so rigid.) The lignin itself begins to be transformed once the log has been buried under further sediments.

Lignin is a polymer molecule—a long chain of identical units. Each unit consists of ten carbons and four oxygen atoms: a six-carbon ring with a three-carbon chain attached to one side and a single carbon attached to another, and two oxygens each on the ring and side chain. The oxygens play a key role in the coalification of lignin.

Hatcher analyzed samples of successive stages of coalification: peat, brown coal, lignite, subbituminous coal, and bituminous coal. When peat becomes brown coal, he found, the lignin skeleton is simply rearranged into a more compact, stable configuration. As the heat and pressure increase, though, they begin plucking oxygen atoms from the lignin units. In lignite, one oxygen has been pulled from the side chain; only three oxygens remain. Subbituminous coal has lost the second side-chain oxygen and one from the ring, leaving only one oxygen per ring. Finally, bituminous coal loses the last oxygen from every other ring on the lignin strand. While the coal is losing oxygen it is also having water squeezed out of it, which makes it more compact. But it is the progressive loss of oxygen that makes the coal more energy-rich: the less oxygen the coal contains, the more it can be oxidized and the more energy it releases when burned.

This process could never unfold in such an organized way, Hatcher thinks, with specific oxygen atoms being lost at predictable times, if strands of lignin in wood cells were arranged in the way fuel scientists have always thought—each long strand randomly clumped and knotted like yarn. That's what synthetic lignin, made in a test tube, looks like; but real lignin, according to Hatcher, must be more orderly. The problem is that it's hard to look at real lignin while it is embedded in cellulose.

Instead, Hatcher and his colleague Jean-Loup Faulon used a computer to calculate the forces among the atoms of a lignin strand, and thus how the strand is most likely to arrange itself in three-dimensional space. The picture they got was orderly indeed: it was a helix, without clumps or knots. Not only did the computer-generated lignin helix nestle neatly into the space between computer-generated cellulose fibers, but also it required less energy to form than a knotted strand of lignin. Both observations support the view, says Hatcher, that real plants probably make lignin helices.

But so far he has no direct evidence of that, and his computer model has been greeted with skepticism. "We're running up against a log of fixed minds on this concept," he complains. "It's almost dogma." Hatcher is still looking for a way to prove the dogma wrong.

**Questions**:

1. What causes a plant to turn into coal as it gets buried in the Earth?
2. What chemical change occurs in lignin molecules as they turn to coal?
3. What structural form might lignin molecules occur in?

Answers on page 301

22

*Each year, geologic hazards kill more than 100,000 people. They cause billions of dollars in damage and loss from the world's economy. This is a much greater loss of life and money than occurs from pollution, global warming, and many other highly publicized environmental problems of our own making. There are many kinds of geologic hazards. Hazards from materials include swelling soils, acid drainage, radon, and asbestos. Hazards from processes include earthquakes, volcanoes, landslides, floods, and coastal hazards. Many of the human and monetary losses from these hazards could be reduced if the expertise of geologists was used more often. Unfortunately, this expertise is often not used, or it is ignored. For example, there are dozens of cases where geologists warned people of imminent volcanic eruptions, only to have thousands killed because they would not leave the vicinity. Perhaps the most tragic economic cases are the repeated warnings of geologists against development and building construction along ocean coastlines, only to see billions of dollars in damage from storms and hurricanes. The Great Flood of 1993 along the Mississippi River was enormously costly because people insist on living within the confines of the floodplain, which geologists warn is certain to flood periodically.*

# Geology, Geologists and Geologic Hazards

Geologic hazards annually take more than 100,000 lives and wrench billions of dollars from the world's economy. It has been demonstrated repeatedly that involvement of professional geologists can minimize hazards and reduce losses.

by The American Institute of Professional Geologists

**Geology** is a science of *materials, processes,* and *change through immense expanses of time* (**deep time**). Materials include the solids, liquids and gases of the planet, and the processes include the physical and chemical transformations that occur on and within the planet. Geology includes the history of life and the interaction between living organisms, including humans, and the planet on which they live. In particular, geology considers the rates of change, the nature of change, and the ordering of events of change through the planet's history.

**Geologists** are those who are trained in and work in any branch of the geological sciences. The American Institute of Professional Geologists considers acquisition of a baccalaureate degree with 36 semester credits of courses in geology as the minimum educational qualifications needed to be a professional geologist. Many states now require licensing, certification or registration of geologists who practice as professionals in areas of geology that relate to public safety and welfare. Geologists who work with groundwater, waste disposal sites, engineering foundation assessments and assessment of geologic hazards are practicing in specialty areas related to public welfare.

The costs of geologic ignorance are staggering. These losses are mostly preventable but will occur whenever necessary geological evaluation is omitted or performed poorly by someone without suitable training. Citizens who associate geology with only the study of dinosaurs, the production of oil and gas, or prospecting for precious metals are not likely to think of consulting a geologist when property is purchased, land use zoning decisions are made, environmental legislation is passed, or building codes are drafted. Yet these are everyday areas of life in which expertise of geologists can prevent severe losses.

Geologists are essential for making regional plans to deal with major geologic hazards. The data that geologists provide about the history and immediate nature of any major geologic hazard establish the basis for risk assessment, emergency preparedness, land-use planning and public awareness programs. The principal tasks of professional geologists in regional projects are (1) to determine what hazards are likely to exist in the area under study; (2) to investigate them thoroughly enough to provide all the necessary data required to characterize the hazard and its anticipated impacts; (3) to compile and communicate the pertinent information in a concise and comprehensive manner to other team members (such as engineers, planners, government representatives, and land owners); and (4) to ensure that the hazards are addressed in any final project plan.

Geologists are also indispensable investigators in local site evaluations. Site evaluations precede many projects which may range in magnitude from a single-family home to a nuclear power plant. Geologists work productively with site landowners, civil engineers, architects, contractors and regulatory personnel. The geologist will answer at least two important **Questions:** *"How will the geologic conditions affect the success of the proposed use of the site?"* (Example—Will a home built here likely complete its expected useful life?) and *"How might the developed site eventually cause adverse impact through its geologic setting?"* (Example—If the storage tank sited here eventually leaks, is it likely to contaminate groundwater supplies over a wide area?)

**Hazardous geologic materials** include **swelling soils**, which expand in the presence of water and can exert pressures sufficient to shatter pavements and building foundations; **toxic minerals**, such as certain types of asbestos, that can cause health hazards; and **toxic gases**, which may include radon gas.

**Hazardous geological processes** most familiar to the public are those that occur as rapid events. **Earthquakes** produced by the process of rapid snapping movements along faults, **volcanoes** produced by upward-migrating magma, **landslides** produced by instantaneous failure of rock masses under the stress of gravity, and **floods** produced by a combination of weather events and land use—all produce massive fatalities and make overnight headlines. Other geologic processes, such as **soil creep** (slow downslope movement of soils that often produces disalignment of fence posts or cracked foundations of older buildings), **frost heave** (upheaval of ground due to seasonal freezing of the upper few feet) and land subsidence act more slowly and over wider regions. These slower processes take a toll on the economy. Human interaction can be an important factor in triggering or hastening the natural processes.

**Lack of awareness of events through deep time** induces human complacency which sometimes proves fatal. It is difficult to perceive of natural dangers in any area where we and preceding generations have spent our lives in security and comforting familiarity. This is because most catastrophic geologic hazards do not occur on a timetable that makes them easily perceived by direct experience in a single lifetime. Yet development within a hazardous area inevitably produces consequences for some inhabitants. Hundreds of thousands of unfortunate people who perished in geological catastrophes such as landslides, floods or volcanic eruptions undoubtedly felt safe up until their final moments. In June, 1991, Clark Air Base in the Philippines was evacuated when Mount Pinatubo, a volcano dormant for over 600 years, began to erupt and put property and lives at risk.

Geologic hazards are not trivial or forgiving; in terms of loss of life, geologic hazards can compare with the most severe catastrophes of contemporary society. Single geological events have, in minutes, killed more people than now live in Madison, the capital of Wisconsin, (pop. 176,000). Where urban density increases and land is extensively developed, the potential severity of loss of life and property from geologic hazards increases.

Some geologic hazards, such as radon gas and asbestos, have only recently been recognized as hazards. Their eventual costs to our society and the degree of danger these actually pose remain, at present, questions without clear answers. Yet substantial sums of money are already being spent to remedy perceived hazards. Some of this expenditure may prove needless. The magnitude of risk must be accurately known in order to tell if costs to remedy a hazard are worth the expenditures.

We are often faced with the decision about whether we can wisely live in areas where geologic forces may actively oppose otherwise pleasant living conditions. An informed decision can be based on several choices.

**(1) *Avoid an area where known hazards exist.*** Avoidance or abandonment of a large area is usually neither practical nor necessary. The accurate mapping of geologic hazards delineates those very specific areas which should be avoided for particular kinds of development. Otherwise hazardous sites may make excellent green belt space or parks. Towns such as Janesville, Wisconsin, have created beautiful parks in areas zoned as floodplains, thus avoiding placing expensive structures where flooding will cause damage.

**(2) *Evaluate the potential risk of a hazard, if activated.*** Risks can never be entirely eliminated, and the process of reducing risk requires expenditures of effort and money. Assuming, without study, that a hazard will not be serious is insufficient. Life then proceeds as though the hazard were not present at all. "It can't happen here" expresses the view that is responsible for some of the greatest losses. Yet it is equally important not to expend major amounts of society's resources to remedy a hazard for which the risk is actually trivial.

**(3) *Minimize the effect of the hazards by engineering design and appropriate zoning.*** Civil engineers who have learned to work with geologists as team members can be solid and effective contributors to minimizing effects of geologic hazards. More structures today fail as a result of incorrectly assessing (or ignoring) the geological conditions at the site than fail due to errors in engineering design. This fact has led many jurisdictions to mandate that geological site assessments be performed by a qualified geologist. Taking geological conditions into account when writing building codes can have a profound benefit. The December, 1988, earthquake in northwest Armenia that killed 25,000 people was smaller in magnitude (about 40% smaller) than the October, 1989, Loma Prieta earthquake in California. The latter actually occurred in an area of higher population density but produced just 67 fatalities. Good construction and design practice in California was rewarded by preservation of lives and property.

In some colleges, civil engineers are not required to take any course, in geology, and there is a need to make geology, taught by qualified geologists, a part of the education of far more civil and environmental engineers. Engineers should be cognizant of the benefits of geological assessment and be able to communicate well with professional geologists. It is no more practical for an engineer without considerable formal training and experience in geology to perform geological site investigations than it is for someone without considerable formal training in engineering to practice engineering.

California, in 1968, became the first American state to require professional geological investigations of construction sites and has reaped proven benefits for that decision. Since then many states have enacted legislation to insure that qualified geologists perform critical site evaluations of the geology beneath prospective structures such as housing developments and landfills. Most of these laws were enacted after 1980.

Zoning ordinances and building codes that are based on sound information and that are conscientiously enforced are the most effective legal documents for minimizing destruction from geologic hazards. After a severe flood, citizens have often been relocated back to the same site with funding by a sympathetic government. This is an example of "living with a geologic hazard" in the illogical sense. A less costly alternative might be to zone most floodplains out of residential use and to financially encourage communities or neighborhoods that suffer repeated damage to relocate to more suitable ground. When damage or injuries occur from a geologic hazard in a residential area, the "solution" is often a lawsuit brought against a developer. The problem has not truly been remedied; the costs of the mistake have simply been transferred to a more luckless party—the future purchasers of liability and homeowners' insurance at higher premiums. A solution would be a map that clearly delineates those hazardous areas where residential development is forbidden. A suitable alternative would be a statute requiring site assessment by a qualified geologist before an area can be developed. Sound land use that takes geology into account can prevent unreasonable insurance premiums, litigation, and repeated government disaster assistance payments for the same mistakes.

**(4)** ***Develop a network of insurance and contingency plans to cover potential loss or damage from hazards.*** Planners and homeowners need not be geologists, but it is useful to them to be able to recognize the geological conditions of the area in which they live, and to realize when they need the services of a geologist. A major proportion of earthquake damage is not covered by insurance. Despite public awareness about earthquakes in California, the 1987 Whittier quake produced 358 million dollars worth of damage of which only 30 million dollars [was] covered by insurance. Many property owners really do not realize that they may be uninsured against pertinent geological hazards.

For the property owner, especially the prospective homeowner, a geological site assessment may answer the following:

***Is the site in an area where landslides, earthquakes, volcanoes or floods have occurred during historic time? Has the area had past underground mining or a history of production from wells? Did the land ever have a previous use that might have utilized underground workings or storage tanks that might now be buried? Does the site rest on fill, and is the quality of the fill and the ground beneath it known? Are there swelling soils in the area? Have geologic hazards damaged structures elsewhere in the same rock and soil formations which underlie the site in question? Has the home ever been checked for radon? If the home is on a domestic well, has the water quality been recently checked? Is the property on the flood plain of a stream? Is the property adjoining a body of water such as a lake or ocean where there have been severe shoreline erosion problems after infrequent (such as 20-year or 50-year) storms?***

The answers to these questions are not always obvious. Individuals who assumed that they lived in areas that never flooded have often been surprised with great uninsured losses. A start in the home owner's assessment may be contact with the state geological survey or a local consulting geologist.

Insurance agents are not always familiar with local geological hazards. After risks have been assessed, the individual can then consult with insurance professionals (agents, brokers, salespersons) to learn which firms offer coverage that would include pertinent risks. ***(Note: homeowners' policies do not routinely cover geologic hazards such as subsidence or earthquake damage.)*** Consulting with the state insurance boards and commissioners can assist one in finding insurers who provide pertinent coverage.

Local governments should

make plans for zoning and for contingency measures such as evacuations with involvement from a professional geologist. The first line of help for local governments lies in their own state geological surveys. Geologists there are employed for service to the public. While they cannot, by law, normally serve as consultants, they can provide much of the available information that is known about the site or region in question and can direct the inquirer to other additional resources. In many cases, the information which the state geological survey can supply will be sufficient. Geologic maps and reports from public and private agencies are most useful in the hands of those trained to interpret them. Significant evidence that reveals a potential geologic hazard may be present in the reports and maps, but it may not be evident to the user with only an introductory background in geology. If significant risks of hazards are thought to exist, then consultation with a professional geologist may be warranted.

**Summary**

Geologic hazards annually take more than 100,000 lives and wrench billions of dollars from the world's economy. Such hazards can be divided into hazards that result dominantly from particular earth materials or from particular earth processes. Most of these losses are avoidable, provided that the public at large makes use of state-of-art geologic knowledge in planning and development. It has been demonstrated repeatedly that involvement of professional geologists can minimize hazards and reduce losses. A public largely ignorant of geology cannot usually perceive the need for geologists in many environmental, engineering or even domestic projects. Currently, most of our public is uneducated about geology, partly because few schools require any geology and partly because there is a lack of rigor in qualifications to teach earth science. This results in a populace prone to making expensive mistakes, particularly in the area of public policy. ❑

**Questions**:

1. What was the first state to require professional geological investigations of construction sites? When was this?
2. What is "deep time"? Are catastrophes more common in deep time than in a human lifespan?
3. What does the Whittier quake illustrate about how well people are often insured for disasters? What percentage of damage was covered by insurance?

Answers on page 301

# 23

*Landslides in the U.S. annually cause over 1.5 billion dollars in losses and 25–50 deaths. Most of the economic cost occurs from highway and road damage. The toll is much greater in other nations, especially those with a large amount of mountainous terrain. In some cases, the death toll of a single slide can be thousands and even hundreds of thousands of people. Geologists can minimize economic and human costs by identifying geologically unstable areas that should be avoided or at least properly engineered. Indications of such instability include evidence of recent slides in an area, clay-rich soils, planes of weakness in the rocks, frequent earthquakes, and slope undercutting. Humans can greatly enhance these natural factors by poor engineering, such as "oversteepening" a slope.*

# Landslides and Avalanches

Each year landslides in the U.S. produce over 1.5 billion dollars in losses and about 25 to 50 deaths. Avoiding high-risk areas, or planning to include measures for stabilization, is a wise choice for builders and financiers. Where good geologic and slope stability maps are available, chances for successful construction and development are greatly enhanced.

by The American Institute of Professional Geologists

**Landslide Occurrences**

A **landslide** consists of earth materials that move by gravitational force. The motion involves downslope movement as well as lateral spread or scatter of materials. Landslides can even involve some upslope movement, as in the case in which a massive slide runs downslope across a valley and sends debris up the opposite valley wall. The term "landslide" applies to rapid movement of earth materials, although landslide movements often begin with slow creep. An **avalanche** is a similar mass movement of snow and ice. Avalanches take about 20 lives each year in the U.S.

Downslope movement of soil and rock is a result of natural conditions on our planet's surface. The constant stress of gravity and the gradual weakening of earth materials through long-term chemical and physical weathering processes insure that, through geologic time, downslope movement is inevitable. However, when we talk about downslope movement as a hazard, we usually say that a slope is hazardous only when the downslope movement is likely to be rapid enough to threaten man-made structures. Many landslides occur in unpopulated areas and cause neither hazard nor risk. Landslides can involve volumes from as small as one falling bolder to massive movement of millions of cubic yards of material. The largest known landslide on earth occurred prehistorically in Iran, and was 14 km (9 miles) wide and 19 km (12 miles) long.

**Dangers of Landslides**

Slope failures result in an annual cost to society of about 1.5 billion dollars. Damages to highways alone from slope failure in the United States approach 1 billion dollars annually. Annual damage to buildings and building sites in the United States is about half a billion dollars. At Malibu, California, a single landslide at Big Rock Mesa recently generated costs of about 92 million dollars in damages and liabilities.

Very few landslides occur in areas where they cannot at all be anticipated. Small flows and slides that take only a few lives are common and account for an average estimated loss of 25 lives in the United States. Worldwide, landslides account for an average of 600 lives each year. Although steep slopes are an obvious topographic condition that should alert anyone to potential landslides, human inability to take seriously those hazards that occur infrequently can increase loss of life. Even professional engineers and regulatory officials who are aware of significant past landslide activity at a site can still be tempted to proceed with projects through unwarranted faith in the belief that "It won't happen here."

It is not possible to prevent all loss of life from small "freak accident" rockfalls, but it is possible to prevent or reduce many massive fatalities. This is because rapid slope failures occur as a result of the presence of specific types of earth materials and particular conditions and processes. Areas with conditions that indicate potential for massive slides can and should be avoided for major projects and developments.

**Indications of Natural Instability**

Slides can be anticipated through the following criteria.

**1. History of past landslides in the nearby area in the same stratigraphic units as the site in question.** For instance, prehistoric wedge failures had been mapped in the Kootenai River valley of Montana, and wedge failures there plagued the site of the Libby dam after construction began in 1967. The shales of the Pierre Formation are well known by geologists in the area of Denver, Colorado, as materials in which many slope failures occur. Therefore, a geologic map that shows where this stratigraphic unit is exposed at the ground surface reveals potentially dangerous areas. Some recent landslides are easily recognized from the ground even by the untrained eye. Old landslides with the potential to move again are likely to be noticed only by experts who employ aerial photographs and remote sensing techniques. If the subtle evidence is ignored, disaster is invited.

**2. Soil types that are rich in silt-to-clay-sized material, particularly soils that are rich in swelling clays.** Usually movement occurs in these soils in spring when the soil is very saturated from thaw and snow melt. However, water is not always required to produce movement. About 200,000 lives were lost in 1920 in Kansu, China, when dry **loess** soils (fine soils developed on fine wind-blown material) when into fluid motion during an earthquake, destroyed dwellings and engulfed the inhabitants. Loose volcanic materials often can absorb so much water that they flow quickly down even gentle slopes. A mudflow, along with gases, destroyed the Roman city of Herculaneum at the base of Mt. Vesuvius in 79 AD. Formations such as the Rissa Clay of Norway and the Leda Clay of Ontario, Canada, are particularly notorious for slope failure by flowage during liquefaction. The clays were deposited during late glacial periods in the seas that occupied these areas at that time. Where fresh ground water removed salts that were deposited with the clays in sea water, the clays became particularly unstable.

**3. Downslope orientation of planes of weakness in bedrock.** Translational movements occur where **bedding planes** (the planes between layers of sedimentary rocks) point in a downslope direction. Orientation of **schistosity** (alignment of platy and rod-shaped minerals in metamorphic rocks) in a downslope direction or orientation of joints parallel to the slope direction also furnishes planar weaknesses that are conducive to slides. The famous Gros Ventre landslide of Wyoming, slides in the Cretaceous rocks near Rapid City, South Dakota, and the landslide into the Vaiont Reservoir, Italy, all occurred in rocks containing weaknesses in planes that dipped steeply downslope.

Evidence for past landslides abounded in the valley used for the impoundment of the Vaiont Reservoir in Italy. Yet, engineers ignored the implications and constructed the reservoir. The catastrophe of 1963 occurred when a huge landslide shot into the reservoir and sent a wave of water over 300 feet (100 m) high across the reservoir and over the dam. The dam withstood the immense wave that overtopped it, and its strength was a credit to its designer. However, 3000 people caught in the path of the onrushing flood wave perished. Excellence in engineering design proved to be no substitute for a geological site investigation.

**4. Slope undercutting.** Landslides are particularly common along stream banks, reservoir shorelines, and large lake and seacoasts. The removal of supporting material by currents and waves at the base of a slope produces countless small slides each year. Particularly good examples are found in the soft glacial sediments along the shores of the Great Lakes of the U.S. and Canada.

**5. Earthquake tremors.** A tremor, even a fairly mild one, can provide the *coup de grace* to a slope which has been resting for decades in a state of marginal instability. Nearly 30 campers were killed when the 1959 landslide in the Madison Canyon of Montana was triggered by an earthquake. Over 130 million tons of debris covered Canada's Highway 3 when two small earthquakes triggered the Hope Mountain slide in British Columbia in 1965. During the 1964 Alaskan earthquake huge blocks of the shoreline slid beneath the ocean at Valdez, Alaska, when rotational slumping took place in materials below sea level. Similar rotational slumping in soft materials damaged large sections of Anchorage.

**6. Sensitive slopes subject to intense rainfall.** When the right geological conditions exist, periods of intense rainfall can trigger the movement of unstable slopes. In one night in January, 1967, 600 people lost their lives in Brazil when a three-hour cloudburst converted green hills into mudflows and attractive stream valleys into flood torrents. In the southern and central Appalachian Mountains, periods of increased frequency of slides often coincide with severe local summer cloudbursts and thunderstorms. More widespread regional rains associated with hurricanes that move inland trigger landslides over larger areas. Studies of the Canadian Rockies also reveal a definite link between rainstorms and rockfalls. Landslides are particularly abundant during the rainy season along the West Coast of North America.

**Human Causes of Rapid Slope Failures**

People are very capable of causing catastrophic slope failures. Designers who are ignorant about geological materials and processes may unwittingly produce conditions at a site that are the same conditions that promote natural failures.

All loose materials such as sand or soil have a natural **angle of repose**, which is the maximum angle that can be measured on a slope when the material is stable and at rest. When any earth material is maneuvered into large piles that have side slopes that are steeper than the angle of repose of the material, the situation invites a landslide.

Mine wastes may be susceptible to absorption of water and gradual deterioration of shear strength. Sometimes these materials are stacked in huge piles with side slopes that are even steeper than the natural angle of repose (about 40°) of stable materials. When the wastes weaken, the huge piles can fail in a landslide. This occurred in Aberfan, Wales, in 1966 when 144 people, 116 of them children, died in a flow of fine-grained mine waste. Six years later, coal mine wastes which had been used as a makeshift dam in a valley failed, and resulted in over 100 deaths and 50 million dollars damage at Buffalo Hollow, West Virginia, when the failure sent a 5-meter (15 ft.) wall of water and mud down the Buffalo Creek Valley.

Around major cities, accumulated waste is often stacked in large landfills. Some old landfills are piled high near their angle of repose. Like mine wastes, these present sources of possible danger from failure, especially in sites subject to earthquake tremors.

An unstable slope can be set into motion by removing supporting material from the toe of the slope, a situation that often results during construction on or at the base of a slope. Movement can also result from loading the slope from above. Loading from above occurs when people build a structure such as a building, a storage tank, or a highway upon materials that cannot remain stable under the additional load.

Human development of areas alters the natural drainage and increases runoff. Storm drains, roof gutters, septic tanks, or leaking water mains can focus enough water into a sensitive slope to generate movement, particularly where intensive housing development exists in several tiers on a single slope. Drainage control is very important in such areas.

Unlike natural lakes, flood-control reservoirs in steep valleys are often subject to seasonal water level variations that can exceed 100 feet. When the reservoirs are lowered quickly after a high water period (usually spring), the water level in the reservoir drops faster than water can exit from the pores and capillaries of the saturated rocks and soils. This leaves waterlogged soils resting tenuously on steep banks, and these soils easily slump into the reservoir. As this material is removed, the soils that lie on slopes above have no toe support down-slope. Sliding then develops further upslope and contributes to abnormal sediment filling rates in the reservoir.

**Landslides and the Concept of Deep Time**

**Mass wasting**, the downslope movement of materials, is the most prevalent of all geological processes. Through this process, materials are eventually carried into streams or onto glaciers. Thereafter, the material is transported (eroded) away by water, ice or wind. Most mass wasting takes place as creep, which is so slow that it cannot be discerned by direct observation. Instead, creep becomes evident only after some years through changes such as misaligned utility poles, out-of-plumb old houses, and bulges in stone retainer walls. As a hazard, creep is so slow that it is harmless and probably represents the more gradual of geological processes. However, not all downslope movement is gradual. Occasionally sections of rock and soil move quickly. A melting snowman is a good analogy of change. Most of the snowman disappears at nearly imperceptible rates by slow, gradual melting, but occasionally a piece breaks and falls away. The piece is rarely seen falling because a fall occurs in only a tiny fraction of the time taken for melting. The falling events are analogous to the type of mass wasting characterized by rock falls or toppling blocks. The frequent "Falling Rock" sign seen on our highways is a reminder of the perpetual problem of falls, but we rarely see a rock actually in motion toppling onto a highway.

Some slope failures occur seasonally. In many areas in the early spring, one sees small slumps that have occurred along the banks of streams, in pastures, or along slopes adjacent to the highway. While it cannot be predicted exactly when a slump will occur in spring, we can anticipate that more slumps will occur in seasons when the soils are wet with water from recently melted snow.

Some landslides occur on the irregular timetables of events such as earthquakes and storms which trigger movement. These triggering events cannot be predicted, but the particular geological conditions in an area which will make it prone to movement can be mapped, thus clearly showing where the higher landslide risks will be for homes and other structures during a future storm or earthquake. Such maps can help planners, lenders, homeowners and insurers.

Massive sliding or flowing failures that involve thousands of tons of material and that move at speeds well in excess of 100 miles-per-hour are the most infrequent of mass wasting events. Such massive movements usually leave scars and other tell-tale signs on the landscape that enable these areas to be recognized as slide-prone many centuries after the event occurs. These events cannot be predicted, but many can be anticipated from evidence of older prehistoric slides, and the area can be avoided for dense development or construction of sensitive facilities such as dams or power plants. The landslide which destroyed the Vaiont Reservoir in Italy should have been anticipated because landslide scars abounded in the valley, and the geological conditions were obviously ideal for massive failures.

Through deep time, there have been a few huge landslides (called ***sturzstroms***) involving billions of cubic yards of material. Movement of one of these has never been witnessed by humans, but they are evident from the rock record in places like the Shasta terrain of California, and they are clearly evident from high-altitude space-probe photographs of the surface of planet Mars.

In summary, mass wasting through deep time is characterized by (1) gradual movement as creep, (2) fairly regular seasonal times of increased landslide activity, and (3) irregular punctuations when massive slide events are triggered by major regional rainstorms or earthquakes. Times of major slide events cannot be predicted. Generally, the larger the landslide, the more infrequently it occurs over time.

**Roles of the Geologist in Minimizing Slope Stability Hazards**

**In field evaluation**. The more an area is susceptible to landsliding, and the more humans encroach into it, the greater the necessity for a complete geologic investigation.

The involvement of professional geologists should be absolutely mandatory in the site characterization for any structure whose failure may endanger human lives or significant amounts of property. No structure is safer than its geological setting.

Geologic examinations of slopes require evaluation of at least the following categories:

1. **The evaluation of topography—relief, steepness and shape of slope**
2. **The types and conditions of the bedrocks that underlie the slope**
3. **The type and thickness of soils present**
4. **The angle and direction of bedding planes or rock fabric**
5. **The frequency and the direction of any discontinuities (such as joints), the extent of the discontinuities, and the kind of infilling material present in them**
6. **The amount and type of vegetation on the slope**
7. **The sources of moisture and the moisture-retaining properties of the materials**
8. **The nature of surface and subsurface drainage, and any human-made drainage**
9. **A determination of the earthquake history of the area**
10. **Evidence of past slides, flows or rockfalls**
11. **An evaluation of possible volume of earth materials susceptible to failure, the possible styles in which they may fail, and the area that may be affected by failure.**

**In public service.** At the regional level, geologists serve as data gatherers, compilers and organizers of basic information that will become available for future use in development of the areas. Geologists employed by state geological surveys and the U.S. Geological Survey provide a tremendous service by constructing geological and slope stability maps. Mapping is based on field study of the soils and rock formations and is aided by remote sensing methods that include satellite and high-altitude photography and computer technology. Geologists' published maps are used by engineers, contractors, developers and homeowners.

**In research.** The quantitative estimation of slope safety is a common task performed by engineering geologists and geotechnical engineers. Calculating a **"factor of safety"** by estimating the ratio of those forces that tend to drive it downslope is a requisite of most engineering design projects that involve any construction on a sloping terrain. The more that is known about the nature of the geological materials and processes present at a site, the more likely a site will be developed without later regret. Research provides better ways to investigate sites and evaluate data. The mining and civil engineering professions benefit particularly from slope stability research.

A vigorous research program on the topic of massive earth movements has been ongoing in Russia for several years. Study of the way landslides move aids in the design of methods of blasting that move materials efficiently during mining and major construction projects. Benefits of such research serve the public whenever applied by engineers.

Computers provide continued research opportunities to model slope failures and rockfalls, and new programs remove the tedium from calculations that formerly were a necessary evil in conventional factor of safety calculations.

In this age of electronic and biological technology, it might seem to the lay reader that most answers to problems that involve something as apparently primitive as falling rock would have been answered, but such is not the case. In particular, the mechanisms that cause rocks to move surprisingly great lateral distances during rock avalanches are still not well understood. Slope stability and rock movement remain very suitable areas for research in our age of high technology.

**In education.** Geology is the most essential of all sciences for understanding what constitutes good land use. It is important that citizens understand slope hazards because so many home sites and developments are on slopes. Slope hazards should be covered in introductory geology and earth science courses, and such courses should be taught only by teachers with extensive formal education in geology.

Geologists in some universities are involved in teaching engineering geology and environmental geology to future designers. Geology should be part of the education of any civil or environmental engineer, environmental scientist or architect. Development and construction projects that are designed using the contributions of professional geologists have proven to be more successful in reducing damages from landslide hazards than those in which the geology is omitted or is performed by someone without credentials in geology. Yet only a few graduating civil engineers actually take a course in engineering geology, and only half take any geology courses.

Geologists are also involved in public education through establishing scientific areas where the public can investigate landslides, often by guided tour. One of the most expensive slides in history occurred near Thistle, Utah, in 1983, and over 200 million dollars was required for the relocation of a major highway and railroad. The site is now used as an educational resource and a classic area for scientific study. So also are the sites of the Madison Canyon landslide, Montana, and the Gros Ventre landslide, Wyoming. ❑

## Questions:

1. What is a "factor of safety"?

2. Which of these contributes to the likelihood of landslides: schistosity, loess soils, slope less than the angle of repose? Which does not?

3. Which is more common through deep time: creep or rapid landslides? Are large landslides more common than small ones?

Answers on page 301

24

*Despite decades of warnings from geologists and other scientists, real estate development along coastlines has continued. Orrin Pilkey, for example, is a geologist who has often publicized that the beach is a dynamic place. There is constant change, ranging from sand migration to sea level fluctuations. While many of us are aware of small-scale change on a daily basis, there is little recognition of the drastic changes that can occur over a few years. Even an inch in the rise of sea level can cause an encroachment of the sea of many feet. Hurricanes and other weather events can cause drastic changes on a still shorter time scale. Houses many feet from shore can be swept away, and large areas of beach can be removed. Construction of groins and other structures to reduce erosion in one area almost always leads to further erosion elsewhere. Even where attempts to control erosion work, such as by pumping sand to the beach from offshore, they are enormously expensive. Such economic costs are causing developers and local governments to realize what geologists have been saying all along: You cannot develop a beach by changing the natural geological and oceanographic dynamics. It is too costly in terms of lives and money. Instead, you must take the natural dynamics into consideration in developing a coastline; this often means relocation away from the shoreline.*

# This Beach Boy Sings a Song Developers Don't Want to Hear

## Orrin Pilkey sounds like a broken record when he says our coastlines can't be tamed, but people are finally beginning to listen

by Steve Kemper

A coastal newspaper, under the headline "Orrin . . . A Plague on Our Land," once gave Orrin H. Pilkey jr. a higher score than Chicken Little in the category of "Shrillest continuing doomsday performance, foretelling imminent ruin, destruction and disaster" and recommended deporting him to Greenland. On another occasion Pilkey so enraged the mayor and town council of Folly Beach, South Carolina, that they passed a special resolution of condemnation, calling his views "insulting, uninformed and radical both in content and intent."

Academic geologists rarely inspire much passion among the general public or attract much attention from politicians. Pilkey is a trouble-making, blunt-spoken exception. A genial 58-year-old who delivers fighting words with a smile, he directs the Program for the Study of Developed Shorelines at Duke University, where he is James B. Duke Professor of Geology. He is also the point man for scientists and environmentalists who believe that because of the "rush to the shore" that has deposited so many millions of people on the nation's coasts, our beaches and barrier islands are being developed to death, from the Great Lakes to the continental coastlines.

Pilkey's message begins with the geological fact that gives his best-known book its title: *The Beaches Are Moving: The Drowning of America's Shoreline*. Pilkey says that beaches exist in a dynamic equilibrium, responding to storms or a rise in sea levels by slowly migrating—retreating toward the land or, in the case of barrier islands, rolling over on themselves. When people interrupt this movement by compacting and confining sand with roads and buildings, says Pilkey, and further disrupt the flow of sand with groins, jetties and seawalls, they demolish nature's way of maintaining beaches and guarantee their destruction. Coastal geologists had been saying these things for years in professional journals without much notice. Pilkey took the science public, in a torrent of combative op-ed pieces, legislative testimony, books and public meetings.

"Most scientists put themselves on a cloud and think they're above dealing with the public," says Stanley Riggs, professor of geology at East Carolina University and occasionally Pilkey's coauthor, "but Orrin enjoys it. He also likes a fight. Not only is he built like a little bulldog, he takes hold of something as if it were a rag doll and just keeps shaking it until there's nothing left. He's done more on a national basis than any hundred of the rest of us."

What puts Pilkey into a hornet's nest of furious homeowners, businesspeople and politicians are his policy recommendations. Retreat from the beach. Prohibit new development. Outlaw seawalls and groins. If an eroding

***Reprinted from Smithsonian, October 1992, with permission of the author. Steve Kemper is a writer living in Connecticut.***

beach threatens a building, force the owner to move it or lose it. If a storm wallops a beachfront house, don't let the owner rebuild, or at least don't provide incentives for shoreline development with federal guarantees of low insurance rates. We can save shorefront homes and buildings for the benefit of a privileged few, Pilkey tells everyone from the U.S. Congress to the North Carolina Beach Buggy Association, or we can save the beaches for everybody.

When he began delivering this message a dozen years ago, the few people who paid attention called Pilkey crazy. Some people still do. But today, influenced largely by his studies and the glare of publicity that he has focused on beaches, North Carolina, South Carolina, Maine and Massachusetts prohibit open-ocean seawalls, and every state except Virginia makes it difficult for anyone who wants to armor a beach. Even Pilkey's archenemy, the U.S. Army Corps of Engineers, now uses "hard solutions" against beach erosion only as a last resort. Congress is considering legislation that reforms the National Flood Insurance Program to limit protection of property in eroding coastal areas.

Pilkey has made beaches a hot topic, but he says he's just warming up. "Coastal engineers are going to be even madder at me before long. Now we're going after some of the main principles and design assumptions behind beach replenishment, seawalls, and so on. I hope that in 10 or 15 years I will have fostered a revolution in coastal engineering."

Pilkey came to Duke in 1965 after three years at the University of Georgia's Marine Institute and has visited such places as Alaska, Portugal, Colombia and Korea to do research and advise governments on coastal issues. His office is a mess. Most of the time he can root around among the journals, slides, articles, correspondence and empty coffee cups to find what he wants—say, his latest contentious letter to the commander of the Corps of Engineers, or the old clip-on tie that he keeps around "for crises." Shelves hold some of the two-dozen books he has written or edited, including the 17 on various coastal states for the series "Living With the Shore." On one wall hangs the Francis Shepard Medal, the highest honor given internationally for marine geology, bestowed in 1987 for his work on turbidity currents and the way they sculpt the deepwater tablelands called abyssal plains.

"That was exciting stuff," he says of the research that occupied him for 20 years and made his academic reputation, "but nobody cares because it doesn't affect human beings." Several things, including constant seasickness, nudged him from the deeps toward the shore. In 1975, to provide his students with a geology guide for field trips, he, his father Orrin sr. and his brother Walter (both engineers) wrote a simple handbook called *How To Live With An Island*. A North Carolina official liked it and printed it.

"The response was unbelievable," says Pilkey. "I was amazed that with this little effort I could have this much impact." Four years later he coauthored *The Beaches Are Moving*, which created another stir and pushed him closer to shore. By the mid-'80s he had become a full-time coastal gadfly. He says the work sparks him because he "can use principles of geology on real world problems and really do something for mankind."

Pilkey was grounded in 1982 during a field trip with a group of high school students. "I accidentally dove 30 feet from the mast into 6 feet of water and broke my neck. But it paid off," he adds, wagging a finger. "All of those students still remember that dive." The injuries continue to nag him and ended his hobby of running marathons. A tidal delta has been accreting at his beltline every since, but on field trips he churns down the beach at a pace that leaves many of his students straggling behind.

He lives on 13 acres in Hillsborough, North Carolina, with his wife, Sharlene, who calls him Pilkey. Their menagerie of animals once included turkeys, peacocks, guinea hens, geese, ducks, chickens, dogs and a goat that attacked Volkswagen Rabbits. Everything but the chickens and dogs is gone now, along with the Pilkeys' five children. In the backyard, a hawser tied to the chimney runs down to a 700-pound anchor.

One night last February, despite blowing snow and bitter cold, more than 200 people showed up in Chatham, Massachusetts, at the elbow of Cape Cod, to hear Pilkey speak. Five years ago a winter storm sliced through the barrier island in front of the town, opening an inlet that exposed the inner shoreline directly to ocean waves. Since then the harbor has begun to choke with sand, threatening the small fishing fleet. Waves and storms have taken big bites out of the shoreline. Ten houses have fallen into the sea or been demolished, and 30 more are under siege.

People want to build seawalls to protect their property, but the town rejects most applications, citing regulations. Owners can appeal, but waves and sand sometimes move more quickly than bureaucracies. One homeowner wrapped his house in red tape just days before the ocean ate it. People are worried about their property, their town, their beaches, their income from tourism—their future. They want solutions.

Enter Pilkey. At 5 foot 4 and 220 pounds he is built like a grenade. Bearded, in old gray jeans and a hooded sweatshirt, he looks more like the fishermen in the audience than like the tweed-and-turtleneck townsfolk. He dresses the same way to teach a class, dig on a beach or speak at a chamber of commerce luncheon.

His first slide is of the Morris Island lighthouse, once a beacon on land, now an island some 400 yards into the open ocean outside Charleston, South Carolina. He shows them St. Catherines Island, Georgia, where trees undercut by surf are falling onto the beach. "But there's no erosion problem here because no one has built anything on the beach." He flashes a picture of construction workers fishing from an unfinished high-rise condo in Garden City, South Carolina. "One thing we've learned is that you shouldn't be able to fish from your condo window."

He explains that since beach migration and rising sea levels are inexorable, coastal communities have three choices: "hard" stabilization (for example, groins and seawalls); replenishment of the beach by bringing in more sand; or relocation and retreat. "There's no doubt that seawalls degrade beaches," says Pilkey, flashing slides of fully armored beaches in New Jersey, Puget Sound and Japan, where they have been scoured away by waves smashing against seawalls.

By contrast, replenishment doesn't destroy beaches but is expensive and even more temporary. Pilkey thinks it makes the most sense in urban settings—New York City's Coney Island, Miami Beach, Rio de Janeiro. For smaller communities, the economics are more questionable. Ten years ago, Ocean City, New Jersey, paid $5.2 million to replenish a beach; six months later, a third of it was gone. This year, at an initial cost of $39 million, the process is starting all over again. Ocean City, Maryland, got a $14.2 million beach in 1988, spent another $25 million in 1990 and '91 for sand and a seawall, lost its protective dune and parts of its beach last winter, and spent $12 million for more sand this past summer. Still, Pilkey tells his Chatham audience that they may want to consider replenishment, at least in preference to seawalls.

The third possibility, relocation and retreat, is Pilkey's favorite for reasons that mix common sense, science and esthetics. "On the shore, nature always bats last," he says, so coastal communities should begin planning now for orderly retreat. If they choose to fight on the beaches, their retreat will be just as inevitable but far more frantic and expensive. "If the expectations of sea level are realistic," he says, "in the future we won't be worrying about places like Chatham and Nags Head but about Charleston and New York and Galveston."

Nine houses in Chatham were directly opposite the breach. Only two remain—the two with seawalls. John Whelan's house is one of them. "I heard what Mr. Pilkey said about seawalls," he says. "I hate to think that I'm hurting my neighbors, but I didn't feel I had a choice."

"It's an understandably selfish reaction," says Richard Miller, chairman of Chatham's Waterways Committee, whose beachfront house is also threatened. "People say, 'I don't care what the results are, I'm going to save this damn house.' But my wife and I have made the philosophical decision not to revet. We will either move it or lose it."

*It Was Time to Haul Out the Sandbags*

Alfred Nelson's summer home, down the beach from Whelan's, is assessed at $1.2 million. Water has crept to within 19 feet of it. Nelson's petition for a seawall was denied three times by the Chatham Conservation Commission. Last year he spent $30,000 on sand and vegetation, but a storm whisked it all away in half an hour, leaving four feet of salt water in his basement and $60,000 in damages. He appealed to the state's attorney general, who in early summer allowed him to fortify his beach temporarily with emergency sandbags, pending a final decision in his suit against the state. So far he has spent about $70,000 on sandbags. If the court denies his request for a permanent seawall, he expects his house to wash away soon after the sea claims his sandbags. "So what do I think of people like Pilkey?" he asks. "They're taking our property, and we have a right to protect it."

Nelson is one of several people suing the state. The number of lawsuits surely will rise with the advancing waterline. Yet the powerful appeal of beachfront property ensures that people will continue to manifest what Whelan calls the "greater-fool phenomenon." There will always be buyers, delighted by the lovely views and surf, who don't see a dynamic system pushing back the shore without regard for property lines.

On another day, Pilkey is barreling down Virginia Beach in Virginia, with his arms outstretched as if to embrace the coast. He is leading 35 students, two reporters, and visiting professors from Portugal, Korea, and China on one of his frequent field trips along the Atlantic shore. This one will go south from here to Ocracoke, in the Cape Hatteras National Seashore on North Carolina's Outer Banks. "We will see many of the spectacular atrocities," Pilkey promises.

Near the top of his list is Sandbridge, a wealthy residential section of Virginia Beach. The shorefront residents here own most of the beach to the low-tide mark, which means that Pilkey's parade is engaged in trespassing. He shrugs. "You can't see American beaches without trespassing."

From the beach, Sandbridge looks like a fortress. About 150 of the approximately 200 homes on the retreating shoreline are defended by long stretches of bulkhead, much of it sheets of steel six to ten feet high. Sandbridge's beach is washing away because bulkheads prevent the sand from moving up onto the shore. To stay dry at high tide in summer, sunbathers position their lounge chairs sideways at the base of the bulkheads. Waves have undercut the houses without bulkheads, exposing pipes and septic tanks, and causing stairways to jut into space eight feet above the sand. New houses are going up on the beach. "Look," says Pilkey, pointing to a three-story castle 100 feet from the high-tide line. "They're just washing the labels off the windows. That is insane."

At the end of the walk, Pilkey's group bumps into George Owens, owner of a beachfront house and president of the Sandbridge Oceanfront Property Owners Association. Like most people in Middle Atlantic beach towns, Owens recognizes Pilkey and has seen his PBS television special about beaches-on-the-move.

"It's depressing to sit there and agree with you, hearing the waves slap up," he tells Pilkey. "I don't want to be like New Jersey with the entire coast armored, but I do want to live on the ocean." He doesn't think the narrowing beach is related to the bulkheads and cites some coastal-engineering reports.

Up to now the beachfront owners in Sandbridge have paid for their own protective measures—$8 million over the past ten years. Now Owens says they desperately need a proposed $6 million to $9 million replenishment project that would be funded by Virginia Beach and U.S. taxpayers. "Sandbridge generates over $3 million in tax revenues each year," says Owens. "Shouldn't you protect that?"

This is the sort of situation that winds Pilkey up. "Those people caused the problem by putting up bulkheads. As long as the bulkheads are there, the beach will have to be replenished every few years, and those people should pay for every cent of it themselves." As for engineering claims that seawalls are innocent, Pilkey retorts: "Ludicrous!"—a favorite adjective when discussing coastal engineering.

At 7 the next morning in Nags Head, Pilkey is already wet. "We've got a real storm," he says, raising his arms like an exultant boxer. "Come out and see science in action. It's beautiful out there." It's a gray day of driving rain, with wind gusts on the beach sandblasting the skin off faces and ears.

Tsung-yi Lin, Pilkey's graduate student from Taiwan, is setting up traps to measure how much sand the wind is moving. Lin explains that this can be affected by the beach's vegetation and topography; the sand's grain size, wetness or dryness; and the wind's speed, direction and duration. These variables combine differently every hour of every day. Yet, Pilkey says, engineers rarely consider the fact that wind carries large amounts of sand onto and off the beach. Lin and Pilkey think this is a serious omission.

Pilkey is convinced that the movement of sand—whether on the beach, in the surf zone or on the near shore—is far too complex and unpredictable to be captured in mathematical equations. Yet such equations are the basis of coastal engineering, which is used to justify and fund beach projects. For example, Pilkey says the Corps of Engineers overestimates the life spans of replenished beaches and hence underestimates their long-term costs. The Corps says that its replenished beach at Sea Bright, New Jersey, will need more fill in six years; Pilkey says one to three years. For Myrtle Beach, the Corps says eight years; Pilkey, three to five.

The Corps disputes Pilkey's numbers and findings. "Orrin has an idea of the coast and how it operates, and it's become a religion with him," says Jim Houston, director of the Corps' Coastal Engineering Research Center in Vicksburg, Mississippi, and author of several recent articles criticizing some of Pilkey's methodology. "You have to be careful you're not using your religion to do science."

Pilkey bristles at charges of sloppy science. But the larger point is that whenever coastal geologists and coastal engineers talk about the shore, they seem to be describing different places. This complicates the public debate, since both sides can easily line up experts whose conclusions are totally at odds.

Pilkey certainly has ample proof that engineered "solutions" often worsen the problem or cause new ones. Thirty years ago the Corps of Engineers spent ten years and $32 million to turn Florida's flood-prone, serpentine Kissimmee River into a neat canal. That led to suffocation of the ecosystem in the surrounding 75-square-mile river basin. Now the Corps intends to put the kinks back in the river, at a cost of $370 million over 15 years. In South Carolina, the jetty that keeps Charleston Harbor open is starving downshore beach communities of sand. And the Corps' groins that have preserved the beaches on the eastern end of Westhampton, Long Island, have caused severe erosion at the western end, leaving a number of Westhampton houses in the surf. The property owners victimized by these examples of governmental mistakes, however, do not get any sympathy from Pilkey. "If we began to pay people for all the Corps' mistakes on the shoreline," he says, "it would destroy the federal budget."

This attitude separates Pilkey even from some of his allies. Says Joseph T. Kelley, a professor at the University of Maine, the state's marine geologist and a good friend of Pilkey's, "I agree with most everything he says. But I work for the state government, and when people's homes get washed away, they're at my front door the next day crying. I don't think Orrin has ever had to deal with people like that. They're real people feeling real pain. So I differ in how I handle the subject, though not in the conclusions."

Pilkey's parade moves down the beach, past a solid line of houses and hotels, to South Nags Head. In North Carolina, sandbags may be used as a temporary measure, but the state prohibits seawalls and other hard solutions, and many people here who formerly lived in what was the second row of houses now have an unobstructed ocean view, courtesy of a recent storm. A few condemned houses totter on crooked stilts at the tide line, testimony to the fact that armoring or no, the sea is going to go where it wishes. "Here you see the results of North Carolina's anti-armoring policy," Pilkey tells his students, gesturing at the gaps in the neighborhood and the doomed houses.

That night, Pilkey runs into Willie Etheridge, an old foe who heads the family fishing business. A burly, powerful man with a ruddy face, Etheridge berates Pilkey once again for opposing the Corps' plan to stabilize nearby Oregon Inlet with massive jetties. The Corps has already built one structure on the southern side of the inlet. Until the Bonner Bridge spanned it in the mid-'60s, the inlet was steadily moving south. To maintain a channel for fishing boats, the Corps must thwart the inlet's ceaseless urge to move on. At the moment, that costs more than $8 million a year for dredging. Pilkey and other scientists believe that the Corps' plan for jetties, which has been rejected several times but is under review again, would cause severe damage to beaches downshore. This would be an economic boondoggle, Pilkey says, costing millions for maintenance and beach repair. Etheridge doesn't believe it and asserts that other jetties on the Atlantic coast "are all working today with no damage."

"Look at Florida," counters Pilkey. "The most jettied place in the country, and every one of them is causing erosion. Every single one, without exception. A jetty at Oregon Inlet is too big a risk."

"My father says, 'Dr. Pilkey's name should be Dr. Maybe, Dr. Could Be or Dr. Possibly,'" says Etheridge.

Pilkey winces but nods. "You have to be that way sometimes, as a scientist."

"Let me tell you a story," says Etheridge. "I was talking to a fisherman the other day and he said, 'Every time I go through Oregon Inlet my throat gets so dry I'm afraid if I don't have a cough drop in my mouth I won't be able to call for help if anything happens.' If we had that jetty, the inlet would be safe, and he wouldn't have to worry about it." He stares at Pilkey, then shakes his head. "I once heard Dr. Pilkey asked whether he would take all the houses off the shore if he could, and leave it the way it was before—and he said yes."

"I think you're probably quoting me accurately there," says Pilkey.

Pilkey chews on the conversation for the rest of the trip. "The chief coastal engineer of the Corps in North Carolina once said in a public meeting on Oregon Inlet, after I and other scientists had already talked, 'You've been hearing a lot of probablys and maybes. The Army Corps doesn't give you probablys, we give you facts'—the implication being that they know exactly how the shore behaves. Pure nonsense."

The pretense of certainty is one of Pilkey's main gripes about coastal engineers, exemplified for him by the fact that they don't use error bars when estimating the longevity and costs of replenishing beaches. "Yes, scientists should have error bars," says the Corps' Houston, "but if you're voting for something in Congress, you don't want to know plus or minus how much it's going to cost, you want a figure." Counters Pilkey, "If science can't make a prediction, science *shouldn't* make a prediction."

Not far south of Oregon Inlet, Pilkey stops to fume about a new wall of sandbags running for a mile or so between the ocean and Highway 12, the only road onto and off Hatteras Island. Until recent years the road was behind an artificial dune. Wind and waves have gradually shaved the dune away, and now winter storms regularly breach it in places, flooding the highway and leaving behind heaps of sand. North Carolina wants to save the road. Pilkey, however, takes the island's perspective. He says that it is responding to the threat of rising sea level by overwashing in preparation for migration, and that the road should be moved to the back of the island or replaced with a ferry service.

"Why are we making one of our last natural barrier islands into an engineering project?" he asks his students, though this natural island featured a man-made dune to the highway's east and has a series of man-made duck ponds to its west. Pilkey says that eight communities to the south depend on the highway as an escape route during hurricanes, so they might disagree with him. "But where does this lead?" he asks, gesturing toward the sandbags. "Are we going to have a wall all the way down the Outer Banks?"

Next stop, Cape Hatteras lighthouse. In 1870 it was 1,500 feet behind high tide, now it is fewer than 200. It is protected by a series of groins and a mountain range of sandbags, dotted today with sunbathers who don't seem to mind that the beach is nearly gone. Pilkey is excited because the National Park Service has decided to move the lighthouse inland which, he says, sets a good example of using retreat for coastal management.

But what about densely populated coastal communities? Does Pilkey expect Ocean City, Maryland, or Sea Bright, New Jersey, to abandon their houses and high-rise hotels, their income from tourists and taxes?

"No, but they do need to face up to what it's going to cost," he answers. "Once they know it's going to cost $2 billion to replenish their beach for the next 50 years, then they can decide whether it's worth it, and they can put into effect things that could move them back in 30 years. Sea Bright has long since spent more than the value of the community to save the community. At what point does it become sort of stupid?"

People in most coastal communities are likely to respond much as Robert Linvill, mayor of Folly Beach, South Carolina, does: "Pilkey is advocating abandoning the homes we've had here for many years. The hell with that."

But Beth Millemann, executive director of the Coast Alliance, a national conservation groups. finds Pilkey anything but unreasonable. "If he was saying these things about people who wanted to build on the edge of a volcano, nobody would question it. But since he's saying it about a subject with a lot of money at stake, some people call him crazy."

Meanwhile, Pilkey is off on another field trip, this time to unpopulated Shackleford Banks in North Carolina's Cape Lookout National Seashore. "It's the most beautiful island in the world," he says, standing on top of a dune, "but the wild horses on it are killing the estuarine side by eating all the spartina. Now if only we could bring in some cougars," he says, rubbing his beard with a mischievous grin, thinking about beaches, looking for a fight. ❑

**Questions**:

1. Pilkey says that coastal communities have three choices to cope with the inevitable rise of sea level and beach migration. What are the three choices?

2. What are two problems with beach replenishment? Is this the favorite solution to beach development problems, according to Pilkey? If not, what is?

3. Give examples of how engineered "solutions" often worsen or cause problems.

Answers on page 301

*Next to the Earth, Mars may have had the most complex geologic history of all the planets in our solar system. Although it lacks plate tectonic processes, it contains evidence of extensive volcanism, including massive volcanoes. Most fascinating to many people is the evidence of surface water. There is evidence that massive floods have occurred and that lakes and rivers may have existed, even if only for brief periods of time. This has brought up the possibility that Mars may have once had a warmer climate that would have allowed liquid water to exist at the surface. Additional evidence suggests that ground ice exists at shallow depths in the highland regions. Recent findings only increase our fascination with this planet and may one day convince enough people that this is a place we should visit.*

# Geology of Mars

by Michael H. Carr, Ruslan O. Kuzmin, and Philippe L. Masson

*Mars has had a diverse geologic history. An ancient, heavily cratered surface preserves evidence of events of the planet's earliest history. Volcanic activity has continued until the recent geologic past and has resulted in the formation of huge shield volcanoes and extensive lava plains. Most of the volcanic activity was concentrated in Elysium and in the much larger Tharsis region, which is at the center of a 4,000-km-diameter, 10-km-high bulge on the planet's surface. Arrayed around the bulge are radial fractures that may be largely responsible for the immense canyons on the east flank of the bulge.*

*Massive floods have erupted periodically onto the surface of the planet, and most of the oldest surfaces are dissected by networks of branching valleys. The networks seemingly were formed by the slow erosion of running water and suggest warmer climates in the past. Evidence for ground ice is common at high latitudes. Although wind-deposited materials are common, the cumulative effect of wind erosion is small. The lack of plate tectonics and the ineffectiveness of wind, water, and ice in reducing relief have led both to the production of features of enormous size compared to those on Earth and to their nearly perfect preservation.*

**Introduction**

Mars, the fourth planet from the Sun, has long been of special interest among the planets because of the potential for eventual human exploration and the possibility of indigenous life. The modern era of Mars exploration started on July 16, 1965, when the Mariner 4 spacecraft flew by the planet. Since that time, the USA and the USSR have sent numerous fly-bys, orbiters, and landers to Mars, the latest being the Viking orbiters and landers, which operated at the planet between 1976 and 1980, and the Soviet Phobos 2 spacecraft, which went into orbit around the planet in 1989. As a result of these missions, we have photographed from orbit almost all the surface at a resolution of 200 m/pixel and small fractions of the surface at resolutions as low as 8 m/pixel. In addition, we have closeup views of the surface and direct chemical measurements at the two Viking landing sites. The various missions have revealed a geologically diverse planet that has huge volcanoes, enormous canyons, extensive dune fields, and numerous channels seemingly eroded by running water (for comprehensive summaries, see Baker, 1982; Carr, 1981). The abundant evidence for water erosion is particularly intriguing as the present climatic conditions are such that liquid water cannot exist at the surface. Water erosion suggests that the planet may have been warmer in the past and more hospitable to life.

Mars has a diameter of 6,788 km, which is intermediate between that of the Earth (12,756 km) and the Moon (3,476 km). The atmosphere is composed of 95.3 percent $CO_2$, 2.7 percent $N_2$, and 1.6 percent argon. The average atmospheric pressure is 7 millibars, but the pressure varies during the year because about 30 percent of the $CO_2$ in the atmosphere condenses on the winter pole and sublimes back into the atmosphere in summer. Surface temperatures range from 150°K at the winter pole to a daily average of 215°K at the equator. At low latitudes, the diurnal temperatures range from about 170°K to 290°K. Under these conditions, liquid water is unstable everywhere, and the planet has a permafrost zone that is several hundred meters thick at the

*Reprinted from Episodes, v. 16 nos. 1 and 2, March & June 1993*

equator and kilometers thick at the poles. At low latitudes (<40°), water ice is unstable at the surface and at all depths below the surface. It will sublime into the atmosphere at rates that are dependent on the permeability of the overlying rocks. Therefore, near-surface materials at low latitudes should have lost all their unbound water. At latitudes 40–80°, ice is stable at depths greater than about a meter below the surface. At the poles, water ice is stable at the surface and has been detected at the north pole, where it is exposed when the seasonal $CO_2$ cap sublimes in summer. The atmosphere contains very little water. On average, if all the water were to precipitate out of the atmosphere, it would form a layer only 1 micrometer deep. Nevertheless, the atmosphere is close to saturation for nighttime temperatures, and water-ice clouds are common. The opacity of the atmosphere varies considerably. During southern summers, dust storms are common; some reach global proportions and effectively obscure the surface from view for several months.

We have closeup views of the surface only at the two Viking landing sites. The Viking 1 lander is located at 22.5° N., 48° W. in a sparsely cratered area that has wrinkle ridges and, from the orbiter, looks like a lunar mare. The surface has a rolling topography and is strewn with rocks in the centimeter to meter size range. On and between the rocks is fine-grained material, which also is present as drifts ranging in size up to 10 m across. The surface of the drift material is partly cemented to form a duricrust, which indicates that the more volatile and soluble components have migrated within the soil profile. Several small areas, seemingly free of the drift material, appear to be bedrock exposures. Although volcanic, the blocks probably have been ejected from impact craters. They are more common than on lunar maria partly because the atmosphere protects the surface from erosive effects of micrometeorite bombardment. The second Viking lander is located at 48° N., 225.6° W., seemingly on a debris flow from a large impact crater, Mie, 170 km to the east. The view from the lander shows a flat plain that is almost featureless except for numerous rocks, mostly in the 10–20 cm size range, that are distributed uniformly over the entire scene.

**Highlands and Plains**

The surface of Mars can be divided into two main components: ancient cratered highlands, covering most of the southern hemisphere, and low-lying plains, which are mostly at high northern latitudes. Superimposed on these two components are the high-standing volcanic provinces of Tharsis and Elysium.

The cratered highlands cover almost two-thirds of the planet. They are mostly at altitudes of 1–3 km above the datum, in contrast to the northern plains, which are mostly 1–2 km below the datum. The cause of the division between the highlands and plains is unclear, but it may be the result of a very large impact at the end of accretion (Wilhelms and Squyres, 1984). The density of impact craters in the martian highlands is comparable to that of the lunar highlands. The surface clearly dates back to the very earliest history of the planet when impact rates were high. On the Moon, the transition from very high impact rates to rates comparable to the present took place at about 3.8 Giga-annum, and the transition probably took place at the same time on Mars. The martian highlands differ from the lunar highlands in three main ways. The first difference is that sparsely cratered plains are more common between the larger craters in the martian highlands. These plains may be volcanic, as indicated by the presence in a few places of flow fronts and wrinkle ridges that resemble those on the lunar maria. On the other hand, many of the plains could be of impact or sedimentary origin. The second difference between the lunar and martian highlands concerns the impact ejecta. The ejecta around lunar craters generally has a coarse, hummocky texture at the rim crest and grades outward into a finer, randomly hummocky or radial texture. On Mars, the ejecta around craters 5–100 km in diameter is arrayed commonly in discrete lobes, each lobe being outlined by a low ridge. This is true of almost all martian impact craters in this size range, irrespective of location. Two reasons have been suggested for the characteristic martian ejecta patterns. The first suggestion is that impact craters above a certain size penetrate the permafrost zone and eject water-laden, or ice-laden, materials that tend to flow across the surface following ejection from the crater (Carr and others, 1977). The second suggestion, based on wind-tunnel experiments, is that the interaction between the ejecta and the atmosphere causes the flowlike patterns (Schultz and Gault, 1984). The third difference between the lunar and martian highlands is the presence, within the martian highlands, of numerous networks of branching valleys. At low latitudes, these are almost everywhere within the highlands. They superficially resemble terrestrial river valleys and are believed to be formed by running water. They will be discussed more fully below in the section on "Water Erosion."

The plains are located mostly in the northern hemisphere. The number of craters superimposed on them varies substantially and indicates that they continued to form throughout the history of the planet. The plains are diverse in origin. The most unambiguous in origin are those on which numerous flow fronts are visible. They clearly formed from lava flows superimposed one on another, and they are most common around the volcanic centers of Tharsis and Elysium. On other plains, such as Lunae Planum, flows are rare, but wrinkle ridges like those on the Moon are common. These also are assumed to be volcanic. However, the vast majority of the low-lying northern plains lack obvious volcanic features. Instead, they are curiously textured and fractured. Many of their characteristics have been attributed to the action of ground ice or to their location at the ends of large flood features, where lakes must have formed and sediments must have been deposited. In some areas, particularly around the north pole, dune fields are visible. In yet other areas are features that have been attributed to the interaction of volcanism and ground ice. Thus, the plains appear to be complex in origin; they formed variously by volcanism and different forms of sedimentation and then subsequently were modified by tectonism and by wind, water, and ice.

**Volcanism**

The most prominent volcanoes are in two regions, Tharsis and Elysium. Tharsis is at the center of a bulge in the planet's surface, the bulge being over 4,000 km across and 10 km high at the center. A similar bulge centered on Elysium is about 2,000 km across and 5 km high. Three large volcanoes are close to the summit of the bulge, and Olympus Mons, the tallest volcano on the planet, is on the northwest flank. All these volcanoes are enormous by terrestrial standards. Olympus Mons is 550 km across and 27 km high, and the three others have comparable dimensions. Lava flows and lava channels are clearly visible on their flanks, and each has a large, complex summit caldera. They all appear to have formed by the eruption of fluid lava with very little pyroclastic activity. The large size of the volcanoes has been attributed to the lack of plate tectonics on Mars and to the greater depths of their magma source, as compared to the depths for terrestrial volcanoes (Carr, 1973). The small number of superimposed impact craters on their flanks indicates that their surfaces are relatively young, although the volcanoes may have been growing throughout much of Mars' history. To the north of Tharsis is Alba Patera, the largest volcano in areal extent on the planet. It is roughly 1,500 km across but only a few kilometers high. Flows are visible on parts of its flank, but elsewhere, the volcano flanks are dissected by numerous branching channels. The easily eroded, channeled deposits have been interpreted as ash (Wilson and Mouginis-Mark, 1987). Densely dissected deposits on other volcanoes such as Ceraunius Tholus in Tharsis, Hecates Tholus in Elysium, and Tyrrhena Patera in the southern highlands also have been interpreted as ash. Thus, Mars seems to have experienced both the Hawaiian style of volcanism, involving the effusion of fluid lava, and more violent, pyroclastic eruptions, which have resulted in the deposition of extensive ash deposits.

Elysium Mons appears to be a shield volcano that formed largely from fluid lava. However, huge channels start at the periphery of the volcano and extend northward down the regional slope for hundreds of kilometers. The channels have streamlined forms and enclose teardrop-shaped islands. Similar large channels start adjacent to Hadriaca Patera on the rim of the large impact basin Hellas. They start at the volcano and extend for hundreds of kilometers down into the Hellas basin. All these channels show characteristics that result from large floods. They are thought to have formed by the massive release of water following melting of ground ice by the volcanoes. Numerous other features in the Elysium area and elsewhere also have been interpreted to be the result of volcano-ice interactions (Squyres and others, 1987).

**Tectonism**

The most widespread indicators of surface deformation are normal faults, indicating extension, and wrinkle ridges, indicating compression. The most obvious deformational features are those associated with the Tharsis bulge. Around the bulge is a vast system of radial grabens that affect about one-third of the planet's surface. The grabens are particularly prominent north of Tharsis, where many are diverted around the volcano Alba Patera and form a fracture ring. Circumferential wrinkle ridges also are present in places, particularly on the east side of the bulge in Lunae Planum. Both the fractures and the compressional ridges are believed to be the result of stresses caused in the lithosphere by the presence of the Tharsis bulge. No comparably extensive system of deformational features occurs around Elysium, but fractures are located in other places where the crust has been differentially loaded, as around large impact basins, such as Hellas and Isidis, and around large volcanoes, such as Elysium Mons and Pavonis Mons.

The vast canyons on the east flanks of the Tharsis bulge are the most spectacular result of crustal deformation. The canyons extend from the summit of the Tharsis bulge eastward for 4,000 km, where they merge with the chaotic terrain and large channels south of the Chryse basin. In the central section, where several canyons merge, they form a depression that is 600 km across and several kilometers deep. Although the origin of the canyons is poorly understood, faulting clearly played a major role (Masson,

1977, 1985). The canyons are aligned along the Tharsis radial faults, and many of the canyon walls are straight cliffs or have triangularly faceted spurs, which clearly indicate faulting. In one location, a large impact crater is present on the canyon floor, as though the floor were a downfaulted section of the surrounding surface. Other processes also were involved in shaping the canyons. Parts of the walls have collapsed in huge landslides; other sections of the walls are deeply gullied. Fluvial sculpture is particularly common in the eastern sections. Faulting may have created most of the initial relief, and then the relief enabled other processes, such as mast wasting and fluvial action, to take place. Creation of massive fault scarps also may have exposed aquifers in the canyon walls and allowed ground water to leak into the canyons and thereby form temporary lakes.

**Water Erosion**

One of the most puzzling aspects of martian geology is the role that water has played in the evolution of the planet. We have seen that liquid water is unstable at the surface under present climatic conditions, yet we see abundant evidence of water erosion. The most intriguing features are large, dry valleys, interpreted as having formed by large floods. Many of the valleys start in areas of what has been termed "chaotic terrain," in which the ground seemingly has collapsed to form a surface of jostled and tilted blocks that are 1–2 km below the surrounding terrain. The areas of chaotic terrain range in size up to several hundred kilometers across. The largest areas are in the Margaritifer Sinus region east of the canyons and south of the Chryse basin. In this area, large, dry valleys emerge from the chaotic terrain and extend northward down the regional slope for over 1,000 km. Several large channels to the north and east of the canyons converge on the Chryse basin and then continue farther north, where they merge into the low-lying northern plains. The valleys emerge full-size and have few, if any, tributaries. They have streamlined walls and scoured floors, and they commonly contain teardrop-shaped islands. All these characteristics suggest that they are the result of large floods, rather than the result of slow erosion by running water. Although most of the flood channels are around the Chryse basin, they are found elsewhere. Those near Elysium and Hellas have been mentioned already, and others occur in Memnonia and western Amazonis. Impact craters superimposed on the flood channels suggest that they have a wide range of ages.

The floods were enormous. The largest known terrestrial floods are those that cut the channeled scablands in eastern Washington State in the USA in the late Pleistocene. These are estimated to have had peak discharges of about $10^7$ cubic meters per second ($m^3/s$), as compared with the average discharge of $10^5$ $m^3/s$ for the USA's Mississippi River today. Discharges for the martian floods are estimated to have been as high as $10^9$ $m^3/s$ (Baker, 1982). The amount of water that flowed through the channels around the Chryse basin is estimated to be the equivalent of 45 m spread over the whole planet (Carr and others, 1987). The cause of the large floods is unclear, and they may not all be of the same origin. We already have seen that the floods in Elysium and Hellas were, in some way, connected with volcanism. Volcanism may have resulted in the melting of ground ice, the migration of ground water beneath the permafrost, and ultimately, the catastrophic breakout of water. Similarly, volcanism in Tharsis may have resulted in the outward migration of ground water to the lower lying regions around the edge of the Tharsis bulge, where repeated breakouts took place. Sediments within the large equatorial canyons suggest that the canyons contained lakes at one time, probably as a result of ground-water flow out from under the surrounding plateau. Catastrophic release of water from these lakes may have caused some of the large channels that connect with the canyons to the east. After the floods were over, large lakes must have been left at the ends of the channels. The fate of this water would have depended on the latitude of the lakes. Many of the Chryse channels and those in Elysium and Hellas end at high latitudes. If climatic conditions then were similar to present conditions, these lakes would have frozen and possibly formed permanent ice deposits, which may account for some of the peculiarities of the low-lying high-latitude plains. Other channels, such as Mangala Vallis, end at low latitudes. In these cases, the water in the terminal lakes would have frozen and then sublimed into the atmosphere, to be frozen out ultimately at the poles.

Other fluvial features appear to be the result of slow erosion by running water. Networks of branching valleys are found throughout the heavily cratered terrain and in a few places on younger surfaces. They resemble terrestrial river valleys in that they have tributaries and increase in size downstream, although only rarely can a channel be observed within the valley. The valleys generally are short compared with terrestrial river systems, most being less than a few hundred kilometers in length, so rarely does one valley system dominate the drainage over a large area. The most plausible explanation for the valleys is that they formed by the slow erosion of running water. The open nature of some networks, the alcovelike terminations of tributaries, and the range of junction angles between branches all suggest ground-water sapping (Pieri, 1980). Other networks, however, lack these characteristics and more resemble valleys formed by surface runoff.

The origin of the valley networks is controversial. Their branching pattern and small size indicate that they were formed by the slow erosion of running water, but the precise mode of formation and the climatic conditions required are both unknown. Under present climatic conditions, small streams on Mars would freeze rapidly so that the source of water for the larger rivers downstream would be cut off. Most of the valley networks are in the cratered highlands, the oldest part of the planet's surface. The valley networks suggest, therefore, that early Mars was much warmer than now, and liquid water could flow across the surface. Greenhouse calculations indicate that, in order for surface temperatures on Mars to be above freezing, a $CO_2$ atmosphere of 1–3 bars thick is required (Pollack, 1979). If such a thick atmosphere were present, it would be inherently unstable because the $CO_2$ would react with the surface rocks in the presence of liquid water and would form carbonates. It has been suggested, therefore, that Mars had an early, thick atmosphere that kept Mars warm, but that early in the planet's history, this atmosphere largely dissipated as a result of carbonate formation, and the planet assumed today's cold, hostile conditions. This simple model is questionable, however, in view of the presence of younger valley networks, such as those on the volcano Alba Patera. Some of these younger valleys have been attributed to local hydrothermal recycling of water, but Baker and others (1991) claim that the younger valleys are also the result of climate change. They think that Mars has experienced repeated, massive changes in global climates and that these changes were triggered by the large floods. Another problem is that massive carbonate deposits have not been detected. If the valley networks are the result of global climate change, then the 1–3 bars of $CO_2$ should be in carbonate deposits younger than the valley networks themselves, but remote sensing has failed to reveal them.

**Ground Ice**

Several features of the martian surface are explained most plausibly as resulting from the presence of ground ice (Kuzmin, 1983; Lucchitta, 1981). We have seen that ice is unstable at all depths below the surface at low latitudes, but it is stable at depths deeper than a few meters at high latitudes. Although ice is unstable at low latitudes, it may be found at depths of a few meters to several hundred meters because of the slow rate of diffusion of water vapor away from the ice, through the overlying materials, into the atmosphere. Kuzmin (1980) attempted to measure the depth to the top of the ground-ice layer by studying the morphology of impact craters, on the assumption that petallike ejecta patterns were due to the presence of ground ice or ground water. He found that, in any given region, the petallike patterns were only around craters larger than a certain minimum size. The size decreased from 6–8 km in diameter at the equator to 2–4 km at 40° latitude. From this, he inferred that ice is common at depths greater than about 300 m at the equator and at depths greater than about 100 m at 40° latitude.

Debris flows also may indicate the presence of ground ice. In the 35–50°-latitude band in both hemispheres, debris flows commonly are seen at the bases of cliffs. These are convex-upward flows that extend about 20 km away from the bases of the cliffs. Such features are rare, if present at all, at low latitudes. The simplest explanation is that at low latitudes where cliffs form, talus simply accumulates on the cliff slope and so inhibits further erosion. At high latitudes, however, because of the presence of ground ice, ice becomes incorporated in the talus and thereby enables the talus to flow away from the cliff, which exposes the slope to further erosion. Debris flows are particularly common in regions of what has been termed "fretted terrain," in which flat-floored valleys, filled with debris flows, extend from low-lying plains far into the cratered uplands. The formation of these valleys appears to be connected in some way with the formation of the debris flows. A general softening of the terrain at high latitudes also has been attributed to ground ice. At low latitudes, many features, such as crater-rim crests, are crisply preserved, whereas poleward of about 40° latitude, similar features are founded or softened in appearance. Squyres and Carr (1986) attributed the general softening at high latitudes to ice-abetted creep of the near-surface materials.

**Poles**

At each pole, and extending outward to about the 80°-latitude circle, is a thick stack of layered sediments. The sediments are at least 1–2 km thick in the north and at least 4–6 km thick in the south. Incised into the smooth, upper surface of the deposits are numerous valleys and low escarpments. These curl out from the pole in a counterclockwise direction in the north and in a predominantly clockwise direction in the south. Between the valleys, which are roughly equally spaced and 50 km apart, the surface of the deposits is very smooth and almost free of craters. For most of the year, the layered deposits are covered with $CO_2$ frost, but in summer, they become partly defrosted. The layering is seen best as a finely layered, horizontal banding on defrosted slopes. The deposits are believed to be composed of dust and ice and the layering to be caused by different proportions of the two components. The scarcity of impact craters indicates a relatively young age, although the age of the deposits could still be on the order of hundreds of millions of years.

The layering suggests cyclic sedimentation, and the origin of the deposits may be connected in some way with the obliquity cycle. The obliquity is the angle between the equatorial plane of a planet and the orbital plane. The obliquity of Mars oscillates between roughly 15° and 35° on a 1.2-million-year cycle. At the highest obliquity, twice as much insolation falls on the poles as at the lowest obliquity. As a consequence, the capacity of the high-latitude regolith to hold adsorbed $CO_2$, the size of the $CO_2$ cap, and the atmospheric pressure all may change as the obliquity changes. These changes could affect global wind regimes, dust storm activity, and sedimentation rates at the poles and thereby cause cyclic sedimentation. Layering also would be caused by any event that resulted in large amounts of water vapor's being introduced into the atmosphere. Such possible events are volcanic eruptions, large floods, and cometary impacts. These all would result in the deposition of an ice-rich layer at the poles.

**Conclusions**

Mars, like the Earth, has had a diverse geologic history. An ancient, heavily cratered surface preserves evidence of events in the planet's earliest history. Volcanic activity has continued throughout the planet's history, possibly to the present, and has resulted in the formation of extensive lava plains and large shield volcanoes. Most volcanic activity has been in two provinces, Tharsis and Elysium, which are at the centers of large bulges in the planet's surface. The Tharsis bulge caused fracturing over about one-third of the planet's surface, and this fracturing may be largely responsible for the formation of enormous canyons down the east flank of the bulge. Massive floods of water have flowed periodically across the surface, and some of these may have been triggered somehow by volcanism. In addition, dissection of some surfaces by networks of small branching valleys, particularly in the ancient cratered highlands, indicates slow erosion by running water and possibly massive, global climate changes. The morphology of impact craters and the presence of numerous features suggesting ground ice together indicate that ground ice is abundant at shallow depths at high latitudes and at greater depths at low latitudes. Thus, Mars is a planet on which most of the geologic processes familiar to us here on Earth have operated. The two planets are, however, very different. The lack of plate tectonics on Mars has led to greater stability of the surface and to the development of enormous volcanoes and canyons. In addition, the ineffectiveness of water erosion in eliminating topographic relief has led to the almost perfect preservation of features widely ranging in age. Despite the excellent photographic coverage and nearly perfect preservation, the origin of many of the features remains obscure. We look forward to Mars Observer, Mars 94, and Mars 96 to help us unlock more of the planet's secrets. ❑

**References**


Baker, V.R., 1982, The channels of Mars: Austin, Texas, USA, University of Texas Press, 198 p.

Baker, V.R., Strom, R.G., Gulick, V.C., Kargel, J.S., Komatsu, G., and Kale, V.S., 1991, Ancient oceans, ice sheets and the hydrological cycle on Mars: Nature (London), v. 352, p. 589–594.

Carr, M.H., 1973, Volcanism on Mars: Journal of Geophysical Research, v. 78, p. 4049–4062.

———1981, The surface of Mars: New Haven, Connecticut, USA, Yale University Press, 232 p.

Carr, M.H., Crumpler, L.S., Cutts, J.A., Greeley, R., Guest, J.E., and Masursky, H., 1977, Martian impact craters and emplacement of ejecta by surface flow: Journal of Geophysical Research, v. 82, p. 4055–4065.

Carr, M.H., Wu, S.S.C., Jordan, R., and Schafer, F.J., 1987, Volumes of channels, canyons and chaos in the circum-Chryse region of Mars: Lunar and Planetary Science XVIII, p. 155–156.

Kuzmin, R.O., 1980, Morphology of fresh martian craters as an indicator of the depth of the upper boundary of the ice-bearing permafrost: A photogeologic study: Lunar and Planetary Science XI, p. 595–596.

———1983, Kriolitosfera Marsa: Moscow, Nauka Publishers, 142 p. [In Russian.]

Lucchitta, B.L., 1981, Mars and Earth: Comparison of cold-climate features: Icarus, v. 45, p.264–303.

Masson, P., 1977, Structural pattern analysis of the Noctis Labyrinthus-Valles Marineris regions of Mars: Icarus, v. 30, p. 49–62.

———1985, Origin and evolution of the Valles Marineris region of Mars: Advances in Space Research, v. 5, p. 83–92.

Pieri, D.C., 1980, Martian valleys: Morphology, distribution, age and origin: Science, v. 210, p. 895–897.

Pollack, J.B., 1979, Climate change on the terrestrial planets: Icarus, v. 37, p. 479–553.

Schultz, P.H., and Gault, D.E., 1984, On the formation of contiguous ramparts around martian impact craters: Lunar and Planetary Science XV, p. 732–733.

Squyres, S.W., and Carr, M.H., 1986, Geomorphic evidence for the distribution of ground ice on Mars: Science, v. 231, p. 249–252.

Squyres, S.W., Wilhelms, D.E., and Moosman, A.C., 1987, Large scale volcano-ground ice interactions on Mars: Icarus, v. 70, p. 385–408.

Wilhelms, D.E., and Squyres, S.W., 1984, The martian hemispheric dichotomy may be due to a giant impact: Nature (London), v. 309, p. 138–140.

Wilson, L., and Mouginis-Mark, P.J., 1987, Volcanic input into the atmosphere from Alba Patera, Mars: Nature (London), v. 330, p. 354–357.

**Questions**:

1. What may have caused the division of Mars's surface into the highlands and plains?

2. How might volcanism influence flooding on Mars?

3. How do the valley networks resemble Earth's river valleys, and how are they different?

Answers on page 301

*Venus is a hot, dry planet, victim of a runaway greenhouse. In 1970 a Russian probe, Venera 7, landed on the surface of Venus and quit working after 23 minutes. The previous probes were crushed by the dense atmosphere before they even made it to the surface. Could this planet ever have contained liquid water? Possibly. New evidence suggests that early in its history Venus may have had an ocean up to 25 meters deep. This ocean may have existed for up to a billion years, which may have been long enough for life to have evolved. After about a billion years, greenhouse gases would have built up in the atmosphere, heating up the surface to a temperature where water could no longer exist in the liquid state. If life did evolve on Venus, it didn't last long.*

# New Evidence of Ancient Sea on Venus

by R. Cowen

Born in the same part of the solar system as our own planet, Venus has a mass, chemical composition, and size similar to Earth's. But the planet known as Earth's twin differs in at least one important respect. Venus is as dry as a bone. So could the two planets truly have a common origin?

A new analysis of spacecraft data suggests that in the distant past, Venus was all wet. The planet may have had an ocean as deep as 25 meters, according to a reexamination of data gathered by the Pioneer Venus satellite, which burned up in the Venusian atmosphere last fall after a 14-year mission (SN: 10/17/92, p.263). The ocean on Venus might have lasted long enough—about a billion years—to support primitive life, says Thomas M. Donahue of the University of Michigan in Ann Arbor. He reported the findings last week at a press conference in Pasadena, Calif.

Donahue and his colleagues base their report on the chemical evidence that water molecules leave behind when they split apart and leave the atmosphere of a planet. This chemical signature comes from the abundance of two atoms—hydrogen and its less abundant isotope deuterium, which has twice hydrogen's mass. From 1978 through 1980, Pioneer Venus recorded the ratio of deuterium to hydrogen in the planet's upper atmosphere.

The craft's early measurements revealed that this deuterium-to-hydrogen ratio is at least 150 times greater on Venus than in any other known place in the solar system. That unusual ratio presumably came about over billions of years during which atmospheric conditions on Venus prompted ionized hydrogen to escape, while gravity kept the heavier deuterium on the planet. Thus, Venus once had at least 150 times as much hydrogen as it does now. And since hydrogen readily bonds with oxygen to produce water, this suggests that Venus once had a minimum of 150 times as much water as it does now.

Those early data would make any ocean on the young Venus only 0.5 meter deep, notes Donahue. But in reexamining the data, two of his collaborators—Richard E. Hartle and Joseph M. Grebowsky of NASA's Goddard Space Flight Center in Greenbelt, Md.—calculated that deuterium on Venus might exit more easily, compared with hydrogen, than estimated.

If Donahue's team is correct, then Venus once had much more deuterium than previously calculated. This means that the planet must also have had more hydrogen in the past, in order to come up with the ratio of deuterium to hydrogen measured by Pioneer Venus. In fact, Donahue says, hydrogen was about 3.5 times more abundant than believed. And since more hydrogen implies more water, he asserts that Venus may once have had an ocean 8 to 25 meters deep. "The data indicate that Venus was a pretty wet planet," he says.

Donahue cautions that Venus' early water supply might have been steam, not liquid. But scientists generally believe that the sun's luminosity was 30 percent lower several billion years ago, and it may have been cool enough on Venus to permit an ocean. Later on, as carbon dioxide and other greenhouse gases—including water vapor itself—rapidly accumulated in the planet's atmosphere, the surface heated up and the proposed ocean disappeared.

A comparison of data collected by Pioneer Venus during its first few and last few years in orbit shows that the Venusian ionosphere is much lower and less dense when the sun is near a minimum in its 11-year sunspot cycle, Donahue notes. This indicates that if Venus had an ocean, it probably did not exist beyond the first billion or so years of the planet's existence, he says.

Climate modeler James Kasting of Pennsylvania State University in University Park says the new analysis hasn't yet convinced him that Venus once had a deep ocean. "It's not clear to me that we are understanding all the processes important for hydrogen escape; it's messier than we used to think," Kasting says.

Victor R. Baker of the University of Arizona in Tucson, who has proposed that Mars once harbored an ocean, says the Venus findings support the view that all of the inner solar system planets were formed from the collision of similar material and once had an abundance of water. ❑

**Questions**:

1. How does the ratio of deuterium to hydrogen indicate that water was once present on Venus?

2. What happened to the missing hydrogen?

3. Extra hydrogen by itself would not mean that liquid water was ever present on Venus. The surface of Venus must have been cooler. What two factors would have given Venus a cooler surface during its first billion years?

Answers on page 301

# 27

*One of the biggest scientific events of 1994 was the impact of the Shoemaker-Levy comet with Jupiter. It produced a dazzling display leaving scars on Jupiter's surface, some of which were larger in diameter than the Earth. But the impact also left some uncertainties as to what actually happened. Large effects were seen at the surface of Jupiter but the effects did not appear to penetrate deeply into the planet's atmosphere. This raised the question of how big and how coherent the impacting fragments were. If they were as large and coherent as originally thought, they should have penetrated deeper into the atmosphere. If they were already broken into small swarms, it was thought that they would not have produced such large surface effects. It may take scientists some time to sort out the data and determine exactly what happened. Whatever the results, no one appears to be disappointed in the event itself.*

## Shoemaker-Levy Dazzles, Bewilders

Astronomers' first opportunity to watch two solar system bodies collide produced the hoped-for fireworks, but relief has turned to puzzlement: Just what happened?
by Richard A. Kerr

Observers poised at their telescopes for the collision between Jupiter and the first fragment of comet Shoemaker-Levy 9 were an anxious bunch. Getting ready for the start of the show on 16 July, they wondered whether there would be anything to see when bits of the disrupted comet dove into Jupiter's atmosphere at 200,000 kilometers per hour. No one knew how big the fragments were; estimates ranged from less than half a kilometer to almost ten times that size. And just days before the impacts began, some astronomers had warned that the 20 or so fragments might not even be solid; they might simply be loose swarms of fragments so small they could vanish without a trace. In the words of a headline in *Nature* that week, "The Big Fizzle is coming."

Guess again. The Shoemaker-Levy show, which went on for 6 days, looked almost as big and dazzling as the most optimistic predictions. "It's like seeing a supernova go off," exclaimed one astronomer who had flown at 12 kilometers over the South Pacific to get a view. "We're all running around like giddy kids," reported another at McDonald Observatory in Texas. A co-discoverer of the comet, Eugene Shoemaker of Lowell Observatory in Flagstaff, Arizona, pronounced himself well pleased with his namesake: "Nature has outdone herself; we're elated." To observers marveling over the first fireballs and dark bruises in Jupiter's atmosphere, it seemed clear that the comet's fragments must have been big, solid projectiles, plunging deep into the atmosphere. But as the week wore on and astronomers analyzed the impact sites, doubts set in.

By the time the curtain had come down on impact week, some researchers were arguing that the impact debris and scars didn't look as if they came from deeply penetrating wounds. And that has left astronomers struggling to explain how impacts that produced such dazzling displays for observers three quarters of a billion kilometers away could have failed to stir Jupiter itself to any great depth. Although planetary scientists had hoped that the impact effects might provide clues about Jupiter's interior—its internal structure and composition, for instance—such questions are on hold for the moment, as astronomers ponder what it was they saw.

At least they have data to ponder, which is something that seemed in doubt as the first fragment closed in on Jupiter. Most comet specialists had assumed that the 21 diffuse clouds of dust and debris seen in telescopes hid 21 solid fragments that would produce visible effects unless they were at the very small end of the size estimates. But 2 days before the first impact, comet specialist Paul Weissman of the Jet Propulsion Laboratory argued in a *Nature* News and Views article that the impacts would be a bust because each nucleus was no more than a loose swarm of small pieces, "like bees buzzing around a hive."

Weissman and some others believe comets are merely clumps of millions of house-sized "dirty snowballs." Ordinarily, these comet bits are held together by little more than their own feeble gravity. So when Shoemaker-Levy made its fatal pass just above the cloud tops of Jupiter 2 years ago, Weissman argued, the planet's powerful gravity pulled the comet apart into puny pieces, loosely gathered 21 swarms. In the days and hours before impact, argued Weissman, astronomer Terrence Rettig of the University of Notre Dame, and others, Jupiter's gravity would stretch those swarms into elongated streams. The streams would pepper the planet "like machine gun bullets lacing into a moving target," burning up in the upper atmosphere as meteors do rather than plunging deep into the planet and exploding to produce a visible fireball.

Most researchers still held out for solid bodies, arguing that even if Weissman were right about the comet's makeup, the swarms would have reassembled themselves over the 2 years since the cometary breakup. And the first impacts seemed to bring a dramatic vindication for the majority. Fragment A, which looked modest-sized in telescopes, sent a plume more than 1000 kilometers above the planet. A day later, the plume from fragment G rose to several thousand kilometers and left a "black eye" of debris 25,000 kilometers across. By midweek, researchers who had simulated the effects of impacters of various sizes in computer models were in agreement: The plume heights and brightnesses and debris fallout patterns implied that the largest fragments were solid bodies 2 to 3 kilometers in diameter. "Everything we calculated is very close to what was observed," said Thomas Ahrens of the California Institute of Technology, one of the modelers.

Later in the week, however, observations and theory began to diverge as astronomers wrestled with the question of how deep the cometary fragments had penetrated into Jupiter's atmosphere. The models had predicted that hefty iceballs, like those favored by the "solid impacter" group, would plunge through Jupiter's uppermost clouds, made of ammonia, and on into the atmosphere's little-known deeper reaches, finally exploding at depths of 100 kilometers and more. Planetary scientists had predicted a cloud layer of ammonium hydrosulfide beneath that. Stirred up by the deep explosions, these compounds would surely appear in the visible impact scars, they thought.

And, just as predicted, the first compound reported by spectroscopists monitoring the darkened impact sites through the Hubble Space Telescope was ammonia, presumably from the uppermost cloud layer. Then sulfur turned up, possibly from the comet's "dirt" or the middle cloud deck. But clear signs of water never showed up during impact week, sowing seeds of doubt about the penetrating power of Shoemaker-Levy's fragments.

Those doubts grew on the day of the last impact, when planetary meteorologist Andrew Ingersoll of Caltech, who had assumed the explosions were deep, jumped ship. At the daily press conference, Ingersoll showed a diagram of a fragment plunging through the water cloud and said, "I don't believe it anymore. I think the comet did not go through the water cloud." What had changed his mind was two dark rings that the Hubble Space Telescope had caught spreading from the site of the G fragment impact.

To judge by the expansion speeds of the two rings, the larger and more energetic of them was in the Jovian stratosphere, above all the clouds, while the weaker one was deeper, within the clouds. Ingersoll assumed the rings were waves of some sort, made visible as their changing pressure triggered chemical changes in the atmosphere. The evidence that the strongest wave was spreading above the cloud deck convinced him that the fragment itself must have exploded above the clouds, leaving just enough residual energy to stir up the ammonia layer.

These signs of shallow penetration left the comet watchers with a puzzle: How can fragments potent enough to produce the dazzling fireballs have failed to make it very far into the atmosphere? One possibility is that the fragments were swarms after all. Swarms, as Weissman had pointed out, would be stopped much higher in the atmosphere than would solid fragments of the same mass; their unexpectedly showy deaths, he now argues, might be a sign that they had not spread out quite as far as he had assumed. Alternatively, Shoemaker and others suggest, weak but coherent fragments might have started to break up under Jupiter's gravity during their final hours, blunting their penetrating power without detracting from the display.

Then there's the possibility that the comet and its fragments were solid but small, at the low end of the size estimates made before the impact (*Science*, 1 July, p.31). Mordecai-Mark MacLow of the University of Chicago, for one, argues that it doesn't necessarily take a big comet chunk to make a big fireball. A half-kilometer ice ball might not make it to the water clouds, he says, but it would be able to put more of its energy into a display visible from Earth. "My models showed these huge [plume] clouds even for half-kilometer objects," he says.

MacLow's proposal will likely prove controversial, like other efforts to make sense of the impacts. But a little discord might have been expected. Hundreds of observers were watching an event unprecedented in planetary science, exploiting every wavelength from radio through visible to ultraviolet and sharing observations in real time. The impacts themselves were actually taking place just out of view, on Jupiter's backside. And some of the most revealing data have yet to come in, such as observations from the Galileo spacecraft, which had a direct view of the impacts. Eventually, however, planetary scientists expect to sort out the confusion—and secure a unique scientific legacy from the dying comet. ❑

**Questions**:

1. Based on the surface effects of the impacts, what was the estimated size of the largest fragments?

2. What is Paul Weissman's idea of the makeup of the comet fragments?

3. What does Eugene Shoemaker think about the makeup of the comet fragments?

Answers on page 301

# Part Four

# Resources and Pollution

*In September, 1994, the United Nations held an international conference to discuss the world over-population problem. Predictions are that, by the year 2050, world population will be between 7.8 to 12.5 billion, compared to 5.6 billion in 1994. The "Cairo document" sets out a plan for countries to invest at least 20% of public money into the social sector to help stem population growth. If richer donor countries raised population spending from 1% to 4%, the UN projects that world population could be slowed to attain the 7.8 billion mark by 2050. Unfortunately, the political and social resistance to such measures are enormous, including religious and other ideological convictions. Some good news is that the rate of global population increase has declined from over 2% per year in the 1960s to 1.6%.*

# Population: The View from Cairo

Next month, an international congress will be held in Egypt's capital to debate a plan to slow world population growth. The plan has broad political support, but experts dispute how effective it will be

by Wade Roush

To get a sense of the next century's population picture, take a look at what is happening in Cairo today. Home to about 9.5 million people, this mega-city is growing by 200,000 people per year. Three million Cairenes lack sewers, half a million live in rooftop huts, and another half-million dwell among the tombs of the "City of the Dead" in Cairo's eastern section. The city's older districts hold as many as 62,000 residents per square kilometer—a population density twice Manhattan's.

For delegates crowding into Cairo for the United Nations International Conference on Population and Development on 5 to 13 September, the stakes in the world effort to stem rapid population growth could hardly be more visible. By the year 2000, UN demographers project, there will be 21 cities of 10 million or more, all but four in developing countries. By 2050 the world's current population of 5.6 billion will have grown to between 7.8 billion and 12.5 billion. The range of variation in these UN projections—a reflection of uncertainty over the most likely pace of fertility decline—is as large as the world's total population in 1984, the date of the last UN world population conference in Mexico City.

"The rapid pace of population growth is everybody's urgent business," says Timothy Wirth, U.S. undersecretary of state for global affairs and chief U.S. negotiator at Cairo. The sense of urgency is shared by members of most of the world's scientific establishments: Last October 60 science academies, led by the U.S. National Academy of Sciences and Britain's Royal Society, held an unprecedented joint meeting on world population that ended with a statement proclaiming that "Humanity is approaching a crisis point with respect to the interlocking issues of population, environment, and development." And a draft Programme of Action, developed by the UN Population Fund and expected to be approved at the Cairo conference, warns: "The decisions that the international community takes over the next several years, whether leading to action or inaction. will have profound implications for the quality of life for all people, including generations not yet born, and perhaps for the planet itself."

The Cairo document—drafted at a series of preparatory meetings held over the past 2 years with extensive input from women's groups and other nongovernmental organizations—sets out a plan for donor countries and developing countries themselves to invest at least 20% of public expenditures in the social sector. Special emphasis is placed on a range of population, health, and education programs that would improve the status and health of women. It also calls for donor spending on population assistance, currently running at about $1 billion a year, to increase from 1.4% percent of official development assistance to 4%. These measures, the document states, "would result in world population growth at levels close to the United Nations low [projection]" of a global population of 7.8 billion by the year 2050.

Like previous efforts to draft population policies, this one has proved controversial. Diplomatic sparring between the Vatican and other delegations over the draft document's emphasis on birth control and its support of a woman's right to safe abortion have dominated pre-Cairo publicity. But a quieter, and ultimately more significant, scientific debate has been going on over how effective the plan is likely to be in reducing fertility rates.

Researchers remain far from a consensus on the overall emphasis of the draft plan and on the benefits that can be expected from the three major mechanisms it espouses for reducing fertility—improved access to modern contraceptives, reduced infant and child mortality, and expanded school enrollment for females. One reason for the lack of consensus is that, as political scientist Steven Sinding, director of population sciences for the Rockefeller Foundation, puts it: "There are still major gaps in our understanding of fertility change and what causes it."

**The search for an acceptable policy**

Nobody is arguing that it will be easy to hold population growth to the low end of the UN's projections. Although the rate of population increase peaked in the 1960s at just over 2% per year and has since fallen to 1.6%—thanks largely to a fertility drop in the developing nations from an average of six children per woman to below four—growth rates throughout much of the developing world are still high. The population "doubling time" at current rates is just 24 years in Africa and 35 in Asia and Latin America, compared to 98 ycars in North America and 1025 in Europe. Moreover, unprecedented numbers of women will be entering their reproductive years in the next two decades, so that even if fertility rates were miraculously reduced to the so-called "replacement level" of 2.1 children per woman by 1995, global population would still climb to about 7.7 billion in 2050, according to UN projections.

What, if any, steps should be taken to bring down these growth rates has long been a source of contention. Twenty years ago, at the first UN intergovernmental conference on population in Bucharest, representatives of developing nations argued that rapid industrialization would provide the solution to the population problem. Arguing that "development is the best contraceptive," they insisted that a massive redistribution of wealth from North to South must precede population stabilization.

Underlying this argument was the now 60-year-old theory of the demographic transition. When high-fertility, high-mortality societies first modernize, the theory holds, improvements in standards of living, public health, and medical technology bring mortality rates down while birth rates remain high. Population grows markedly, as it did in Europe during the Industrial Revolution. Eventually, however, cultural and economic changes associated with urbanization and industrialization—for example, the increasing net economic cost of raising children—bring birth rates down as well, first among upper- and middle-income classes and then among workers. This completes the transition, leaving fertility at replacement level or lower—as has happened in North America, Japan, and much of Europe.

But even before the Bucharest conference, demographers had noted some odd misfirings in the mechanism supposedly linking economic development to declining fertility. In a few industrializing nations like Brazil and Mexico, fertility remained stubbornly high, while other nations such as Colombia and Sri Lanka saw birth rates plummet without significant industrial development.

Over the past two decades demographers have also taken a closer look at fertility declines in the industrializing West and have found a hodgepodge of historical patterns. In nineteenth-century France, Germany, and Sweden, lower birth rates actually preceded mortality declines, and in Australia (as in Bangladesh and Kenya today) fertility declined across all socio-economic classes at about the same time, rather than trickling from the wealthiest down to the poorest members of society. "The fertility transitions in Europe followed patterns in many places that were only loosely related to industrialization, and the same thing is true in much of the developing world," says Sinding.

Against the backdrop of this reassessment of demographic transition theory, the second UN population conference, held in Mexico City in 1984, put more emphasis on increasing access to modern family planning technologies. The conference was, however dominated by a rancorous dispute over the Reagan Administration's advocacy of market-oriented models of development. Chief U.S. negotiator James Buckley declared, for example, that rapid population growth is a "neutral factor" in the economic health of developing countries. And the family-planning goals were undercut by the U.S. government's announcement that it would cut off funding for international groups providing abortion counseling.

The Cairo Programme of Action takes a different tack from the ones that emerged from the Bucharest and Mexico City gatherings. It supplants population planning's old focus on contraception and industrial development with a new emphasis on the connections between population, poverty, inequality, environmental decay, and the need for "sustainable" development. The education of women, for example, is advocated both as a means of deflecting pressures for large families and as a spur to income-generating activities.

And this time around, the United States is in full support. The Clinton Administration last year restored funding for population programs that had been cut in the Reagan era and has consolidated population planning in Wirth's new State Department office. Population programs, says Wirth, are increasingly being viewed as "the basic wedge into development programs" rather than the other way around.

"You have three things going together—poverty, high fertility, and environmental degradation—affecting the production basis of rural life," explains Partha Dasgupta, an economist at Cambridge University in England. Dasgupta's studies of rural households in India and sub-Saharan Africa find that the depletion of natural resources like water and fuel wood creates a need for extra hands around the house, hence more births, hence even fewer resources to go around. Programs to interrupt the spiral by conserving resources and providing women with cash earnings therefore benefit the environment, the economy, and population stability all at the same time. Says Dasgupta, "The public policy needs that stare you in the face are precisely the things that you might think were reasonable even if you weren't worried about population."

This thrust is strongly supported by international women's groups. Former member of Congress Bella Abzug, speaking on behalf of the Women's Environment and Development Organization—which helped shape the Programme of Action through a series of critical papers and extensive lobbying with individual governments—argues, for example, that "family planning and fertility rates cannot be seen as abstractions in themselves."

Some members of the population research establishment are, however, worried by the document's failure to address population goals more directly. Charles Westoff, a demographer at Princeton's Office of Population Research, notes that the plan of action eschews quantitative fertility targets, which are seen by Abzug and other feminists as coercive. "You search [the Cairo document] in vain for an explicit statement that in certain parts of the world, such as sub-Saharan Africa, women are having a lot more children, and want to have more children, than is commensurate with replacement fertility," Westoff complains. Adds economist and demographer Paul Demeny, a senior associate at the Population Council in New York and editor of the journal *Population and Development Review*: "I think that articulating population programs simply as yet another need-satisfying welfare program, without involving the rationale that prompted these programs in the first place—governments' concern about the harm caused by too-rapid aggregate population growth—greatly weakens the argument that these programs should be a high priority." The real question, many researchers say, is how much each of the three major recommendations outlined in the Cairo plan can affect demographic trends.

**Contraception: Unmet needs**

Between 1970 and 1990, world contraceptive use increased from 30% of couples to 55%, and average family size fell from 4.9 children to 3.5, halfway to replacement level, according to Sharon Camp, former senior vice president of Population Action International. Fertility declined most sharply in countries that instituted strong, government-sponsored family planning programs in the 1960s and '70s, including China, Botswana, Kenya, Zimbabwe, Morocco, and the now-prosperous Eastern and Southeastern Asian nations of Singapore, South Korea, Taiwan, and Thailand. Demographer John Bongaarts, vice president and director of research at the Population Council, estimates that 40% of the world's fertility decline is attributable to better contraceptive access. "There's no doubt that the investment in family planning has paid off," Bongaarts says.

Sinding agrees: "Clearly [contraception] made the difference between the century it took fertility to decline in some of the now-industrialized countries and the 15 years it's taken in places like Taiwan and Korea," he says. With a confidence bred of this apparent success, family-planning advocates like Sinding argue that substantial further progress can be made toward replacement-level fertility by meeting the remaining "unmet need" for reliable birth-control methods. A 1991 survey conducted by Westoff and Luis Hernando Ochoa for the Maryland-based Demographic and Health Surveys found that approximately 120 million married women worldwide do not wish to have any more children but are not currently using any modern contraceptive method. The Cairo plan calls for the elimination of this unmet need by the year 2015.

A study published in the March issue of *Population and Development Review*, however, has touched off a spirited dispute over the actual extent of unmet need for contraception. Lant Pritchett, a senior economist at the World Bank, compared actual fertility rates in 53 developing countries with various measures of people's desired family sizes. His regression analyses found that approximately 90% of the differences in actual fertility between countries was attributable to differences in desired fertility and that the number of unwanted births is no lower in regions where people have greater access to family planning services. High fertility in regions like sub-Saharan Africa, Pritchett concludes, is explained almost completely by a high desire for children, not by any shortage of low-cost contraceptives.

Pritchett also challenges the most widely accepted version of unmet need. Westoff and Ochoa, he says, "decide who needs contraception and then argue that anyone who does not use it has unmet need." If women who avoid contraception for reasons other than cost or availability are removed from the unmet need category, Pritchett calculates, then the fertility reduction achievable through expanded access to family planning adds up to less than half a birth per woman—a small improvement in African countries like Niger, Côte d'Ivoire, or Uganda, where the average woman has 7.4 children. Pritchett insists that he is not opposed to the expansion of family-planning programs, which he says provide many benefits for women. "The main point of my paper is that the demographic impact is likely to be small."

Westoff acknowledges that his measures of "unmet need" include women who say they intentionally avoid using contraceptives, but he argues that any woman who says she wants no more children should be counted as having a real need. He agrees with Pritchett, however, that access is often not the issue. "Meeting unmet need is more complex than simply providing contraceptive methods. The obstacles tend to be things like lack of information, concern about side effects, religious or other fatalistic attitudes, women concerned about how their husbands feel, and those kinds of things. These require education and information."

Whatever the actual number of couples who want contraceptives but lack money or easy access, says Wirth, the next leap toward replacement-level fertility will probably be more difficult than the last. "The easy part of the unmet need has been done" through family-planning programs focused on urban areas and lower-middle-class and middle-class individuals, Wirth says. "The much tougher part of unmet need is ahead. That's the very poor in cities, and that's the very rural."

**Reducing childhood mortality**

Although mortality rates among infants and young children have been significantly reduced everywhere since World War II, a child's chances of survival past the age of 4 are still much better if he or she is born in an industrialized nation (97% in the United States, compared to 71% in Ethiopia, 76% in India, and 80% in Kenya). Many researchers believe—and the Cairo plan reiterates—that reducing fertility rates will be impossible without further reductions in infant and child mortality, because only parents who are confident that their children will grow to support them in old age can attain their desired family size without "overshooting." As Julius Nyrere, the former president of Tanzania, has said, "The most powerful contraceptive is the knowledge that your children will survive."

But "we don't really know as much as the [Cairo] document pretends we know about infant and child mortality," says Sinding. "The Cairo document says that among a wide menu of social policies one might invest in, this is a particularly important one because of the effect on fertility. I would like to believe that, but the evidence is not strong." A recent World Bank study of the neighboring West African countries of Côte d'Ivoire and Ghana, for example, found no statistically significant link between child mortality and fertility within individual households. The study did find a link at the community level; in Ghana, one fewer child is born for every five who escape an early death, an effect similar to that found in some Asian and Latin American countries. But the study pointed to "conceptual and statistical problems" that make it difficult to prove that low levels of child mortality actually slow population growth.

Paul Schultz, an economist at Yale University's Center for Economic Growth and one author of the World Bank study, says he believes "there's nowhere in the world where you get fertility coming down until you get child mortality under some degree of control." But he adds that it is very difficult to sort out the confounding factors in the relationship between the two. "Child mortality is heavily shaped by women's education and resources in the family. It's very hard to infer whether [the link between child mortality and fertility] is causal or whether additional factors are influencing both. If you're a scientist you have to say the linkage is there but we can't prove it, and that's frustrating."

**Women's education**

George Moffett, a diplomatic correspondent for the *Christian Science Monitor*, writes in his new book *Critical Masses: The Global Population Challenge* that "countries in which education is least accessible to girls have the highest fertility rates in the world, while those that provide the greatest access have the lowest." In India's Kerala state, for example, an 87% literacy rate among women is widely credited with helping to push the fertility rate down to 2.3 children per woman, among the lowest in the developing world. And almost no one disputes the finding that increases in school enrollment ratios among females have a greater negative effect on fertility than do equivalent increases among males.

"The link between greater education and fertility is very clear and powerful and invariant," says Sinding. "It is as good a bet for public policy as social science is generally able to provide."

But just as with child mortality, the mechanisms linking education to changes in desired and actual fertility are uncertain. Educated women may want to have smaller families in order to stretch the resources available for their own children's schooling, or their education may give them profitable alternative uses for their own time so that the cost of motherhood goes up. They may marry and begin bearing children later and have fewer children as a result, or they may be more effective users of family-planning methods and better protectors of their children's health. "People have not really tried to test which of those pathways is the main pathway, and there is some debate about which programs are the most effective," says Elizabeth King, a senior World Bank economist.

Indeed, high levels of female education do not always result in lower fertility. Martha Ainsworth, another World Bank economist who has recently completed a study of education, contraceptive use, and fertility in 14 countries in sub-Saharan Africa, found that although 60% of women in Ghana are educated—half of them at the secondary level—the effect on fertility there is no greater than in Senegal, Togo, Mali, Niger, and other countries where women have much less schooling. It is often difficult to measure any effect in regions where so few women overall have schooling, Ainsworth says. "Sometimes I can't disentangle how much of what I'm measuring is the effect of selectivity," that is, the fact that females who seek education are more likely ahead of time and want smaller families and to use contraception, she says.

In spite of these uncertainties, most population researchers believe that if the Cairo plan is implemented, it may gradually shift desired family size downward. "Cairo is a good thing even if it's not being guided by the scientific evidence," concludes Yale's Schultz. "This is merely a linking of various lobby groups that have an interest and a confluence of logic. It's not necessarily empirically documented, but it's plausible in certain parts of the world."

Michael Teitelbaum, a demographer at the Alfred P. Sloan Foundation, adds that he has "never run into a single policy issue that has unanimous scientific support." The important point, he says, is that U.S. policy makers have "re-established a leadership position" on population issues and that, through Wirth's new Global Affairs office at the State Department, they are "trying their best to make the connections" between foreign policy and demographic trends.

As the delegates gather in Cairo next week they would do well to bear in mind one fact: During the 9 days they will spend debating the plan of action, the world's population will grow by some 2.1 million. ❑

**Questions**:

1. What are the population "doubling times" for Africa and Europe?

2. How many cities on Earth will have over 10 million people by the year 2000?

3. Between 1970 and 1990, world contraceptive use increased to ___% and average family size fell to ___ children.

Answers on page 301

# 29

*The 1980s may have been the last decade when civilization could anticipate a future of assured economic growth and certain improvement in the general human condition. By one measurement after another, the global pattern of continual growth and improvement, experienced since World War II, is apparently coming to an end. The main causes involve exponential changes that indicate an increasingly rapid pace: human population growth, resource use, and pollution are but a few global indicators increasing exponentially. The result is a stabilizing, or decline, of the average quality of life in the world. For example, availability of many foods is declining for the first time since World War II. These reflect the "diminishing returns" that are characteristic of environmental resources, such as many marine fish, that are reaching the limits of exploitability. But it is important not to become overwhelmed by "gloom and doom" information. It has been said that it is easier to create the future than to predict it. Each of us can contribute to this creation in whatever way we choose. Conversely, if we "tune out" unpleasant information, that will not make the problems disappear.*

## A Decade of Discontinuity

**The 1980s may have been the last decade in which humankind could anticipate a future of ever-increasing productivity on all fronts. By one measure after another, the boom we have experienced since mid-century is coming to an end.**

by Lester R. Brown

When the history of the late 20th century is written, the 1990s will be seen as a decade of discontinuity—a time when familiar trends that had seemed likely to go on forever, like smooth straight roads climbing toward an ever-receding horizon, came to abrupt bends or junctures and began descending abruptly. The world's production of steel, for example, had risen almost as reliably each year as the sun rises in the morning. The amount of coal extracted had risen almost uninterruptedly ever since the Industrial Revolution began. Since the middle of this century, the harvest of grain had grown even faster than population, steadily increasing the amount available both for direct consumption and for conversion into livestock products. The oceanic fish catch, likewise, had more than quadrupled during this period, doubling the consumption of seafood per person.

These rising curves were seen as basic measures of human progress; we *expected* them to rise. But now, within just a few years, these trends have reversed—and with consequences we have yet to grasp. Meanwhile, other trends that were going nowhere, or at most rising slowly, are suddenly soaring.

That such basic agricultural and industrial outputs should begin to decline, while population continues to grow, has engendered disquieting doubts about the future. These reversals, and others likely to follow, are dwarfing the discontinuities that occurred during the 1970s in the wake of the 1973 rise in oil prices. At that time, an overnight tripling of oil prices boosted energy prices across the board, slowed the growth in automobile production, and spurred investment in energy-efficient technologies, creating a whole new industry.

The discontinuities of the 1990s are far more profound, originating not with a handful of national political leaders as with the OPEC ministers of the 1970s, but in the collision between expanding human numbers and needs on the one hand and the constraints of the earth's natural systems on the other. Among these constraints are the capacity of the oceans to yield seafood, of grasslands to produce beef and mutton, of the hydrological cycle to produce fresh water, of crops to use fertilizer, of the atmosphere to absorb CFCs, carbon dioxide, and other greenhouse gases, of people to breathe polluted air, and of forests to withstand acid rain.

Though we may not have noticed them, these constraints drew dramatically closer between 1950 and 1990, as the global economy expanded nearly fivefold. Expansion on this scale inevitably put excessive pressure on the earth's natural systems, upsetting the natural balances that had lent some stability to historical economic trends. The trends were driven, in part, by unprecedented population growth. Those of us born before 1950 have seen world population double. In 1950, 37 million people were added to the world's population. Last year, it was 91 million.

### Against the Grain

The production of grain, perhaps the most basic economic measure of human well-being, increased 2.6 fold from 1950 to 1984. Expanding at nearly 3 percent per year, it outstripped population growth, raising per capita grain

*Lester R. Brown. "A Decade of Discontinuity," World Watch Magazine; July/August 1993, Worldwatch Institute, Washington, D.C. Reprinted with permission.*

consumption by 40 percent over the 34-year period, improving nutrition and boosting consumption of livestock products—meat, milk, eggs, and cheese—throughout the world.

That period came to an end, ironically, around the time the United States withdrew its funding from the United Nations Population Fund. During the eight years since 1984, world grain output has expanded perhaps one percent per year. In per capita terms, this means grain production has shifted from its steady rise over the previous 34 years to a *decline* of one percent per year since then—a particularly troubling change both because grain is a basic source of human sustenance and because of the likely difficulty in reversing it.

This faltering of basic foodstuffs was triggered by other, earlier discontinuities of growth—in the supply of cropland, irrigation water, and agricultural technologies. Cropland, measured in terms of grain harvested area, expanded more or less continuously from the beginning of agriculture until 1981. The spread of agriculture, initially from valley to valley and eventually from continent to continent, had come to a halt. Since 1981, it has not increased. Gains of cropland in some countries have been offset by losses in others, as land is converted to nonfarm uses and abandoned because of erosion.

Irrigation, which set the stage for the emergence of early civilization, expanded gradually over a span of at least 5,000 years. After the middle of this century, the growth in irrigated area accelerated, averaging nearly 3 percent per year until 1978. Around that time, however, as the number of prime dam construction sites diminished and underground aquifers were depleted by overpumping, the growth of irrigated area fell behind that of population. Faced with a steady shrinkage of cropland area per person from mid-century onward, the world's farmers since 1978 have faced a shrinking irrigated area per person as well.

Although there was little new land to plow from mid-century onward, the world's farmers were able to achieve the largest expansion of food output in history by dramatically raising land productivity. The engine of growth was fertilizer use, which increased ninefold in three decades—from 14 million tons in 1950 to 126 million tons in 1984—before starting to slow.

In 1990, the rise in fertilizer use—what had been one of the most predictable trends in the world economy—was abruptly reversed. It has fallen some 10 percent during the three years since the 1989 peak of 146 million tons. Economic reforms in the former Soviet Union, which removed heavy fertilizer subsidies, account for most of the decline. Letting fertilizer prices move up to world market levels, combined with weakened demand for farm products, dropped fertilizer use in the former Soviet Union by exactly half between 1988 and 1992. This was an anomalous decline, from which there should eventually be at least a partial recovery.

More broadly, however, growth in world fertilizer use has slowed simply because existing grain varieties in the United States, Western Europe, and Japan cannot economically use much more fertilizer. U.S. farmers, matching applications more precisely to crop needs, actually used nearly one-tenth less fertilizer from 1990 to 1992 than they did a decade earlier. Using more fertilizer in agriculturally advanced countries does not have much effect on production with available varieties.

The backlog of unused agricultural technology that began to expand rapidly in the mid-19th century now appears to be diminishing. In 1847, German agricultural chemist Justus von Leibig discovered that all the nutrients removed by plants could be returned to the soil in their pure form. A decade later, Gregor Mendel discovered the basic principles of genetics, setting the stage for the eventual development of high-yielding, fertilizer-responsive crop varieties. As the geographic frontiers of agricultural expansion disappeared in the mid-20th century, the adoption of high-yielding varieties and rapid growth in fertilizer use boosted land productivity dramatically. In the 1960s, an array of advanced technologies for both wheat and rice producers was introduced into the Third World—giving rise to a growth in grain output that was more rapid than anything that had occurred earlier, even in the industrial countries.

Although it cannot be precisely charted, the backlog of unused agricultural technology must have peaked at least a decade ago. Most of the known means of raising food output are in wide use. The highest-yielding rice variety available to farmers in Asia in 1993 was released in 1966—more than a quarter-century ago. Today, the more progressive farmers are peering over the shoulders of agricultural scientists looking for new help in boosting production, only to find that not much is forthcoming. Agricultural scientists are worried that the rapid advance in technology characterizing the middle decades of this century may not be sustainable.

**Less Meat *and* Less Fish**

The growth in meat production, like that of grain, is slowing. Between 1950 and 1987, world meat production increased from 46 million tons to 161 million tons—boosting the amount per person from 18 kilograms in 1980 to 32 kilograms (about 70 pounds) in 1987. Since then, however, it has not increased at all. The one percent decline in per capita production in 1992 may be the beginning of a gradual world decline in per capita meat production, another major discontinuity.

Underlying this slowdown in overall meat production is a rather dramatic slowdown in the production of beef and mutton, resulting from the inability of grasslands to support more cattle and sheep. From 1950 to 1990, world beef output increased 2.5-fold. Now, with grasslands almost fully used—or overused—on every continent, this growth may be nearing an end. From 1990 to 1992, per capita beef production for the world fell 6 percent.

The supply of fish, like that of meat, no longer keeps pace with increases in human numbers. Here, too, there has been a reversal of the historic trend. Between 1950 and 1989, the global catch expanded from 22 million tons to 100 million tons. The per capita seafood supply increased from 9 to 19 kilograms during this period. Since 1989, the catch has actually declined slightly, totalling an estimated 97 million tons in 1992. United Nations marine biologists believe that the oceans have reached their limit and may not be able to sustain a yield of more than 100 million tons per year.

Throughout this century, it has been possible to increase the fish take by sending out more ships, using more sophisticated fishing technologies, and going, literally, to the farthest reaches of the ocean. That expansion has now come to an end. The world's ocean catch per capita declined 7 percent from 1989 until 1992, and is likely to continue declining as long as population continues to grow. As a result, seafood prices are rising steadily.

Getting more animal protein, whether it be in the form of beef or farm-raised fish, now depends on feeding grain and soybean meal. Those desiring to maintain animal protein intake now compete with those trying to consume more grain directly.

**Fossil Fuels: The Beginning of the End**

While biological constraints are forcing discontinuities in agriculture and oceanic fisheries, it is atmospheric constraints—the mounting risks associated with pollution and global warming—that are altering energy trends. Throughout the world energy economy, there are signs that a major restructuring is imminent. On the broadest level, this will entail a shifting of investment from fossil fuels and nuclear power toward renewables—and toward greater energy efficiency in every human activity.

We cannot yet see the end of the fossil fuel age, but we can see the beginning of its decline. World oil production peaked in 1979. Output in 1992 was four percent below that historical high. World coal production dropped in 1990, in 1991, and again in 1992 (partly because of the recession), interrupting a growth trend that had spanned two centuries. If strong global warming policies are implemented, this could be the beginning of a long-term decline in coal dependence.

Of the three fossil fuels, only natural gas is expanding output rapidly and is assured of substantial future growth. Gas burns cleanly and produces less carbon dioxide than the others, and is therefore less likely to be constrained by stricter environmental policies. While oil production has fallen since 1979, gas production has risen by one-third.

With oil, it was the higher price that initially arrested growth. More recently, it has been the pall of automotive air pollution in cities like Los Angeles, Mexico City, and Rome that has slowed the once-unrestrained growth in motor vehicle use and, therefore, in oil use. With coal, it was neither supply nor price (the world has at least a few centuries of coal reserves left), but the effects of air pollution on human health, of acid rain on forests and crops, and of rising $CO_2$ concentrations on the earth's climate that have sent the industry into decline. Several industrial countries have committed themselves to reducing carbon emissions. Germany, for example, plans to cut carbon emissions 25 percent by 2005. Switzerland is shooting for a 10 percent cut by 2000, and Australia for 20 percent by 2005. Others, including the United States, may soon join them.

With the beginning of the end of the fossil fuel age in sight, what then will be used to power the world economy? Fifteen years ago, many would have said, with little hesitation, that nuclear power will. Once widely thought to be the energy source of the future, it has failed to live up to its promise (the problems of waste disposal and safety have proved expensive and intractable) and is being challenged on economic grounds in most of the countries where it is produced.

Nuclear generating capacity reached its historical peak in 1990. Though it has declined only slightly since then, it now seems unlikely that there will be much, if any, additional growth in nuclear generating capacity during this decade—and perhaps ever.

**The Winds of Change**

Even as the nuclear and fossil fuel industries have faltered, three new technologies that harness energy directly or indirectly from the sun to produce electricity—solar thermal power plants, photovoltaic cells, and wind generators—are surging. In wind power, particularly, breakthroughs in turbine technology are setting the stage for rapid expansion in the years ahead. Wind electricity generated in California already produces enough electricity to satisfy the residential needs of San Francisco and Washington, D.C. Indeed, it now seems likely that during the 1990s, the growth in wind generating capacity will exceed that in nuclear generating capacity. Three countries—Denmark, the Netherlands, and Germany—have plans to develop a minimum of a thousand megawatts of wind generating capacity by 2005. China aims to reach the same goal by 2000. Given the rapid advances in the efficiency of wind generating machines and the falling costs of wind generated electricity, the growth in wind power over the remainder of this decade could dwarf even current expectations.

The potential for wind power far exceeds that of hydropower, which currently supplies the world with one-fifth of its electricity. England and Scotland alone have enough wind generating potential to satisfy half of Europe's electricity needs. Two U.S. states—Montana and Texas—each have enough wind to satisfy the whole country's electricity needs. The upper Midwest (the Dakotas east through Ohio) could supply the country's electricity without siting any wind turbines in either densely populated or environmentally sensitive areas. And wind resource assessments by the government of China have documented 472,000 megawatts of wind generating potential, enough to raise China's electricity supply threefold.

For Third World villages not yet connected to a grid, a more practical source is photovoltaic arrays, which may already have a competitive advantage. With the World Bank beginning to support this technology, costs will fall fast, making photovoltaic cells even more competitive. Wind, photovoltaic cells, and solar thermal power plants all promise inexpensive electricity as the technologies continue to advance and as the economies of scale expand. Over the longer term, cheap solar electricity in various forms will permit the conversion of electricity into hydrogen, which will offer an efficient means of energy transportation and storage.

Technological advances that increase the *efficiency* of energy use are in some ways even more dramatic than the advances in harnessing solar and wind resources. Striking gains have been made in the energy efficiency of electric lighting, electric motors, the thermal efficiency of windows, and cogenerating technologies that produce both electricity and heat. One of the most dramatic, as recently noted in *World Watch* (May/June 1993), is the new compact fluorescent light bulb—which can supply the same amount of light as an incandescent bulb while using only one-fourth as much electricity. The 134 million compact fluorescent bulbs sold worldwide in 1992 saved enough electricity to close 10 large coal-fired power plants.

The discontinuities that have wreaked havoc with once-reliable trends are not random, but reflect an escalating awareness of the need to transform the global economy into one that is sustainable. They reflect the unavoidable reality that we have entered an era in which satisfying the needs of the 91 million people being added each year depends on reducing consumption among those already here. At this rate, by the year 2010, this growth will amount to a net addition equal to nearly 200 cities the size of New York, or 100 countries the size of Iraq—dramatically reducing the per capita availability of cropland and irrigation water. At some point, as people begin to grasp the implications of this new reality, population policy will become a central concern of national governments.

**Economic Entropy**

Whether in basic foodstuffs and fresh water, or in overall economic output, the decade of discontinuity has begun. Growth in the world economy reached its historical high at 5.2 percent per year during the 1960s. It then slowed to 3.4 percent per year in the 1970s, and 2.9 percent in the 1980s. Despite this slowdown, the per capita output of goods and services rose as overall economic growth stayed ahead of population growth. Now that, too, may be reversing.

From 1990 to 1992, the world economy expanded at 0.6 percent per year. If the International Monetary Fund's recent projection of 2.2 percent in world economic growth for 1993 materializes, we will find ourselves three years into this decade with an income per person nearly 2 percent lower than it was when the decade began. Even using an economic accounting system that overstates progress because it omits environmental degradation and the depletion of natural capital, living standards are falling.

Evidence is accumulating that the world economy is not growing as easily in the 1990s as it once did. The conventional economic wisdom concerning the recession of the early 1990s attributes it to economic mismanagement in the advanced industrial countries (particularly the United States, Germany, and Japan) and to the disruption associated with economic reform in the centrally planned economies. These are obviously the dominant forces slowing world economic growth, but they are not the only ones. As noted above, growth in the fishing industry, which supplies much of the world's animal protein, may have stopped. Growth in the production of beef, mutton, and other livestock products from the world's rangelands may also be close to an end. The world grain harvest shows little prospect of being able to keep pace with population, much less to eliminate hunger. And scarcities of fresh water are limiting economic expansion in many countries. With constraints emerging in these primary economic sectors—sectors on which much of the Third World depends—we may be moving into an era of slower economic growth overall.

The popular question of "growth or no growth" now seems largely irrelevant. A more fundamental question is how to satisfy the basic needs of the world's people without further disrupting or destroying the economy's support systems. The real challenge for the 1990s is that of deciding how the basic needs of all people can be satisfied without jeopardizing the prospects of future generations.

Of all the discontinuities that have become apparent in the past few years, however, it is an upward shift in the population growth trend itself that may be most disturbing. The progress in slowing human population growth so evident in the 1970s has stalled—with alarming implications for the long-term population trajectory. Throughout the 1960s and 1970s, declining fertility held out hope for getting the brakes on population growth before it began to undermine living standards. The 1980s, however, turned out to be a lost decade, one in which the United States not only abdicated its leadership role, but also withdrew all financial support from the U.N. Population Fund and the International Planned Parenthood Federation. This deprived millions of couples in the Third World of access to the family planning services needed to control the number or timing of their children.

The concern that population growth could undermine living standards has become a reality in this decade of discontinuity. There is now a distinct possibility that the grain supply per person will be lower at the end of this decade than at the beginning, that the amount of seafood per person will be substantially less, and that the amount of meat per person will also be far less than it is today.

The absence of any technology to reestablish the rapid growth in food production that existed from 1950 to 1984 is a matter of deepening concern. In early 1992, the U.S. National Academy of Sciences and the Royal Society of London together issued a report that warned: "If current predictions of population growth prove accurate and patterns of human activity on the planet remain unchanged, science and technology may not be able to prevent either irreversible degradation of the environment or continued poverty for much of the world."

Later in the year, the Union of Concerned Scientists issued a statement signed by nearly 1,600 of the world's leading scientists, including 96 Nobel Prize recipients, noting that the continuation of destructive human activities "may so alter the living world that it will be unable to sustain life in the manner that we know." The statement warned: "A great change in our stewardship of the earth and the life on it is required, if vast human misery is to be avoided and our global home on this planet is not to be irretrievably mutilated."

The discontinuities reshaping the global economy define the challenge facing humanity in the next few years. It is a challenge not to the survival of our species, but to civilization as we know it. The question we can no longer avoid asking is whether our social institutions are capable of quickly slowing and stabilizing population growth without infringing on human rights. Even as that effort gets underway, the same institutions face the complex issue of how to distribute those resources that are no longer expanding, among a population that is continuing to grow by record numbers each year. ❑

## Questions:

1. The author says the discontinuities of the 1990s are caused by a collision between expanding human numbers on the one hand and the constraints of earth's natural systems on the other. What are some of the natural environmental constraints referred to?

2. How much did global fertilizer use increase between 1950 and 1984? What has happened to it since 1989?

3. Which of the three fossil fuels is expanding in use? Which has more potential: wind power or hydropower? How many people are being added to the population of the Earth each year?

Answers on page 301

30

*In 1991, 26 countries generated electricity with nuclear power. It provided over half the electricity for France, Belgium, and Sweden. In contrast, the U.S. produced only about 22% of its electricity with nuclear power. Indeed, fears about nuclear energy have led to much less use of it in the U.S. than predicted in the 1970s. Only about 110 nuclear power plants are now operating, compared to the more than 1,000 plants predicted by this time. Furthermore, very few plants are planned or under construction. There are a number of reasons for this: cheap fossil fuels, excessive time and expense of nuclear plants, mismanagement of existing plants, and especially negative public opinion. Does this mean nuclear power has no future in the U.S.? Not necessarily. For example, some environmentalists point out that nuclear power does not release many of the air pollutants caused by fossil fuels, such as acid rain, smog, and especially greenhouse gases that may be causing global warming. Redesigning reactors would alleviate some of the problems by decreasing the likelihood of meltdowns and increasing efficiency to reduce the amount of nuclear waste.*

# The Future of Nuclear Power

America will choose nuclear power only if demand for electricity accelerates, nuclear costs are contained and global-warming worries grow

by John F. Ahearne

The dawn of what some believed would be the age of nuclear power in America came quietly, at a remote site in the Idaho desert. On December 20, 1951, Experimental Breeder Reactor Number 1 (EBR-1) generated enough electricity to turn on four light bulbs. A decade later, it appeared that nuclear power would make a significant contribution to the future production of electricity in the United States. The Atomic Energy Commission predicted in the mid-1970s that there would be more than 1,000 large nuclear power plants in the United States by the year 2000.

The number of operating nuclear reactors in the U.S. at the turn of the century, however, is likely to be 110—about one-tenth of what was once predicted. Not one reactor has been ordered by a U.S. utility since 1978, and 64 reactors will be more than 20 years old by 2000. Recently, the owners of the Yankee Rowe plant in Massachusetts closed the reactor rather than seek an extension of its license (most plants are licensed for 40 years), which would have required expensive modifications that would have made its electricity cost far more than that from other sources. We have neither grown to depend on nuclear power nor suffered the shortages that were predicted if we did not build nuclear plants.

What happened? Has nuclear power lost all significance in the production of electricity? Not at all. Nuclear power produces electricity throughout the world. In 1991 nuclear power plants provided more than half of the electricity for France (72.7 percent), Belgium (59.3 percent) and Sweden (51.6 percent); in all, 26 countries generated electricity with nuclear power. The United States produced 21.7 percent of its electricity from nuclear power in 1991, placing the U.S. 14th in the world in the percentage of electricity generated from nuclear power plants. In 1989 six American states obtained more than 50 percent of their electricity from nuclear power, led by Vermont with 76 percent. Another nine states obtained more than 25 percent of their electricity from nuclear plants. Because the U.S. is the largest consumer of electricity in the world, it uses more nuclear power than any other nation, even though its percentage is relatively low.

Although nuclear power had a promising beginning and continues to provide significant amounts of electricity, its future in the U.S. looks grim. The reasons for this turn of events are explored in a report by the National Research Council, the research arm of the National Academy of Sciences, published in 1992 and titled, *Nuclear Power: Technical and Institutional Options for the Future.* A troubled past has created doubts about nuclear power's future, and this report concludes that America's nuclear-power industry has suffered from four problems: the growth in the demand for energy slowed, the cost and construction time of nuclear plants increased, the nuclear plants did not achieve the expected levels of operation, and public opinion turned away from nuclear power as a desirable technology.

The early predictions for the impact of nuclear power assumed a combination of factors: a high rate of growth in the demand for energy, adequate financial investment and public trust. These factors failed to materialize. This leads to the

**Reprinted by permission from AMERICAN SCIENTIST, January/February 1993, Journal of Sigma Xi, The Scientific Research Society.**

present question: Does nuclear power have a future in the United States? Some people would argue that the primary issue is reactor safety, and that the United States would build more nuclear plants if the public could be assured that the plants were safe. I do not think that safety is the primary issue. In large part, this decision will be made by the electric utilities, and they will build nuclear power plants only if the demand for energy again shows a high rate of growth and if the plants are less expensive and run more reliably. Otherwise, a nuclear-powered future is unlikely in the United States.

**A Drop in Demand**

Any technology's success depends on the demand for what it produces. In the 1960s and the early 1970s, the demand for electricity grew rapidly. Utility planners believed that the rate of growth would continue, and would require power plants with large generating capacities. Nuclear power plants can generate large quantities of electricity.

The utilities' expectations of demand—and therefore their construction plans—turned out to be far in excess of what was needed. The story of this miscalculation is told in the forecasts of the National Electricity Reliability Council (NERC), which represents the electric utilities and publishes an annual summary of the generating capacity available in the United States as well as the expected growth. The forecast is based on what utility planners believe they will be building in the future.

In 1974 NERC forecast a total generating capacity of about 700,000 megawatts for 1982. By 1978 the NERC projection for 1982 had fallen below 500,000 megawatts, and by 1981 it had dropped below 450,000 megawatts. Analysts in the energy community call a graph of these projections a NERC fan, because the forecast lines spread out like an open oriental fan, showing how utility plans changed constantly as growth in demand dropped year by year. In 1974 electricity demand was expected to grow at an average annual rate of 7.6 percent for the following decade. The growth rate turned out to be 2.9 percent. Growth began to slow down in the early 1970s, after the first "oil shock," when OPEC (the Organization of Petroleum Exporting Countries) demonstrated its power to raise oil prices. Although oil was not the principal fuel for generating electricity, the oil-price increases induced an examination of all energy use.

The United States is, and has been for decades, one of the most energy-consumptive countries, in terms of its energy use per capita. Some of this consumption undoubtedly arises from the size of the country, which creates a substantial fuel bill for transportation. In the early 1970s, however, energy was inexpensive, and the United States gave little thought to efficiency in its use. Few builders of houses incorporated energy-efficient designs, such as larger studs and more insulation in the walls, roof and floor. Factory machinery remained energy inefficient, because the cost of energy was a small item on the balance sheet.

The oil shocks (the second came in 1978) forced substantial revisions in how America used energy. The change becomes evident when one compares energy use to the Gross Domestic Product (GDP), the value of goods and services produced within the United States. Prior to the 1970s energy use and the GDP grew at nearly identical rates. Many argued that a correlation between the two is inevitable, and that a reduction in the use of energy would lead to a reduction in the growth of GDP. Isolated voices, such as those of David Freeman of the Ford Energy Project and Amory Lovins of the Rocky Mountain Institute, challenged this view. Then in 1977 the Carter administration declared that energy use was the most important issue facing the country. Americans acquired an interest in conservation and efficiency that led to a slowdown in the rate at which the demand for energy was growing, although GDP continued to rise. Growth in the demand for electricity remained correlated with GDP, but the rate of growth was modest compared to earlier predictions. Nuclear power utilities were caught with large sums of money invested in plants that were never built. More than $20 billion was invested in more than 100 nuclear projects that were eventually canceled.

Some in the nuclear industry claim that regulatory obstacles and public antagonism, fueled by environmentalists and the media, caused the halt in nuclear power. But that is not the case. In 1988 Ken Davis, a long-time senior executive at Bechtel Corporation, one of the world's largest constructors of nuclear power plants, and later a Deputy Secretary of Energy, wrote that "the hiatus in orders for new nuclear power plants in the U.S. since 1978 [resulted] entirely from the over-ordered situation at that time—there has been no reason for any utility to order a power reactor since." The demand factor must be understood before we can address the question of what would cause nuclear power to be a viable option in the United States. Demand is unrelated to public antagonism. The change in demand growth is merely a manifestation of a change in the use of electricity.

**Excessive Time and Expense**

As utility planners considered how to meet the huge demand for electricity projected in the early 1970s, nuclear power plants had several advantages. In addition to having large generating capacities, the plants use a type of fuel whose cost was relatively low. Oil, although thought to be abundant, was considered an expensive fuel for generating electricity. The choice of fuels, then, was between nuclear and coal. The cost of building a nuclear power plant was substantial, but it was considered predictable, and utility planners believed they would have no trouble recouping the construction costs through the rates charged to their customers.

Like the predictions of demand, these expectations turned out to be incorrect. The cost of building a nuclear plant increased dramatically between the 1970s and 1980s, and became more and more unpredictable. The time required to complete a new plant also grew longer and harder to predict. Furthermore, the utilities began to have trouble recouping the spiraling costs of plant construction from ratepayers.

The trends in cost and construction time show some dramatic shifts. The most notable change is seen when one compares the 1970s with the decade that followed. One might conclude that some event explains the sudden increase in construction costs and time, and there might be such an event in the record: the accident in 1979 at the Three Mile Island nuclear power plant near Harrisburg, Pennsylvania. There is much debate about what caused the great slowdown after Three Mile Island. Many things happened in the years that immediately followed. The Nuclear Regulatory Commission (NRC) spent two years considering what regulatory changes would be necessary as a result of the lessons learned at Three Mile Island. Meanwhile, utility companies slowed down construction. Eventually the NRC issued new requirements that caused utilities to modify the design of some plants under construction; these requirements included increased operator training and instrumentation to monitor the water level in nuclear reactors. Some of the slowdown might be attributed to the uncertain regulatory climate, but it is important to note that utility planners were also noticing the declining projections of demand for generating capacity.

Regardless of the cause, a slowdown in construction did arise. In the United States, from 1970 through 1979, the average construction time—the time from ground breaking to the first generation of electricity—for 63 nuclear power plants was 6.3 years. From 1980 through 1989, the average construction time shot up to 11 years for 47 plants. The average cost for a kilowatt of generating capacity went from $817 for 13 plants going into operation between 1971 and 1974 to $3,100 for 10 plants that went into operation during 1987 and 1988. (All costs are given in 1988 dollars.) These construction costs are known as "overnight costs." The overnight cost represents how much it would have cost to build the plants overnight with no inflation and no interest payments on loans.

Comparing costs from one country to another is not useful because different governments provide different levels of support, and the cost of money is quite different. Construction time, however, is well defined. Prior to 1978, the United States was in line with the rest of the world in construction time; the average time required to build a plant was less than in the United Kingdom and about 10 percent greater than in France. This picture changed drastically after 1978. From 1979 to 1990, it took nearly twice as long, on average, to build a plant in the United States as it did in France, and nearly two-and-a-half times as long as it took in Japan.

The unpredictability of the numbers was as daunting to utility planners as their size. If a utility plans to meet future needs with a nuclear plant, and the plant is not ready when the utility needs more generating capacity, the utility must seek another source, which requires additional spending. But even when the average construction time was a respectable six years, the minimum time was three years and the maximum time exceeded a decade. Later, when the average construction time increased to 11 years, the range was from six to 19 years. It is worth noting that some large plants going into operation between 1980 and 1989, after Three Mile Island, were completed in about six years, a reasonable time by any standard.

In 1988 the most expensive nuclear plant cost 25 percent more than the most expensive coal-fired power plant. The least expensive nuclear plant, on the other hand, cost nearly 30 percent less than the least expensive coal plant. In the early 1970s, the lowest cost and the highest cost for a nuclear plant differed by a factor of 2.5. That factor increased to more than three for the period from 1977 to 1988. The only consistency is that nuclear plants of the same generation and the same size ranged widely in construction costs. A planner could not predict construction cost and time for a given plant design. The reasons for this wide variation were explored by the National Academy panel, but no clear explanation emerged. Some experts have suggested that construction management was the key variable.

State utility commissions also influence the cost of nuclear plants by regulating how costs can be recovered. When a utility commits to building a power plant, its planners expect to include the construction cost in the utility's rate base, the figure used to determine how much the utility can charge for a kilowatt-hour of electricity. In some areas, the inclusion of costs for nuclear plants resulted in a potential 25-percent increase in the rates that could be charged.

In the face of public anger over the rate shock, state utility commissions began questioning the cost of nuclear-plant construction. The commissions introduced prudency reviews, post-construction reviews of power plants to determine which costs may be recovered from ratepayers. There are debates on whether prudency reviews are fair or accurate, but these issues are irrelevant to the purpose of this article. Prudency reviews exist, and reviews of 88 nuclear power plants during the 1980s prevented investor-owned utilities from recovering $14.4 billion in plant costs. This significantly decreased the return on investment, or profitability, of utilities with nuclear plants, because the $14.4 billion had to be taken from the return that investors hoped to receive. By comparison, 11 reviews of coal-fired plants led to the exclusion of only $700 million from the rate base. The average exclusion per nuclear plant was 2.5 times that of coal plants. In addition, many states decided not to allow the cost of construction work in progress to be added to the rate base at all.

**Mismanagement**

Even if a nuclear plant could be built at reasonable cost and on time, its profitability depends on adequate management, and nuclear plants in the United States have had a mixed record. The National Academy report states: "Effective plant management has been identified both by the NRC and by the industry as critical to safe, economical operation of nuclear power plants." Poor management has serious consequences. The NRC shut down the Pilgrim reactor in Massachusetts for four years because of poor management. A finding that operators were asleep in the control room led to a two-year shutdown at Peachbottom in New Jersey. An extended period of poor management by both the board of directors and plant supervisors preceded the referendum that closed Rancho Seco in California. The Tennessee Valley Authority's (TVA's) nuclear-power program, one of the largest in the United States, was shut down completely for more than four years because of questions about its management.

One indicator of management performance is the load factor, the ratio between the electricity that a plant produces and the electricity that it is designed to produce if it runs 100 percent of the time. In the United States, all nuclear power plants are base-load plants. That is, they are designed to operate continuously to provide the electricity required by those customers whose requirements do not change. Extra demand for electricity comes from load-following units, used to supply needs such as air conditioning on summer days. France has so much nuclear power that some of its nuclear power plants are run to follow demand. Through 1987 the average lifetime load factors of nuclear power plants in the major countries differed significantly: 78.2 percent in Canada, 73.6 percent in West Germany, 68 percent in Japan and 60.5 percent in the United States. (I omit France because of the load following, which skews calculations of the load factor.) If nuclear plants in the United States had performed more reliably, and thereby increased their load factor, then rate commissioners and utility planners would have a more positive attitude about this fuel source.

The information given above indicates the great weaknesses that nuclear power has from the perspective of utility planners. Nuclear power plants fail the three questions critical to planning: How much will the plant cost? When will the plant be ready? How well will it run? The usual answers for nuclear power are: It will cost a lot, it will take a long time to build, and it won't run well. Worse yet, the honest answer to each question is: We don't know. Some nuclear power plants are built at reasonable cost; other construction budgets are exorbitant. Some nuclear plants are built in a reasonable period of time; others take incredibly long to build. And although some nuclear power plants in the United States run among the best in the world, many are among the worst.

**Public Opinion**

Public opposition is frequently described as one of the major problems for nuclear power. Bert Wolfe, former head of nuclear operations at General Electric, mentioned two years ago that he took his own, unscientific survey by talking to seatmates on his many airplane flights. The typical response to his questions was: Nuclear power has problems— it costs too much and may be unsafe— and we don't really need it, so why should we use it? It is not clear what the public consensus is regarding nuclear power, but opponents want assurance that the plants are running safely today and that radioactive waste will not create a problem in the future.

Opponents of nuclear power cite several issues of concern. The first argument is that nuclear power costs too much. The rate shock mentioned earlier made a significant contribution to this perspective. The second argument is that nuclear power is not safe. The National Academy report, which incorporated the opinions of people with a wide range of positions on nuclear power, reached several conclusions on safety, including: "The risk to the health of the public from the operation of current reactors in the United States is very small. In this fundamental sense, current reactors are safe. A significant segment of the public has a different perception, and also believes the level of safety can and should be increased." The accidents at Three Mile Island and at Chernobyl, Ukraine, in 1986, made safety a dominant issue among the opponents of nuclear power.

Another problem presented by the opponents of nuclear power is the question of how to get rid of the waste generated by plants. Nuclear plants generate both low-level and high-level radioactive waste. Low-level waste is material whose radioactivity will decay to safe levels in less than 50 years. It can be buried in shallow landfills. It is a broad category, including radioactive material, such as isotopes of nickel and cobalt from corroding pipes, that can appear in the clothing and the tools of the reactor workers. Low-level waste is also generated by processes outside the nuclear power industry. In 1989, for instance, 52 percent of the low-level waste in the United States came from reactors, but the next largest percentage, 35 percent, came from commercial industrial activity.

High-level waste, such as fission products, requires centuries to decay to safe levels of radioactivity, and it most be stored in sealed containers because of the intensity of its ionizing radiation. It has two sources: commercial reactors, whose spent fuel must be disposed of, and the defense complex. The high-level waste from defense facilities exceeds the commercial volume by a factor of about 20, but commercial high-level waste contains more radioactivity.

For years the low-level waste produced in the United States has been sent to Beatty, Nevada; Hanford, Washington; and Barnwell, South Carolina. In 1989 60 percent of the volume of low-level waste and 84 percent of the radioactivity from low-level waste went to Barnwell. The Low-Level Waste Policy Act of 1985 requires that all states either form a compact with other states to select a disposal site or develop a site themselves. This process has foundered, leading to repeated extensions of the cut-off date in the act. The choice of a low-level-waste site runs into the same opposition as the locating of a hazardous-waste-disposal facility or a toxic-chemical incinerator. The problems are best described by the variety of emerging acronyms: LULU (locally unwanted land use), NIMBY (not in my backyard) and NIMTO (not in my term of office). The public near a selected waste site rises in opposition, arguing that the site should be somewhere else.

It is much more difficult to select a high-level-waste disposal site. Low-level waste has the protective coloration of association with medical use (which generates waste from the use of radioactive materials for clinical tests and therapeutic uses), even though the medical community produced less than 0.1 percent of the low-level waste in 1988. A high-level-waste site, on the other hand, is strictly for the disposal of remnants from making nuclear weapons and of the fuel from generating electricity from nuclear power. Congress continues to agonize over selecting high-level-waste disposal sites, and it has painfully passed laws, by slim margins, that establish a process to select a permanent repository for high-level nuclear waste. The last time through this issue, Congress selected Yucca Mountain in Nevada. Many issues surround that selection. Is it a site that will remain geologically stable for thousands of years? Was it chosen largely because it borders the nuclear test site in a highly isolated, barren landscape of a low-population state? The Nevada congressional delegation and the state government continue to fight any examination of their state for a site. Regardless of whether Yucca Mountain is an appropriate site and whether it can meet the regulations of the NRC and the Environmental Protection Agency, the constant opposition from the state government gives the general public the impression that there is something wrong with nuclear power because its "garbage" must be disposed of against the opposition of local residents.

The waste-disposal dilemma, however, is not stopping utilities from building nuclear plants. When I ask utility planners about the waste-disposal problem, they admit that they would like to have it solved, if only to calm some opponents of nuclear power. But when I ask planners whether they would build a nuclear plant if waste disposal were not a problem, they still say, "No." The waste issue may well affect the future of nuclear power, but it is not the deciding issue at present.

Carl Sagan of Cornell University wrote in *Parade* magazine that "... there is one other problem: all nuclear power plants use or generate uranium and plutonium that can be employed to manufacture nuclear weapons." Faced with such statements, the proponents of nuclear power quickly argue that nuclear power plants have not been a source of nuclear-weapons development. But the question is not the record but the perceived threat. The issue, according to the report from the National Academy, is that "many skeptics of nuclear power, including what Alvin Weinberg [former director of the Oak Ridge National Laboratory] has called the 'articulate elites,' believe the potential for nuclear weapons proliferation is a major threat posed by the use of nuclear power."

I usually ignore polling data as a source of public opinion on nuclear power. With the exception of a very small blip around the

Persian Gulf war, polls indicate that the American public believes that energy is not important. Consequently, a poll that considers nuclear power versus coal, for example, is asking about a subject that the public considers unimportant. I do not place much credence on conclusions from most of these polls.

I am, however, interested in the results of two polls taken in early 1992. Cambridge Reports/Research International asked opinion leaders what they think about nuclear power and what they believe the public thinks about nuclear power. Opinion leaders were defined as federal legislators and officials, state legislators and officials, business executives, financial executives, leaders of public-interest groups, representatives of the national and local media, and policy experts. (The data do not indicate the mix of respondents.) The opinion leaders were asked: "Practically speaking, how important a role do you think nuclear energy should play in meeting America's future energy needs?" Bruskin/Goldring then asked randomly selected members of the public a similar question. The answers "very important" or "somewhat important" were chosen by 72 percent of the opinion leaders and 73 percent of the public. The answers "not too important" and "somewhat important" cover 27 percent of the opinion leaders and 22 percent of the public. Then the opinion leaders were asked: "What about the American public: Do you think the majority of Americans would say that nuclear energy should play an important role in meeting America's future energy needs or do you think the majority would say that nuclear energy should not play an important role?" Opinion leaders thought that 25 percent of the public would say "an important role," and that 63 percent of the public would say "not an important role." The opinion leaders believe that most of the public is opposed to nuclear power, but the poll indicates that most of the public is in favor of this power source, at least as a future option. The perception of public opposition to nuclear power may be exaggerated.

**Future Options**

What would have to happen to make nuclear power an attractive option to an electrical utility in the United States? The report from the National Academy lists developments that might promote a return to nuclear power. The primary issue, perhaps, is the growth in demand for electricity. The use of nuclear reactors will be logical if we ever need more generating capacity of a magnitude that can be best met by large power plants. Today, this is not the case in the United States.

Nuclear power would also become more important, according to the report, if environmental policies force a reduction in emissions from plants that generate electricity with coal, oil and natural gas. Coal is now the dominant fuel for generating electricity in America, and the preferred choice for new generating capacity is natural gas. Natural gas is an appealing choice because gas-fired plants can be built in two to three years, and they can be built in sizes as small as 100 to 200 megawatts, making them a good choice for the currently projected demand, which is one or two percent per year for electricity. In addition, natural gas is currently plentiful and inexpensive. Therefore, only harsh emissions policies would turn utilities from coal and natural gas under current demand forecasts.

The remaining issues are largely concerned with public opinion about nuclear power. The report says that if nuclear power is to be an option in the future, existing plants must be operated safely, and this information must be relayed to the public. Norman Rasmussen, the former head of the nuclear-engineering department at the Massachusetts Institute of Technology, voiced a similar opinion, saying that what is needed is, "No more Peachbottoms, no more Pilgrims, no more TVA's." The Academy report goes on to say that public opinion might be improved by involving the public in nuclear-power issues such as planning, siting and oversight. In the opinion of the industry, however, the public already has more than enough opportunities for participation. Many in the nuclear-power industry believe that public participation has blocked low-level-waste sites, canceled the Shoreham reactor on Long Island, New York, and led to extensive delays such as the extra years required to begin operation at the Seabrook plant in New Hampshire. Critics of nuclear power, on the other hand, believe that local participation is necessary for the acceptance of waste-disposal sites as well as nuclear plants.

The other issues cited by the report involve nuclear waste. The report urges that the public be made aware that low-level radiation comes from nature as well as from nuclear-power-plant waste. This issue may be the source of the greatest disagreement between scientists and non-scientists. Radiation is mysterious to many people. It is seen as deadly and produced by sources beyond public control. A scientist might argue that the radiation from nuclear plants and that likely to be produced by waste sites is insignificant in comparison to natural radioactivity radon gas from underground, radioactivity from the stone in buildings and cosmic radiation. Much of the general public, however, remains unconvinced that they should accept any additional radioactivity. Indeed, radiation is hazardous. Several years ago, radioactive material was disposed of in a Brazilian dump where children found it. They later suffered horrible, painful deaths. The firefighters and operators at Chernobyl suffered radiation burns and massive exposures. These events involved real hazards. Nuclear power demands extreme attention to detail for safe operation, but the small amounts of radiation measured outside nuclear plants or those likely to be released from waste sites are, to a nuclear physicist, trivial.

The report goes on to say that the stalemate surrounding the disposal of high-level waste must be resolved. The report does not suggest that Yucca Mountain be forced down Nevada's throat. Rather, surface storage may be acceptable for many decades. Spent fuel is currently stored at nuclear plants in pools of water, but containers could be developed, either at the plants or at government sites, that could be used to store the fuel for decades. It is not necessary to put the fuel underground at this time.

Finally, the report addresses weapons. It states that a nuclear future can only develop if people are assured that "nuclear power would not materially affect the likelihood of nuclear weapons proliferation."

I would add that a nuclear future would require different attitudes among utility planners and members of the financial community. Their attitudes might be changed if the problems of cost, construction time and plant operation could be solved.

**Nuclear Technology**

All nuclear reactors generate electricity by capturing the thermal energy released when a heavy nucleus splits into smaller nuclei. Typically a uranium nucleus breaks apart into two smaller nuclei and at the same time emits two or three neutrons. All of these fission fragments have high kinetic energy, which is converted after collisions with other nuclei into the random motion of thermal energy. The neutrons can be absorbed by other uranium nuclei, inducing them to split in turn and thereby maintaining a chain reaction. The fissile element in most nuclear reactors is an isotope of uranium, $^{235}U$.

The commercial nuclear reactors in the United States are called light-water reactors. They use ordinary, or light, water both as a moderator to slow neutrons and as a coolant to remove heat from the nuclear fuel. The nuclear fuel is pressed into pellets that are encased in rods of a zirconium alloy. Control rods, which absorb neutrons, can be inserted or withdrawn from around the fuel, thereby slowing or increasing the rate of fission. The fuel assembly is placed in a pressure vessel, which is inside a sturdy containment building. A variety of systems are in place to automatically stop the reaction and to provide emergency cooling.

Light-water reactors come in two styles: the boiling-water reactor and the pressurized-water reactor. In a boiling-water reactor, the coolant boils, producing steam that drives a turbine. In the pressurized-water reactor, the coolant is under higher pressure, so that it does not boil. The heat from the coolant is passed via a heat exchanger to a second system that also contains water. The water in this system boils, producing steam to turn a turbine.

**Renovating Reactor Designs**

Several of the problems surrounding nuclear power could be solved through new reactor designs. The nuclear-power industry, however, was somewhat slow to recognize that new designs would be necessary. For many years, the industry claimed that regulatory hassles alone caused the high costs and long construction times common in the United States. The industry's proposal for a solution was to eliminate at least one of the hearings required for federal approval of a new nuclear power plant and to streamline the NRC's review process. Some of this has been done. The Energy Act of 1992 was just signed by President Bush, and it includes "one-step" licensing for nuclear plants. Although this may be too bluntly stated, the nuclear industry wanted to get the NRC out of the way. A new group of enlightened nuclear-power and utility executives, however, recognizes that getting rid of the NRC will not solve the industry's problems. Instead, the industry must modify reactor designs.

The Electric Power Research Institute (EPRI)—a research group funded by the electric utilities in the United States and some foreign countries—has developed a set of criteria for new reactors that focus on the problems of previous generations. These criteria indicate that reactors should be designed to last for 50 years, to have a load factor of 87 percent, to be built in less than four-and-a-half years (even for large, 1,500-megawatt plants), to have an overnight cost of $1,300 per kilowatt of generating capacity and to exceed the safety goals established by the NRC.

EPRI developed these criteria in concert with the nuclear-power industry, and the list is being reviewed by the NRC. The criteria are fine goals, and I suspect that the financial community would raise the funds for a nuclear plant if EPRI's numbers could be guaranteed, and if the growth in demand indicated the need for a new plant. But the nuclear industry looks like the shepherd boy who too often cried "wolf!" The industry claims that it can design nuclear power plants that could be built quickly, cost less and run well, but utility planners, utility commissioners and financiers remain skeptical. They want to see some of these plants built and in operation.

New reactor designs fall into several categories. The first

category includes evolutionary light-water reactors, designed in the United States by Westinghouse, General Electric and Combustion Engineering. These reactors are improved versions of the large plants that these companies built in the past. General Electric's design is the furthest along, and the company is currently building two such plants in Japan. These plants may appear similar to the old designs, but they are significantly simplified and have improved control systems. For example, the advanced boiling-water reactor requires 50 percent fewer welds in the structure that encloses the fuel and its cooling water. Likewise, the advanced pressurized-water reactor requires less piping. Both systems use multiplexing in instrumentation and control systems, reducing the amount of electrical cable needed, and thus reducing construction cost. These systems also use digital controls that provide self-diagnostics during operation, which increases the likelihood of safe operation.

It is possible that General Electric's Japanese plants will be built quickly, at lower cost, and will run well. But this is not likely to influence the financial community because the current generation of Japanese plants were also built quickly, at lower cost, and they run well.

The next category of reactors are smaller, around 600 megawatts, and are widely described as "passively safe." The cooling system relies on natural circulation. A large pool of water is put above the reactor. If an accident causes the reactor's core to get too hot, gravity drives the water into the reactor, and the buoyancy of heated water induces the water to circulate. The Westinghouse AP600—a 600-megawatt plant—is the furthest along in this class. Its design indicates that, compared with a conventional 600-megawatt plant, the AP600 will have 60 percent fewer valves, 35 percent fewer large pumps, 75 percent less piping, 80 percent less control cabling and 80 percent less ductwork. Many questions, however, must be answered about passively safe reactors. They remain "paper reactors," under development through a joint venture of industry and the Department of Energy.

Another category of reactors uses gas as a coolant and graphite as a moderator. The MHTGR (modular high-temperature gas reactor) is cooled with helium, and this design is advocated by General Atomics in the United States. This reactor operates at much higher temperatures than does a light-water reactor, encouraging cogeneration—the production of both electricity and process steam that could be used in manufacturing.

The novel feature of the MHTGR is so-called containment-in-a-pellet. Most current reactors in the United States include a large containment building, usually a very thick, reinforced-concrete structure. If an accident causes melting of the fuel and the release of energy and radiation, the containment isolates the danger. In the MHTGR, the fuel is composed of pellets of uranium, 0.8 millimeter in diameter, that are surrounded by layers of pyrolytic carbon and silicon carbide. The designers predict that the maximum temperature reached in such an accident could not melt the coating. Therefore, in theory, the coating would prevent the release of radiation, even if all of the cooling gas is lost.

The designers argue that the MHTGR does not need a containment building, which adds significant cost to a plant. The design's proponents also argue that the emergency planning zone (EPZ)—the area in which the NRC requires an evacuation plan defined essentially by a 10-mile radius from the plant—could be eliminated. The EPZ requirement has created many difficulties, and it was probably the reason that the Shoreham plant on Long Island never operated. The MHTGR's most ardent advocates argue that this reactor would be so safe that it could be built near a city, without a containment building. The National Academy group, however, was skeptical about the justification of eliminating a containment building.

The breeder reactor is the final design category of interest. The purpose of this reactor is apparent from its name: It breeds fuel as it runs. Such a reactor can be fueled with the common isotope $^{238}U$—the form in which uranium is usually found—and plutonium. The plutonium undergoes fission, much like the $^{235}U$ in a light-water reactor, producing neutrons that can be absorbed by $^{238}U$ to create $^{239}U$, which eventually becomes plutonium. This reactor requires fast neutrons for absorption by $^{238}U$, so there is no moderator. The coolant for a breeder reactor is a liquid metal, currently sodium.

A key turn of events in the history of the breeder reactor was a battle that erupted in the late 1970s between the White House and Congress over plans for the government-operated Clinch River Breeder Reactor (CRBR) in Oak Ridge, Tennessee. When Jimmy Carter was elected, he intended to remove plutonium from the nuclear fuel cycle because of a concern over weapons proliferation. Although James Schlesinger, Carter's Secretary of Energy and former director of the Atomic Energy Commission, found the White House position specious at best, he and his staff concluded that the CRBR was a great waste of money because the breeder reactor would not be needed until well past the year 2000. At the same time, nuclear-power advocates were concerned that the United States was going to run out of uranium. If that were true and the country anticipated a very large nuclear-power program, then the obvious solution was to build breeder reactors.

Schlesinger and his staff, however, concluded that the United States probably has more uranium than it will use. With the expense of the breeder and the ample supply of uranium, the case for the Clinch River reactor disappeared, and it became a costly, unnecessary government program. Congress finally killed the project under President Reagan, a supporter of nuclear power and of the breeder program.

There are other reactor designs, but they are even less likely to affect the market in the United States. For instance, many people may wonder about fusion reactors, but fusion is far too experimental to know when or if it would be an economically attractive option.

Even with new reactor designs, there is no sign of interest among utilities. That will not develop until the demand for electricity increases and reactors are an economical choice—built on time, at reasonable cost and running well. It will be easiest to convince utilities of the last point, running well. Nuclear plants in the United States are doing better, under constant pressure from the NRC to improve operating practices and from the Institute of Nuclear Power Operations to improve all practices. The average load factor in 1991 was 70.2 percent, which is respectable, especially given that the worst performers have not been removed.

I believe that nuclear power will become economically attractive only if non-nuclear sources of electricity face new obstacles. Emission controls are the most likely candidates because of concerns about climate change and greenhouse warming. Nevertheless, a 1991 National Academy report, *Policy Implications of Greenhouse Warming*, says that nuclear power remains an unlikely option. The report states: "Questions about the appropriateness of current technologies and public opposition to nuclear power ... currently make this option difficult to implement. To the extent that concern about greenhouse warming replaces concern about nuclear energy and inherently 'safe' nuclear plants are developed, this option increases in priority ranking." Instead of nuclear power, the report focuses on energy conservation, new renewable technologies and better use of coal.

It has been more than 40 years since the first nuclear reactor powered four light bulbs in the desert of Idaho. At that time, many expected a nuclear-powered America in the 1990s. That will not happen. The current trends in the demand for electricity do not indicate that the growth rate will soon return to what we experienced prior to the 1970s. It remains to be seen whether new reactor designs can decrease the cost and time of construction, and whether they will increase the load factor and reactor safety to the necessary levels. Does nuclear power have a future in the United States? Even though nuclear power *is* currently important, very important in many states, I doubt that it will play a significant role in meeting America's future energy needs. ❑

**Bibliography**

Ahearne, John F. 1989. *Will Nuclear Power Recover in a Greenhouse?* Washington, D.C.: Resources for the Future, Inc.

Carter, Luther J. 1987. *Nuclear Imperatives and Public Trust: Dealing with Radioactive Waste.* Washington, D.C.: Resources for the Future, Inc.

Cohen, Bernard L. 1990. *The Energy Option: An Alternative for the 90's.* New York: Plenum Press.

Flavin, Chris. 1988. *The Case Against Reviving Nuclear Power.* Washington, D.C.: World Watch Institute.

Lewis, H. W. 1990. *Technological Risk.* New York: W. W. Norton and Company.

National Research Council. 1992. *Nuclear Power: Technical and Institutional Options for the Future.* Washington, D.C.: National Academy Press.

Taylor, John J. 1989. Improved and safer nuclear power. *Science* 244:318–325.

## Questions:

1. What does the author conclude about the future of nuclear power in the U.S.?
2. What is a breeder reactor? What two fuels can it use? Why is there concern about plutonium?
3. What is a load factor? What is the U.S. load factor, and what does it say about U.S. nuclear power plant performance reliability?

Answers on page 301

# 31

*No major industrialized nation is more dependent on fossil fuels than the U.S. Fossil fuels, such as coal and petroleum, are nonrenewable resources; it is projected that the next few decades will completely exhaust world supplies of petroleum. The U.S., which derives about 90% of its energy from fossil fuels, is now importing about 50% of its petroleum. This is unsafe, for political and economic reasons. Fossil fuels are also a major contributor to the Greenhouse Effect. For these and many other reasons, there is growing interest in developing renewable, alternative energy sources. These include biomass, hydropower, wind power, and direct solar (photovoltaics). Such alternatives have long been technologically feasible, but are only now becoming economically attractive. Such economic attractiveness can only increase, as fossil fuels become scarcer and their environmental impacts more well known to the public.*

# Renewable Energy: Economic and Environmental Issues

## Solar energy technologies, paired with energy conservation, have the potential to meet a large portion of future US energy needs

by David Pimentel, G. Rodrigues, T. Wang, R. Abrams, K. Goldberg, H. Staecker, E. Ma, L. Brueckner, L. Trovato, C. Chow, U. Govindarajulu, and S. Boerke

The United States faces serious energy shortages in the near future. High energy consumption and the ever-increasing US population will force residents to confront the critical problem of dwindling domestic fossil energy supplies. With only 4.7% of the world's population, the United States consumes approximately 25% of the total fossil fuel used each year throughout the world. The United States now imports about one half of its oil (25% of total fossil fuel) at an annual cost of approximately $65 billion (USBC 1992a). Current US dependence on foreign oil has important economic costs (Gibbons and Blair 1991) and portends future negative effects on national security and the economy.

Domestic fossil fuel reserves are being rapidly depleted, and it would be a major drain on the economy to import 100% of US oil. Within a decade or two US residents will be forced to turn to renewable energy for some of their energy needs. Proven US oil reserves are projected to be exhausted in 10 to 15 years depending on consumption patterns (DOE 1991a, Matare 1989, Pimentel et al. 1994, Worldwatch Institute 1992), and natural gas reserves are expected to last slightly longer. In contrast, coal reserves have been projected to last approximately 100 years, based on current use and available extraction processes (Matare 1989).

The US coal supply, however, could be used up in a much shorter period than the projected 100 years, if one takes into account predicted oil and gas depletion and concurrent population growth (DOE 1991a, Matare 1989). The US population is projected to double to more than one-half billion within the next 60 years (USBC 1992b). How rapidly the coal supply is depleted will depend on energy consumption rates. The rapid depletion of US oil and gas reserves is expected to necessitate increased use of coal. By the year 2010, coal may constitute as much as 40% of total energy use (DOE 1991a). Undoubtedly new technologies will be developed that will make it possible to extract more oil and coal. However, this extra extraction can only be achieved at greater energy and economic costs. When the energy input needed to power these methods approaches the amount of energy mined, extraction will no longer be energy cost-effective (Hall et al. 1986).

Fossil fuel combustion, especially that based on oil and coal, is the major contributor to increasing carbon dioxide concentration in the atmosphere, thereby contributing to probable global warming. This climate change is considered one of the most serious environmental threats throughout the world because of its potential impact on food production and processes vital to a productive environment. Therefore, concerns about carbon dioxide emissions may discourage widespread dependence on coal use and encourage the development and use of renewable energy technologies.

Even if the rate of increase of per capita fossil energy consumption

**Reprinted with permission of the authors from BioScience, September 1994, vol. 44, no. 8, pp. 536-547, copyright 1994. American Institute of Biological Sciences.**

is slowed by conservation measures, rapid population growth is expected to speed fossil energy depletion and intensify global warming. Therefore, the projected availability of all fossil energy reserves probably has been overstated. Substantially reducing US use of fossil fuels through the efficient use of energy and the adoption of solar energy technologies extends the life of fossil fuel resources and could provide the time needed to develop and improve renewable energy technologies.

Renewable energy technologies will introduce new conflicts. For example, a basic parameter controlling renewable energy supplies is the availability of land. At present more than 99% of the US and world food supply comes from the land (FAO 1991). In addition, the harvest of forest resources is presently insufficient to meet US needs and thus the United States imports some of its forest products (USBC 1992a). With approximately 75% of the total US land area exploited for agriculture and forestry, there is relatively little land available for other uses, such as biomass production and solar technologies. Population growth is expected to further exacerbate the demands for land. Therefore, future land conflicts could be intense.

In this article, we analyze the potential of various renewable or solar energy technologies to supply the United States with its future energy needs. Diverse renewable technologies are assessed in terms of their land requirements, environmental benefits and risks, economic costs, and a comparison of their advantages. In addition, we make a projection of the amount of energy that could be supplied by solar energy subject to the constraints of maintaining the food and forest production required by society. Although renewable energy technologies often cause fewer environmental problems than fossil energy systems, they require large amounts of land and therefore compete with agriculture, forestry, and other essential land-use systems in the United States.

**Assessment of renewable energy technologies**

Coal, oil, gas, nuclear, and other mined fuels currently provide most of US energy needs. Renewable energy technologies provide only 8%. The use of solar energy is, however, expected to grow. Renewable energy technologies that have the potential to provide future energy supplies include: biomass systems, hydroelectric systems, hydrogen fuel, wind power, photovoltaics, solar thermal systems, and passive and active heating and cooling systems.

**Biomass energy systems**

At present, forest biomass energy, harvested from natural forests, provides an estimated 3.6 quads ( 1.1 x $10^{18}$ Joules) or 4.2% of the US energy supply. Worldwide, and especially in developing countries, biomass energy is more widely used than in the United States. Only forest biomass will be included in this US assessment, because forest is the most abundant biomass resource and the most concentrated form of biomass. However, some biomass proponents are suggesting the use of grasses, which on productive soils can yield an average of 5t • $ha^{-1}$ • $yr^{-1}$ (Hall et al. 1993, USDA 1992).

Although in the future most biomass probably will be used for space and water heating, we have analyzed its conversion into electricity in order to clarify the comparison with other renewable technologies. An average of 3 tons of (dry) woody biomass can be sustainably harvested per hectare per year with small amounts of nutrient fertilizer inputs (Birdsey 1992). This amount of woody biomass has a gross energy yield of 13.5 million kcal (thermal). The net yield is, however, lower because approximately 33 liters of diesel fuel oil per hectare is expended for cutting and collecting wood and for transportation, assuming an 80-kilometer roundtrip between the forest and the plant. The economic benefits of biomass are maximized when biomass can be used close to where it is harvested.

A city of 100,000 people using the biomass for a sustainable forest (3 tons/ha) for fuel would require approximately 220,000 ha of forest area, based on an electrical demand of 1 billion kWh (860 x 10° kcal = 1 kWh) per year. Nearly 70% of the heat energy produced from burning biomass is lost in the conversion into electricity, similar to losses experienced in coal-fired plants. The area required is about the same as that currently used by 100,000 people for food production, housing, industry, and roadways (USDA 1992).

The energy input/output ratio of this system is calculated to be 1:3. The cost of producing a kilowatt of electricity from woody biomass ranges from 7¢ to 10¢, which is competitive for electricity production that presently has a cost ranging from 3¢ to 13¢ (USBC 1992a). Approximately 3 kcal of thermal energy is required to produce 1 kcal of electricity. Biomass could supply the nation with 5 quads of its total gross energy supplied by the year 2050 with the use of at least 75 million ha (an area larger than Texas, or approximately 8% of the 917 million ha in the United States).

However, several factors limit reliance on woody biomass. Certainly, culturing fast-growing trees in a plantation system located on prime land might increase yields of woody biomass. However, this practice is unrealistic because prime land is essential for food production. Furthermore, such intensely managed systems require additional fossil fuel inputs for heavy machinery, fertilizers, and pesticides, thereby diminishing the net energy available. In addition, Hall et al. (1986) point out that energy is not the highest priority use of trees.

If natural forests are managed for maximal biomass energy production, loss of biodiversity can be expected. Also, the conversion of natural forests into plantations increases soil erosion and water runoff. Continuous soil erosion and degradation would ultimately reduce the overall productivity of the land. Despite serious limitations of plantations, biomass production could be increased using agroforestry technologies designed to protect soil quality and converse biodiversity. In these systems, the energy and economic costs would be significant and therefore might limit the use of this strategy.

The burning of biomass is environmentally more polluting than gas but less polluting that coal. Biomass combustion releases more than 100 different chemical pollutants into the atmosphere (Alfheim and Ramdahl 1986). Wood smoke is reported to contain pollutants known to cause bronchitis, emphysema, and other illnesses. These pollutants include up to 14 carcinogens, 4 cocarcinogens, 6 toxins that damage cilia, and additional mucus-coagulating agents (Alfheim and Ramdahl 1986, DOE 1980). Of special concern are the relatively high concentrations of potentially carcinogenic polycyclic aromatic hydrocarbons (PAHs, organic compounds such as benzo (a) pyrene) and particulates found in biomass smoke (DOE 1980). Sulfur and nitrogen oxides, carbon monoxide, and aldehydes also are released in small though significant quantities and contribute to reduced air quality (DOE 1980). In electric-generating plants, however, as much as 70% of these air pollutants can be removed by installing the appropriate air-pollution control devices in the combustion system.

Because of pollutants, several communities (including Aspen, Colorado) have banned the burning of wood for heating homes. When biomass is burned continuously in the home for heating, its pollutants can be a threat to human health (Lipfert et al. 1988, Smith 1987b).

When biomass in the form of harvested crop residues is used for fuel, the soil is exposed to intense erosion by wind and water (Pimentel et al. 1984). In addition to the serious degradation of valuable agricultural land, the practice of burning crop residues as a fuel removes essential nutrients from the land and requires the application of costly fossil-based fertilizers if yields are to be maintained. However, the soil organic matter, soil biota, and water-holding capacity of the soil cannot be replaced by applying fertilizers. Therefore, we conclude that crop residues should not be removed from the land for a fuel source (Pimentel 1992).

Biomass will continue to be a valuable renewable energy resource in the future, but its expansion will be greatly limited. Its use conflicts with the needs of agricultural and forestry production and contributes to major environmental problems.

**Liquid fuels**

Liquid fuels are indispensable to the US economy (DOE 1991a). Petroleum, essential for the transportation sector as well as the chemical industry, makes up approximately 42% of total US energy consumption. At present, the United States imports about one half of its petroleum and is projected to import nearly 100% within 10 to 15 years (DOE 1991a). Barring radically improved electric battery technologies, a shift from petroleum to alternative liquid and gaseous fuels will have to be made. The analysis in this section is focused on the potential of three liquid fuels: ethanol, methanol, and hydrogen.

**Ethanol**. A wide variety of starch and sugar crops, food processing wastes, and woody materials (Lynd et al. 1991) have been evaluated as raw materials for ethanol production. In the United States, corn appears to be the most feasible biomass feedstock in terms of availability and technology (Pimentel 1991).

The total fossil energy expended to produce 1 liter of ethanol from corn is 10,200 kcal, but note that 1 liter of ethanol has an energy value of only 5130 kcal. Thus, there is an energy imbalance causing a net energy loss. Approximately 53% of the total cost (55¢ per liter) of producing ethanol in a large, modern plant is for the corn raw material (Pimentel 1991). The total energy inputs for producing ethanol using corn can be partially offset when the dried distillers grain produced is fed to livestock. Although the feed value of the dried distillers grain reduces the total energy inputs by 8% to 24%, the energy budget remains negative.

The major energy input in ethanol production, approximately 40% overall, is fuel needed to run the distillation process (Pimentel 1991). This fossil energy input contributes to a negative energy balance and atmospheric pollution. In the production process, special membranes can separate the ethanol from the so-called beer produced by fermentation. The most promising systems rely on distillation to bring the ethanol concentration up to 90%, and selective-membrane processes are used to further raise the ethanol concentration to 99.5% (Maeda and Kai 1991). The energy input for this upgrading is approximately 1280 kcal/liter. In laboratory tests, the total input for producing a liter of ethanol can potentially be reduced from 10,200 to 6200 kcal by using membranes, but even then the energy balance remains negative.

Any benefits from ethanol production, including the corn by-products, are negated by the environmental pollution costs incurred from ethanol production (Pimentel 1991). Intensive corn production in the United States causes serious soil erosion and also requires the further draw-down of groundwater resources. Another environmental problem is caused by the large quantity of stillage or effluent produced. During the fermentation process approximately 13 liters of sewage effluent is produced and placed in the sewage system for each liter of ethanol produced.

Although ethanol has been advertised as reducing air pollution when mixed with gasoline or burned as the only fuel, there is no reduction when the entire population system is considered. Ethanol does release less carbon monoxide and sulfur oxides than gasoline and diesel fuels. However, nitrogen oxides, formaldehydes, other aldehydes, and alcohol—all serious air pollutants—are associated with the burning of ethanol as fuel mixture with or without gasoline (Sillman and Samson 1990). Also, the production and use of ethanol fuel contribute to the increase in atmospheric carbon dioxide and to global warming, because twice as much fossil energy is burned in ethanol production than is produced as ethanol.

Ethanol produced from corn clearly is not a renewable energy source. Its production adds to the depletion of agricultural resources and raises ethical questions at a time when food supplies must increase to meet the basic needs of the rapidly growing world population.

**Methanol.** Methanol is another potential fuel for internal combustion engines (Kohl 1990). Various raw materials can be used for methanol production, including natural gas, coal, wood, and municipal solid wastes. At present, the primary source of methanol is natural gas. The major limitation in using biomass for methanol production is the enormous quantities needed for a plant with suitable economies of scale. A suitably large methanol plant would require at least 1250 tons of dry biomass per day for processing (ACTI 1983). More than 150,000 ha of forest would be needed to supply one plant. Biomass generally is not available in such enormous quantities from extensive forests and at acceptable prices (ACTI 1983).

If methanol from biomass (33 quads) were used as substitute for oil in the United States, from 250 to 430 million ha of land would be needed to supply the raw material. This land area is greater than the 162 million ha of US cropland now in production (USDA 1992). Although methanol production from biomass may be impractical because of the enormous size of the conversion plants (Kohl 1990), it is significantly more efficient than the ethanol production system based on both energy output and economics (Kohl 1990).

Compared to gasoline and diesel fuel, both methanol and ethanol reduce the amount of carbon monoxide and sulfur oxide pollutants produced, however both contribute other major air pollutants such as aldehydes and alcohol. Air pollutants from these fuels worsen the tropospheric ozone problem because of the emissions of nitrogen oxides from the richer mixtures used in the combustion engines (Sillman and Samson 1990).

**Hydrogen.** Gaseous hydrogen, produced by the electrolysis of water, is another alternative to petroleum fuels. Using solar electric technologies for its production, hydrogen has the potential to serve as a renewable gaseous and liquid fuel for transportation vehicles. In addition hydrogen can be used as an energy storage system for electrical solar energy technologies, like photovoltaics (Winter and Nitsch 1988).

The material inputs for a hydrogen production facility are primarily those needed to build a solar electric production facility. The energy required to produce 1 billion kWh of hydrogen is 1.3 billion kWh of electricity (Voigt 1984). If current photovoltaics require 2700 ha/1 billion kWh, then a total area of 3510 ha would be needed to supply the equivalent of 1 billion kWh of hydrogen fuel. Based on US per capita liquid fuel needs, a facility covering approximately 0.15 ha (16,300 ft$^2$) would be needed to produce a year's requirement of liquid hydrogen. In such a facility, the water requirement for electrolytic production of 1 billion kWh/yr equivalent of hydrogen is approximately 300 million liters/yr Voigt 1984).

To consider hydrogen as a substitute for gasoline: 9.5 kg of hydrogen produces energy equivalent to that produced by 25 kg of gasoline. Storing 25 kg of gasoline requires a tank with a mass of 17 kg, whereas the storage of 9.5 kg of hydrogen requires 55 kg (Peschka 1987). Part of the reason for this difference is that the volume of hydrogen fuel is about four times greater than that for the same energy content of gasoline. Although the hydrogen storage vessel is large, hydrogen burns 1.33 times more efficiently than gasoline in automobiles (Bockris and Wass 1988). In tests, a BMW 745i liquid-hydrogen test vehicle with a tank weight of 75 kg, and the energy equivalent of 40 liters (320,000 kcal) of gasoline, had a cruising range in traffic of 400 km or a fuel efficiency of 10 km per liter (24 mpg) (Winter 1986).

At present, commercial hydrogen is more expensive than gasoline. For example, assuming 5¢ per kWh of electricity from a conventional power plant, hydrogen would cost 9¢ per kWh (Bockris and Wass 1988). This cost is equivalent of 67¢/liter of gasoline. Gasoline sells at the pump in the United States for approximately 30¢/liter. However, estimates are that the real cost of burning a liter of gasoline ranges from $1.06 to $1.32, when production, pollution, and other external costs are included (Worldwatch Institute 1989). Therefore, based on these calculations hydrogen fuel may eventually be competitive.

Some of the oxygen gas produced during the electrolysis of water can be used to offset the cost of hydrogen. Also the oxygen can be combined with hydrogen in a fuel cell, like those used in the manned space flights. Hydrogen fuel cells used in rural and suburban areas as electricity sources could help decentralize the power grid, allowing central power facilities to decrease output, save transmission costs, and make mass-produced, economical energy available to industry.

Compared with ethanol, less land (0.15 ha versus 7 ha for ethanol) is required for hydrogen production that uses photovoltaics to produce the needed electricity. The environmental impacts of hydrogen are minimal. The negative impacts that occur during production are all associated with the solar electric technology used in production. Water for the production of hydrogen may be a problem in the arid regions of the United States, but the amount required is relatively small compared with the demand for irrigation water in agriculture. Although hydrogen fuel produces emissions of nitrogen oxides and hydrogen peroxide pollutants, the amounts are about one-third lower than those produced from gasoline engines (Veziroglu and Barbir 1992). Based on this comparative analysis, hydrogen fuel may be a cost-effective alternative to gasoline, especially if the environmental and subsidy costs of gasoline are taken into account.

**Hydroelectric systems**

For centuries, water has been used to provide power for various systems. Today hydropower is widely used to produce electrical energy. In 1988 approximately 870 billion kWh (3 quads or 9.5%) of the United States' electrical energy was produced by hydroelectric plants (FERC 1988, USBC 1992a). Further development and/or rehabilitation of existing dams could produce an additional 48 billion kWh per year. However, most of the best candidate sites already have been fully developed, although some specialists project increasing US hydropower by as much as 100 billion kWh if additional sites are developed (USBC 1992a).

Hydroelectric plants require land for their water-storage reservoirs. An analysis of 50 hydroelectric sites in the United States indicated that an average of 75,000 ha of reservoir area are required per 1 billion kWh/yr produced. However, the size of reservoir per unit of electricity produced varies widely, ranging from 482 ha to 763,000 ha per 1 billion kWh/yr depending upon the hydro head, terrain, and additional uses made of the reservoir. The latter include flood control, storage of water for public and irrigation supplies, and/or recreation (FERC 1984). For the United States the energy input/output ratio was calculated to be 1:48; for Europe an estimate of 1:15 has been reported (Winter et al. 1992).

Based on regional estimates of land use and average annual energy generation, approximately 63 million hectares of the total of 917 million ha of land area in the United States are currently covered with reservoirs. To develop the remaining best candidate sites, assuming land requirements similar to those in past developments, an additional 24 million hectares of land would be needed for water storage.

Reservoirs constructed for hydroelectric plants have the potential to cause major environmental problems. First, the impounded water frequently covers agriculturally productive, alluvial bottom land. This water cover represents a major loss of life and destruction of property. Further, dams alter the existing plant and animal species in the ecosystem (Flavin 1985). For example, coldwater fishes may be replaced by warmwater fishes, frequently blocking fish migration (Hall et al. 1986). However, flow schedules can be altered to ameliorate many of these impacts. Within the reservoirs, fluctuations of water levels alter shorelines and cause downstream erosion and changes in physiochemical factors, as well as the changes in aquatic communities. Beyond the reservoirs, discharge patterns may adversely reduce downstream water quality and biota, displace people, and increase water evaporation losses (Barber 1993). Because of widespread public environmental concerns, there appears to be little potential for greatly expanding either large or small hydroelectric power plants in the future.

**Wind power**

For many centuries, wind power like water power has provided energy to pump water and run mills and other machines. In rural America windmills have been used to generate electricity since the early 1900s.

Modern wind turbine technology has made significant advances over the last 10 years. Today, small wind machines with 5 to 40 kW capacity can supply the normal electrical needs of homes and small industries (Twidell 1987). Medium-size turbines 100kW to 500kW produce most of the commercially generated electricity. At present, the larger, heavier blades required by large turbines upset the desirable ratio between size and weight and create efficiency problems. However, the effectiveness and efficiency of the large wind machines are expected to be improved through additional research and development of lighter weight but stronger components (Clarke 1991). Assuming a 35% operation capacity at a favorable site, the energy input/output ratio of the system is 1:5 for the material used in the construction of medium-size wind machines.

The availability of sites with sufficient wind (at least 20 km/h) limits the widespread development of wind farms. Currently, 70% of the total wind energy (0.01 quad) produced in the United States is generated in California (AWEA 1992). However, an estimated 13% of the contiguous US land area has wind speeds of 22 km/h or higher; this area then would be sufficient to generate approximately 20% of US electricity using current technology (DOE 1992). Promising areas for wind development include the Great Plains and coastal regions.

Another limitation of this energy resource is the number of wind machines that a site can accommodate. For example, at Altamont Pass, California, an average of one turbine per 1.8 ha allows sufficient spacing to produce maximum power (Smith and Ilyin 1991). Based on this figure approximately 11,700 ha of land are needed to supply 1 billion kWh/yr. However, because the turbines themselves only occupy approximately 2% of the area or 230 ha, dual land use is possible. For example, current agricultural land developed for wind power continues to be used in cattle, vegetable, and nursery stock production.

An investigation of the environmental impacts of wind energy production reveals a few hazards. For example, locating the wind turbines in or near the flyways of migrating birds and wildlife refuges may result in birds flying into the supporting structures and rotating blades (Clarke 1991, Kellett 1990). Clarke suggests that wind farms be located at least 300 meters from nature reserves to reduce this risk to birds.

Insects striking turbine blades will probably have only a minor impact on insect populations, except for some endangered species. However, significant insect accumulation on the blades may reduce turbine efficiency (Smith 1987a).

Wind turbines create interference with electromagnetic transmission, and blade noise may be heard up to 1 km away (Kellet 1990). Fortunately, noise and interference with radio and television signals can be eliminated by appropriate blade materials and careful placement of turbines. In addition, blade noise is offset by locating a buffer zone between the turbines and human settlements. New technologies and designs may minimize turbine generator noise.

Under certain circumstances shadow flicker has caused irritation, disorientation, and seizures in humans (Steele 1991). However, as with other environmental impacts, mitigation is usually possible through careful site selection away from homes and offices. This problem slightly limits the land area suitable for wind farms.

Although only a few wind farms supply power to utilities in the United States, future widespread development may be constrained because local people feel that wind farms diminish the aesthetics of the area (Smith 1987a). Some communities have even passed legislation to prevent wind turbines from being installed in residential areas (Village of Cayuga Heights, New York, Ordinance 1989). Likewise areas used for recreational purposes, such as parks, limit the land available for wind power development.

**Photovoltaics**

Photovoltaic cells are likely to provide the nation with a significant portion of its future electrical energy (DeMeo et al. 1991). Photovoltaic cells produce electricity when sunlight excites electrons in the cells. Because the size of the units is flexible and adaptable, photovoltaic cells are ideal for use in homes, industries, and utilities.

Before widespread use, however, improvements are needed in the photovoltaic cells to make them economically competitive. Test photovoltaic cells that consist of silicon solar cells are currently up to 21% efficient in converting sunlight into electricity (Moore 1992). The durability of photovoltaic cells, which is now approximately 20 years, needs to be lengthened and current production cost reduced about fivefold to make them economically feasible. With a major research investment, all of these goals appear possible to achieve (DeMeo et al. 1991).

Currently, production of electricity from photovoltaic cells costs approximately 30¢/kWh, but the price is projected to fall to approximately 10¢/kWh by the end of the decade and perhaps reach as low as 4¢ by the year 2030, provided the needed improvements are made (Flavin and Lenssen 1991). In order to make photovoltaic cells truly competitive, the target cost for modules would have to be approximately 8¢/kWh (DeMeo et al. 1991).

Using photovoltaic modules with an assumed 7.3% efficiency (the current level of commercial units), 1 billion kWh/yr of electricity could be produced on approximately 2700 ha of land, or approximately 0.027 ha per person, based on the present average per capita use of electricity. Thus, total US electrical needs theoretically could be met with photovoltaic cells on 5.4 million ha (0.6% of US land). If 21% efficient cells were used, the total area needed would be greatly reduced. Photovoltaic plants with this level of efficiency are being developed (DeMeo et al. 1991).

The energy input for the structural materials of a photovoltaic system delivering 1 billion kWh is calculated to be approximately 300 kWh/$m^2$. The energy input/

output ratio for production is about 1:9 assuming a life of 20 years.

Locating the photovoltaic cells on the roofs of homes, industries, and other buildings would reduce the need for additional land by approximately 5% (USBC 1992a), as well as reduce the costs of energy transmission. However, photovoltaic systems require back-up with conventional electrical systems, because they function only during daylight hours.

Photovoltaic technology offers several environmental advantages in producing electricity compared with fossil fuel technologies. For example, using present photovoltaic technology, carbon dioxide emissions and other pollutants are negligible.

The major environmental problem associated with photovoltaic systems is the use of toxic chemicals such as cadmium sulfide and gallium arsenide, in their manufacture (Holdren et al. 1980). Because these chemicals are highly toxic and persist in the environment for centuries, disposal of inoperative cells could become a major environmental problem. However, the most promising cells in terms of low cost, mass production, and relatively high efficiency are those being manufactured using silicon. This material makes the cells less expensive and environmentally safer than the heavy metal cells.

**Solar thermal conversion systems**

Solar thermal energy systems collect the sun's radiant energy and convert it into heat. This heat can be used for household and industrial purposes and also to drive a turbine and produce electricity. System complexity ranges from solar ponds to the electric-generating central receivers. We have chosen to analyze electricity in order to facilitate comparison to the other solar energy technologies.

**Solar ponds.** Solar ponds are used to capture solar radiation and store it at temperatures of nearly 100°C. Natural or man-made ponds can be made into solar ponds by creating a salt-concentration gradient made up of layers of increasing concentrations of salt. These layers prevent natural convection from occurring in the pond and enable head collected from solar radiation to be trapped in the bottom brine.

The hot brine from the bottom of the pond is piped out for generating electricity. The steam from the hot brine burns freon into a pressurized vapor, which drives a Rankine engine. This engine was designed specifically for converting low-grade heat into electricity. At present, solar ponds are being used in Israel to generate electricity (Tabor and Doran 1990).

For successful operation, the salt-concentration gradient and the water levels must be maintained. For example, 4000 ha of solar ponds lose approximately 3 billion liters of water per year under the arid conditions of the southwestern United States (Tabor and Doran 1990). In addition, to counteract the water loss and the upward diffusion process of salt in the ponds, the dilute salt water at the surface of the ponds has to be replaced with fresh water. Likewise salt has to be added periodically to the heat-storage zone. Evaporation ponds concentrate the brine, which can then be used for salt replacement in the solar ponds.

Approximately 4000 ha of solar ponds (40 ponds of 100 ha) and a set of evaporation ponds that cover a combined 1200 ha arc needed for the production of 1 billion kWh of electricity needed by 100,000 people in one year. Therefore, a family of three would require approximately 0.2 ha (22,000 sq ft) of solar ponds for its electricity needs. Although the required land area is relatively large, solar ponds have the capacity to store heat energy for days, thus eliminating the need for back-up energy sources from conventional fossil plants. The efficiency of solar ponds in converting solar radiation into heat is estimated to be approximately 1:5. Assuming a 30-year life for a solar pond, the energy input/output ratio is calculated to be 1:4. A 100-hectare (1 $km^2$) solar pond is calculated to produce electricity at a rate of approximately 14¢ per kWh. According to Folchitto (1991), this cost could be reduced in the future.

In several locations in the United States solar ponds are now being used successfully to generate heat directly. The heat energy from the pond can be used to produce processed steam for heating at a cost of only 2¢ to 3.5¢ per kWh (Gommend and Grossman 1988). Solar ponds are most effectively employed in the Southwest and Mid-west.

Some hazards are associated with solar ponds, but most can be prevented with careful management. For instance, it is essential to use plastic liners to make the ponds leak-proof and thereby prevent contamination of the adjacent soil and groundwater with salt. Burrowing animals must be kept away from the ponds by buried screening (Dickson and Yates 1983). In addition, the ponds should be fenced to prevent people and other animals from coming in contact with them. Because some toxic chemicals are used to prevent algae growth on water surface and freon is used in the Rankine engine, methods will have to be devised for safely handling these chemicals (Dickson and Yates 1983).

**Solar receiver systems.** Other solar thermal technologies that concentrate solar radiation for large scale energy production include distributed and central receivers. Distributed receiver technologies use rows of parabolic troughs to focus sunlight on a central-pipe receiver that runs above the troughs. Pressurized water and other fluids are heated in the pipe and are used to generate steam to drive a turbo-generator for electricity production or provide industry with heat energy.

Central receiver plants use computer-controlled, sun-tracking mirrors, or heliostats, to collect and concentrate the sunlight and redirect it toward a receiver located atop a centrally placed tower. In the receiver, the solar energy is captured as heat energy by circulating fluids, such as water or molten salts, that are heated under pressure. These fluids either directly or indirectly generate steam, which is then driven through a conventional turbo-generator to yield electricity. The receiver system may also be designed to generate heat for industry.

Distributed receivers have entered the commercial market before central receivers, because central receivers are more expensive to operate. But, compared to distributed receivers, central receivers have the potential for greater efficiency in electricity production because they are able to achieve higher energy concentrations and higher turbine inlet temperatures (Winter 1991). Central receivers are used in this analysis.

The land requirements for the central receiver technology are approximately 1100 ha to produce 1 billion kWh/yr, assuming peak efficiency, and favorable sunlight conditions like those in the western United States. Proposed systems offer four to six hours of heat storage and may be constructed to include a back-up alternate energy source. The energy input/output ration is calculated to be 1:10. Solar thermal receivers are estimated to produce electricity at approximately 10¢ per kWh, but this cost is expected to be reduced somewhat in the future, making the technology more competitive (Vant-Hull 1992). New technical advances aimed at reducing costs and improving efficiency include designing stretched membrane heliostats, volumetric-air ceramic receivers, and improved overall system designs (Beninga et al. 1991).

Central receiver systems are being tested in Italy, France, Spain, Japan, and the United States (at the 10-megawatt Solar One pilot plant near Barstow, California; Skinrood and Skvarna 1986). Also, Luz's Solar Electric Generating System plants at Barstow use distributed receivers to generate almost 300 MW of commercial electricity, (Jensen et al. 1989).

The potential environmental impacts of solar thermal receivers include: the accidental or emergency release of toxic chemicals used in the heat transfer system (Baechler and Lee 1991); bird collisions with a heliostat and incineration of both birds and insects if they fly into the high temperature portion of the beams; and—if one of the heliostats did not tract properly but focused its high temperature beam on humans, other animals, or flammable materials—burns, retinal damage, and fires (Mihlmester et al. 1980). Flashes of light coming from the heliostats may pose hazards to air and ground traffic (Mihlmester et al. 1980).

Other potential environmental impacts include microclimate alteration, for example reduced temperature and changes in wind speed and evapotranspiration beneath the heliostats or collecting troughs. This alteration may cause shifts in various plant and animal populations. The albedo in solar-collecting fields may be increased from 30% to 56% in desert regions (Mihlmester et al. 1980). An area of 1100 ha is affected by a plant producing 1 billion kWh.

The environmental benefits of receiver systems are significant when compared to fossil fuel electrical generation. Receiver systems cause no problems of acid rain, air pollution, or global warming (Kennedy et al. 1991).

**Passive heating and cooling of buildings**

Approximately 23% (18.4 quads) of the fossil energy consumed yearly in the United States is used for space heating and cooling of buildings and for heating hot water (DOE 1991a). At present only 0.3 quads of energy are being saved by technologies that employ passive and active solar heating and cooling of buildings. Tremendous potential exists for substantial energy savings by increased energy efficiency and by using solar technologies for buildings.

Both new and established homes can be fitted with solar heating and cooling systems. Installing passive solar systems into the design of a new home is generally cheaper than retrofitting an existing home. Including passive solar systems during new home construction usually adds less than 10% to construction costs (Howard and Szoke 1992); a 3-5% added first cost is typical[1]. Based on the cost of construction and the amount of energy saved measured in terms of reduced heating costs, we estimate the cost of passive solar systems to be approximately 3¢ per kWh saved.

Improvements in passive solar technology are making it more effective and less expensive than in the past. In the area of window designs, for example, current research is focused on the development of superwindows and with high-insulating values and smart or electrochromic windows that can respond to electrical current, temperature, or incident sunlight to control the admission of light energy (Warner 1991). Use of transparent insulation materials makes window designs that transmit from 50% to 70% of incident solar energy while at the same time providing insulat-

[1] B.D. Howard, 1992, personal communication. The Alliance to Save Energy, Washington, D.C.

ing values typical of 25 cm of fiber glass insulation (Chahroudi 1992). Such materials have a wide range of solar technology applications beyond windows, including house heating and transparent, insulated collector-storage walls and integrated storage collectors for domestic hot water (Wittwer et al. 1991).

Active solar heating technologies are not likely to play a major role in the heating of buildings. The cost of energy saved is relatively high compared with passive systems and conservation measures.[2]

Solar water heating is also cost-effective. Approximately 3% of all the energy used in the United States is for heating water in homes (DOE 1991a). In addition, many different types of passive and active water heating solar systems are available and are in use throughout the United States. These systems are becoming increasingly affordable and reliable (Wittwer et al. 1991). The cost of purchasing and installing an active solar water heater ranges from $2500 to $6000 in the northern regions and $2000 to $4000 in the southern regions of the nation (DOE/CE 1988).

Although none of the passive heating and cooling technologies require land, they can cause environmental problems. For example, some indirect land-use problems may occur, such as the removal of trees, shading and rights to the sun (Schurr et al. 1979). Glare from collectors and glazing could create hazards to automobile drivers, pedestrians, bicyclists, and airline pilots. Also, when houses are designed to be extremely energy efficient and airtight, indoor air quality becomes a concern because air pollutants may accumulate inside. However, installation of well-designed ventilation systems promotes a healthful exchange of air while reducing heat loss during the winter and heat gain during the summer. If radon is a pollutant present at unsafe levels in the home, various technologies can mitigate the problem (ASTM 1992).

[2] See footnote 1.

**Comparing solar power to coal and nuclear power**

Coal and nuclear power production are included in this analysis to compare conventional sources of electricity generation to various future solar energy technologies. Coal, oil, gas, nuclear, and other mined fuels are used to meet 92% of US energy needs. Coal and nuclear plants combined produce three quarters of US electricity (USBC 1992a).

Energy efficiencies for both coal and nuclear fuels are low due to the thermal law constraint of electric generator designs: coal is approximately 35% efficient and nuclear fuels approximately 33% (West and Kreith 1988). Both coal and nuclear power plants in the future may require additional structural material to meet clean air and safety standards. However, the energetic requirements of such modifications are estimated to be small compared to the energy lost due to conversion inefficiencies.

The dots of producing electricity using coal and nuclear energy are 3¢ and 5¢ per kWh, respectively (EIA 1990). However, the costs of this kind of energy generation are artificially low because they do not include such external costs as damages from acid rain produced from coal and decommissioning costs for the closing of nuclear plants. The Clean Air Act and its amendments may raise coal generation costs, while the new reactor designs, standardization, and streamlined regulations may reduce nuclear generation costs. Government subsidies for nuclear and coal plants also skew the comparison with solar energy technologies (Wolfson 1991).

Clouding the economic costs of fossil energy use are the direct and indirect US subsidies that hide the true cost of energy and keep the costs low, thereby encouraging energy consumption. The energy industry receives a direct subsidy of $424 per household per year (based on an estimated maximum of $36 billion for total federal energy subsidies [ASE 1993]). In addition, the mined-energy industry, like the gasoline industry, does not pay for the environmental and public health costs of fossil energy production and consumption.

The land requirements for fossil fuel and nuclear-based plants are lower than those for solar energy technologies. The land area required for electrical production of 1 billion kWh/year is estimated at 363 ha for coal and 48 ha for nuclear fuels. These figures include the area for the plants and both surface and underground mining operations and waste disposal. The land requirements for coal technologies are low because it uses concentrated fuel sources rather than diffuse solar energy. However, as the quality of fuel ore declines, land requirements for mining will increase. In contrast, efficient reprocessing and the use of nuclear breeder reactors may decrease the land area necessary for nuclear power.

Many environmental problems are associated with both coal and nuclear power generation (Pimentel et al. 1994). For coal, the problems include the substantial damage to land by mining, air pollution, acid rain, global warming, as well as the safe disposal of large quantities of ash (Wolfson 1991). For nuclear power, the environmental hazards consist mainly of radioactive waste that may last for thousands of years, accidents, and the decommissioning of old nuclear plants (Wolfson 1991).

Fossil-fuel electric utilities account for two-thirds of the sulfur dioxide, one-third of the nitrogen dioxide, and one-third of the carbon dioxide emissions in the United States (Kennedy et al. 1991). Removal of carbon dioxide from coal plant emissions could raise costs to 12¢/kWh; a disposal tax on carbon could raise coal electricity costs to 18¢/kWh (Williams et al. 1990).

The occupational and public health risks of both coal and nuclear plants are fairly high, due mainly to the hazards of mining, ore transportation, and subsequent air pollution during the production of electricity. However, there are 22 times as many deaths per unit of energy related to coal than of nuclear energy production because 90,000 times greater volume of coal than nuclear ore is needed to generate an equivalent amount of energy.[3]

Also, and as important, coal produces more diffuse pollutants than nuclear fuels during normal operation of the generating plant. Coal-fired plants produce air pollutants—including sulfur oxides, nitrogen oxide, carbon dioxide, and particulates—that adversely affect air quality and contribute to acid rain. Technologies do exist for removing most of the air pollutants, but their use increases the cost of a new plant by 20-25% (IEA 1987). By comparison, nuclear power produces fewer pollutants than do coal plants (Tester et al. 1991).

**Transition to solar energy and other alternatives**

The first priority of a sustainable US energy program should be for individuals, communities, and industries to conserve fossil energy resources. Other developed countries have proven that high productivity and a high standard of living can be achieved with considerably less energy expenditure compared to that of the United States. Improved energy efficiency in the United States, other developed nations, and even in developing nations would help both extend the world's fossil energy resources and improve the environment (Pimentel et al. 1994).

The supply and demand for fossil and solar energy; the requirements of land for food, fiber, and lumber; and the rapidly growing human population will influence future US options. The growth rate of the US population has been increasing and is now at 1.1% per year (USBC 1992b); at this rate, the present population of 260 million will increase to more than a half billion in just 60 years. The presence of more people will require more land for homes, businesses, and roads. Population density directly influences food production, forest product needs, and energy requirements. Considerably more agricultural and forest land will be needed to provide vital food and forest products, and the drain on all energy resources will increase. Although there is no cropland shortage at present (USDA 1992), problems undoubtedly will develop in the near future in response to the diverse needs of the growing US population.

Solar energy technologies, most of which require land for collection and production, will compete with agriculture and forestry in the United States and worldwide. Therefore, the availability of land is projected to be a limiting factor in the development of solar energy. In the light of this constraint, an optimistic projection is that the current level of nearly 7 quads of solar energy collected and used annually in the United States could be increased to approximately 37 quads (Ogden and Williams 1989, Pimentel et al. 1984). This higher level represents only 43% of the 86 quads of total energy currently consumed in the United States. Producing 37 quads with solar technology would require approximately 173 million ha, or nearly 20% of US land area. At present this amount of land is available, but it may become unavailable due to future population growth and increased resource consumption. If land continues to be available, the amounts of solar energy (including hydropower and wind) that could be produced by the year 2050 are projected to be: 5 quads from biomass, 4 quads from hydropower, 8 quads from wind power, 6 quads from solar thermal systems, 6 quads from passive and active solar heating, and 8 quads from photovoltaics.

Another possible future energy source is fusion energy (Bartlett 1994, Matare 1989). Fusion uses nuclear particles called neutrons to generate heat in a fusion reactor vessel. Nuclear fusion differs from fission in that the production of energy does not depend on continued mining. However, high costs and serious environmental problems are anticipated (Bartlett 1994). The environmental problems include the production of enormous amounts of heat and radioactive material.

The United States could achieve a secure energy future and a satisfactory standard of living for everyone if the human population were to stabilize at an estimated optimum of 200 million (down from today's 260 million) and conservation measures were to lower per capita energy consumption to about half the present level (Pimentel et al. 1994). However, if the US population doubles in 60 years as is more likely, supplies of energy, food, land, and water will become inadequate, and land, forest, and general environmental degradation will escalate (Pimentel et al. 1994, USBC 1992a).

Fossil energy subsidies should be greatly diminished or withdrawn and the savings should be invested to encourage the development and use of solar energy technologies. This policy would increase the rate of adoption of solar energy technologies and lead to a smooth

[3] D. Hammer, 1993, personal communication. Cornell University, Ithaca, NY.

transition from a fossil fuel economy to one based on solar energy. In addition, the nation that becomes a leader in the development of solar energy technologies is likely to capture the world market for this industry.

## Conclusions

This assessment of alternate technologies confirms that solar energy alternatives to fossil fuels have the potential to meet a large portion of future US energy needs, provided that the United States is committed to the development and implementation of solar energy technologies and that energy conservation is practiced. The implementation of solar technologies will also reduce many of the current environmental problems associated with fossil fuel production and use.

An immediate priority is to speed the transition from reliance on nonrenewable energy sources to reliance on renewable, especially solar-based, energy technologies. Various combinations of solar technologies should be developed consistent with characteristics of different geographic regions, taking into account the land and water available and regional energy needs. Combined, biomass energy and hydroelectric energy in the United States currently provide nearly 7 quads of solar energy, and their output could be increased to provide up to 9 quads by the year 2050. The remaining 28 quads of solar renewable energy needed by 2050 is projected to be produced by wind power, photovoltaics, solar thermal energy, and passive solar heating. These technologies should be able to provide energy without interfering with required food and forest production.

If the United States does not commit itself to the transition from fossil to renewable energy during the next decade or two, the economy and national security will be adversely affected. Starting immediately, it is paramount that US residents must work together to conserve energy, land, water, and biological resources. To ensure a reasonable standard of living in the future, there must be a fair balance between human population density and energy, land, water, and biological resources.

## Acknowledgments

We thank the following people for reading an earlier draft of this article, for their many helpful suggestions, and in some cases, for providing additional information: A. Baldwin, Office of Technology Assessment, US Congress; A.A. Bartlett, University of Colorado, Boulder; E. DeMeo, Electric Power Research Institute; H. English, Passive Solar Industries Council; S.L. Frye, Bechtel; M. Giampietro, National Institute of Nutrition, Rome, Italy; J. Goldemberg, Universidade de Sao Paulo, Brazil; C.A.S. Hall, College of Environmental Science and Forestry, SUNY, Syracuse; D. O. Hall, King's College, London, United Kingdom; S. Harris, Oak Harbor, WA; J. Harvey, New York State Energy Research and Development Authority; B.D. Howard, The Alliance to Save Energy; C.V. Kidd, Washington, DC; N. Lenssen, Worldwatch Institute; L.R. Lynd, Dartmouth College; J.M. Nogueira, Universidade de Brasilia, Brazil; M.G. Paoletti, University of Padova, Italy; R. Ristenen, University of Colorado, Boulder; S. Sklar, Solar Energy Industries Association; R. Swisher, American Wind Energy Association; R.W. Tresher, National Renewable Energy Laboratory; L.L. Vant-Hull, University of Houston; Wang Zhaoqian, Zheijan Agricultural University, China; P.B. Weisz, University of Pennsylvania, Philadelphia; Wen Dazhong, Academia Sinica, China; C.J. Winter, Deutsche Forschungsanstalt fur Luft und Raumfahrt, Germany; D.L. Wise, Northeastern University, Boston; and at Cornell University: R. Baker, S. Bukkens, D. Hammer, L. Levitan, S. Linke, and M. Pimentel.

## References cited

Advisory Committee on Technology Innovation (ACTI). 1983. *Alcohol Fuels: Options for Developing Countries*. National Academy Press, Washington, DC.

Alfheim, I., and T. Ramdahl. 1986. *Mutagenic and Carcinogenic Compounds from Energy Generation.* Final Report No. NP-6752963 (NTIS No. DE 86752963). Center for Industriforkning, Oslo, Norway.

Alliance to Save Energy (ASE). 1993. *Federal Energy Subsidies: Energy, Environmental, and Fiscal Impacts.* The Alliance to Save Energy, Washington, DC.

American Society for Testing Materials (ASTM) 1992. Standard guide for radon control options for the design and construction of new low rise residential buildings. Pages 1117-1123 in *Annual Book of American Society for Testing Materials*. E1465-92. ASTM, Philadelphia, PA.

American Wind Energy Association (AWEA). 1991. Wind energy comes of age. *Solar Today* 5: 14-16.

______. 1992. *Wind Technology Status Report*. American Wind Energy Association, Washington, DC.

Baechler, .M. C. and A. D. Lee. 1991. Implications of environmental externalities assessments for solar thermal powerplants. Pages 151-158 in T. R. Mancini, K. Watanabe and D. E. Klett, eds. *Solar Engineering 1991*. American Society of Mechanical Engineering, New York.

Barber, M. 1993. Why more energy? The hidden cost of Canada's cheap power. *Forests, Trees and People Newsletter* No. 19: 26-29.

Bartlett, A. A. 1994. Fusion: An illusion or a practical source of energy? *Clearinghouse Bulletin* 4(1): 1-3, 7.

Beninga, K., R. Davenport, J. Sandubrue, and K. Walcott. 1991. Design and fabrication of a market ready stretched membrane heliostat. Pages 229-234 in T. R. Mancini, K. Watanabe and D. E. Klett, eds. *Solar Engineering 1991*. American Society of Mechanical Engineering, New York .

Birdsey, R. A. 1992. Carbon storage and accumulation in United States forest ecosystems. General Technical Report WO 59. USDA Forest Service, Washington, DC.

Bockris, J. O. M., and J. C. Wass. 1988. About the real economics of massive hydrogen production at 2010 AD. Pages 101-151 in T. N. Veziroglu and A. N. Protsenko, eds. *Hydrogen Energy VII* Pergamon Press, New York.

Chahroudi, D. 1992. Weather panel architecture: a passive solar solution for

cloudy climates. *Solar Today* 6: 17-20.
Clarke. A. 1991. Wind energy progress and potential. *Energy Policy* 19: 742-755.
DeMeo, E. A., F. R. Goodman, T. M. Peterson, and J. C. Schaefer. 1991. *Solar Photovoltaic Power: A US Electric Utility R&D Perspective.* Edited by 2.1. P. S. Conference. IEEE Photovoltaic Specialist Conference Proceedings, New York.
Dickson, Y. L., and B. C. Yates. 1983. *Examination of the Environmental Effects of Salt-Gradient Solar Ponds in the Red River Basin of Texas.* North Texas State University, Denton Institute of Applied Sciences, Denton, TX.
Department of Energy (DOE). 1980. *Health Effects of Residential Wood Combustion: Survey of Knowledge and Research.* DOE, Technology Assessments Division, Washington, DC.
______. 1990. *The Potential of Renewable Energy.* SERI, DOE, Golden, CO.
______. 1991a. *Annual Energy Outlook with Projections to 2010.* DOE, Energy Information Administration, Washington DC.
______. 1991b. *1989 International Energv Annual.* DOE, Washington, DC.
______. 1992. *Wind Energy Program Overview.* National Renewable Energy Laboratory, DOE, Golden, CO.
DOE/Conservation Energy (CE). 1988. *Passive and Active Solar Domestic Hot Water Systems.* FS-119. DOE, Conservation and Renewable Energy Inquiry and Referral Service, Washington, DC.
Energy Information Administration (EIA). 1990. *Electric Plant Cost and Power Production Expenses.* EIA, Washington, DC.
Electric Power Research Institute (EPRI). 1991. Photovoltaic system performance assessment for 1989. EPRI Interim Report 65-7286. EPRI, Los Angeles, CA.
Food and Agriculture Organization (FAO). 1991. *Food Balance Sheets.* FAO, Rome.
Federal Energy Regulatory Commission (FERC). 1984. *Hydroelectric Power Resources of the United States.* FERC, Washington, DC.
______. 1988. *Hydroelectric Power Resources of the United States: Developed and Undeveloped.* FERC, Washington, DC.
Flavin, C. 1985. *Renewable Energy at the Crossroads.* Center for Renewable Resources, Washington, DC.
Flavin, C., and N. Lenssen. 1991. Here comes the sun. Pages 10-18 in L. Brown et al., eds. *State of the World 1991.* Worldwatch Institute, Washington, DC.
Folchitto, S. 1991. Seawater as salt and water source for solar ponds. *Sol. Energy* 46: 343-351.
Gibbons, J. H., and P. D. Blair. 1991. US energy transition: on getting from here to there. *Physics Today* 44: 22-30.
Gommend, K., and G. Grossman. 1988. Process steam generation by temperature boosting of heat from solar ponds. *Solar Ponds* 41: 81-89.
Hall, C. A. S., C. J. Cleveland, and R. L. Kaufmann. 1986. *Energy and Resource Quality: The Ecology of the Economic Process.* Wiley, New York.
Hall, D. O., F. Rosillo-Calle. R. H. Williams, and J. Woods. 1993. Biomass for energy: supply prospects. In T. B. Johansson, H. Kelley, A. K. N. Reddy. and R. H. Williams, eds. *Renewable Energy.* Island Press, Washington, DC.
Holdren, J. P., G. Morris, and I.Mintzer. 1980. Environmental aspects of renewable energy. *Annual Review of Energy 5*: 241 - 291.
Howard, B., and S. S. Szoke. 1992. *Advances in Solar Design Tools.* 5th Thermal Envelope Conference, December 1992. US Dept. of Energy, ASHRAE, Clearwater, FL.
Intemarional Commission on Large Dams (ICLD). 1988. *World Register of Dams.* ICID. Paris.
International Energy Agency (IEA) 1987. *Clean Coal Technology.* OECD Publications, Paris.
____. 1991. *Energy Statistics of OECD Countries.* IEA, Paris.
Jensen, C., H. Price, and D. Kearney. 1989. The SEGS power plants: 1989 performance. Pages 97-102 in A. H. Fanney and K.O. Lund, eds. *Solar Engineering 1989.* American Society of Mechanical Engineering, New York.
Kellet, J. 1990. The environmental impacts of wind energy developments. *Town Planning Review* 61: 139-154.
Kennedy, T., J. Finnell, and D. Kumor. 1991. Considering environmental costs in energy planning: alternative approaches and implementation. Pages 145-150 in T. R. Mancini. K;. Watanabe and D. E. Klett eds. *Solar Engineering* 1991. American Society of Mechanical Engineering, New York.
Kohl, W. L. 1990. *Methanol as an Alternative Fuel Choice: An Assessment.* The Johns Hopkins Foreign Policy Institute, Washington, DC.
Lipfert, F. W., R. G. Malone, .M. L. Dawn, N. R. Mendell, and C. C. Young. 1988. *Statistical Study of the Macroepidemiology of Air Pollution and Total Mortality.* Report No. BNL 52122. US Dept. of Energy, Washington, DC.
Lynd, L. R., J. H. Cushman, R. J. Nichols, and C. E. Wyman. 1991. Fuel ethanol from cellulose biomass. *Science* 251: 1318-1323.
Maeda, Y., and M. Kai. 1991. Recent progress in pervaporation membranes for water/ethanol separation. Pages 391-435 in R. V. M. Huang, ed. *Pervaporation Membrane Separation Processes.* Elsevier, Amsterdam, the Netherlands.
Matare, H. F. 1989. *Energy: Fact and Future.* CRC Press, Boca Raton, FL.
Mihlimester, P. E., J. B. Thomasian, and M. R. Riches. 1980. Environmental and health safety issues. Pages 731-762 in W. C. Dickinson and P. N. Cheremisinoff, eds. *Solar Energy Technology Handbook.* Marcel Dekker, New York.
Moore, T. 1992. High hopes for high-power solar. *EPRI Journal* 17 (December): 1 6-25.
Ogden, J. M., and R. H. Williams. 1989. *Solar Hydrogen: Moving Beyond Fossil Fuels.* World Resources Institute, Washington, DC.
Peschka, W. 1987. The status of handling and storage techniques for liquid hydrogen in motor vehicles. *International Journal of Hydrogen Energy* 12: 753- 64.
Pimentel D. 1991. Ethanol fuels: Energy security, economics, and the environment. *Journal of Agricultural and Environmental Ethics* 4: 1-13.
Pimentel, D. 1992. Competition for land: development, food, and fuel. Pages 325-348 in M. A. Kuliasha, A. Zucker and K. J. Ballew, eds. *Technologies for a Greenhouse-constrained Society.* Lewis, Boca Raton, FL.
Pimentel, D., L. Levitan, J. Heinze, M. Loehr, W. Naegeli, J. Bakker, J. Eder, B. Modelski, and M. Morrow. 1984. Solar energy, land and biota. *SunWorld* 8:70-73, 93-95.
Pimentel, D., M. Herdendorf, S. Eisenfeld, L. Olanden, M. Carroquino, C. Corson, J. McDade, Y. Chung, W. Cannon, J. Roberts, L. Bluman, and J. Gregg. 1994. Achieving a secure energy future: environmental and economic issues. *Ecological Economics* 9:201-219.
Schurr, S.M., J. Darmstadter, H. Perry, W. Ramsay, and M. Russell. 1979. *Energy in America's Future: The Choices Before Us. A Study Prepared for RFF National Energy Strategies.* The Johns Hopkins University Press, Baltimore, MD.
Sillman, S., and P. J. Samson. 1990. Impact of methanol-fueled vehicles on rural and urban ozone concentrations during a region-wide ozone episode in the midwest. Pages 121-137 in W.L. Kohl, ed. *Methanol as an Alternative Fuel Choice: An Assessment.* The Johns Hopkins Foreign Policy Institute, Washington, DC.
Skinrood, A. C., and P. E. Skvarna. 1986. Three years of test and operation at Solar One. Pages 105-122 in M. Becker, ed. *Solar Thermal Receiver Systems.* Springer-Verlag, Berlin.
Smil, V. 1984. On energy and land. *Am. Sci.* 72: 15-21.
Smith, D. R. 1987a. The wind farms of the Altamont Pass area. *Annual Review of Energy* 12: 145-183.
Smith, D. R., and M. A. Ilyin. 1991. Wind and solar energy, costs and value. *The American Society of Mechanical Engineers*

*10th Annual Wind Energy Symposium*: 29-32.

Smith, K. R. 1987b. *Biofuels, Air Pollution and Health: A Global Review*. Plenum Press, New York.

Steele, A. 1991. An environmental impact assessment of the proposal to build a wind farm at Langdon common in the North Pennines, U.K. *The Environmentalist* 11: 195-212.

Tabor, H. Z., and B. Doran. 1990. The Beith Ha'arva 5MWe solar pond power plant (SPPP)-Progress Report. *Solar Energy* 45: 247-253.

Tester, J. W., D. O. Wood, and N. A. Ferrari. 1991. *Energy and the Environment in the 21st Century*. The MIT Press, Cambridge, MA.

Twidell, J. 1987. *A Guide to Small Wind Energy Conversion Systems*. Cambridge University Press, Cambridge, UK.

US Bureau of the Census (USBC). 1992a. *Statistical Abstract of the United States 1992*. US Government Printing Office, Washington, DC.

____. 1992b. *Current Population Reports*. January ed. USBC, Washington, DC.

US Department of Agriculture (USDA).1992. *Agricultural Statistics*. US Government Printing Office, Washington, DC.

Vant-Hull, L. L. 1992. Solar thermal receivers: current status and future promise. *American Solar Energy Society* 7: 13-16

Veziroglu, T. N., and F. Barbir. 1992. Hydrogen: the wonder fuel. *International Journal of Hydrogen Energy* 17: 391-404.

Voigt, C. 1984. Material and energy requirements of solar hydrogen plants. *International Journal of Hydrogen Energy* 9: 491 -500.

Warner, J. L. 1991. Consumer guide to energy saving windows. *Solar Today* 5:10-14.

West, R. E., and F. Kreith. 1988. *Economic Analysis of Solar Thermal Energy Systems*. The MIT Press, Cambridge, MA.

Williams, T. A., D. R. Brown, J. A. Dirks, K. K. Humphreys, and J. L. La Marche. 1990. Potential impacts of $CO_2$, emission standards on the economics of central receiver power systems. Pages 1-6 in J. T. Beard and M. A Ebadian, eds. *Solar Engineering 1990*. American Society of Mechanical Engineering, New York.

Winter, C. J. 1986. Hydrogen energy—Expected engineering break-throughs. Pages 9-29 in T. N. Veziroglu, N. Getoff and P. Weinzierl, eds. *Hydrogen Energy Progress VI*, Pergamon Press, New York.

Winter, C. J. 1991. High temperature solar energy utilization after I5 years R&D: kick-off for rhe third generation technologies. *Solar Energy Materials* 24: 26-39.

Winter, C. J., and J. Nitsch, eds. 1988. *Hydrogen as an Energy Carrier: Technologies, Systems and Economy*. Springer-Verlag, Berlin.

Winter, C. J., W. Meineke, and A. Neumann 1992. Solar thermal power plants: No need for energy raw material_only conversion technologies pose environmental questions. Pages 198l-1986 in M. E Arden, S. M. A. Burley and M. Coleman eds. 1991 *Solar World Congress: Proceedings of the Biennial Congress of the International Solar Energy Society*. Vol 2, part II. Pergamon Press, Oxford, UK.

Wittwer, V., W. Platzer, and M. Rommell. 1991. Transparent insulation: an innovative European technology holds promise as one route to more efficient solar buildings and collectors. *Solar Today* 5: 20-22.

Wolfson, R. 1991. *Nuclear Choices*. MIT Press, Cambridge, MA.

Worldwatch Institute. 1989. *State of the World*. Worldwatch Institute, Washington, DC.

______. 1992. *State of the World*. Worldwatch Institute, Washington, DC.

## Questions:

1. With only ___% of the world's population, the U.S. consumes about ___% of the world's fossil fuel.

2. For photovoltaics to be cost competitive, the cost of modules should be ___/kWh. The cost by the end of the decade is projected to be ___/kWh.

3. About ___% of the fossil energy consumed yearly in the U.S. is used for space heating and cooling of buildings and hot water.

Answers on page 301

*The General Mining Law was passed by the U.S. Congress in 1872 to help resolve the legal chaos over mining claims in the "Wild West." But the law also promoted mining activity by essentially giving the land away. It permits individuals to stake a claim on land while paying the government no money for rent on the land or taxes on profits on the minerals extracted. Furthermore, if valuable minerals are found, the claimholder can purchase the land for $5 per acre or less. Environmentalists consider the law to be an archaic relic of times when resources were plentiful and people scarce. This is not only because it allows land to be readily overtaken and mined, but also because it has no provisions for repairing environmental damage from mining. Mining companies often degrade water quality from heavy metals in mine drainage, for example, but are not required to pay for damages. Mine waste can be very toxic. Abandoned mines often end up on the Superfund listing while the company simply declares bankruptcy and makes the taxpayer pay the cleaning bill, if it gets cleaned up at all. Attempts to reform the law generally involve requiring the mining companies to pay royalties on profits made, since mining federal lands produces $3–4 billion worth of minerals per year. Powerful mining lobbies have prevented such reform in the past, but there are signs of change.*

# 1872 Mining Law: Meet 1993 Reform

## The Last Great Land Giveaway May Soon Be Over

by Jay Letto

One hundred and twenty-two years ago, Ulysses S. Grant was President, Charles Darwin continued work on his evolution theory, George Pullman introduced the "sleeper car," and the General Mining Law of 1872 was enacted. Since then, we've had 24 new presidents, evolution has become a science and the sleeper car has lost out to air travel. But the mining law remains. Passed by Congress primarily to bring law and order to a 19th century "Wild West" shooting itself apart over mining claims, the 1872 Law also sought to lure immigrants and Easterners to settle the vast stretches of public land in the West. In 1993, however, the miners are usually wealthy and the mining companies often foreign-owned, and just about everybody—except mining companies and their powerful lobby—thinks the law has long outstayed its welcome.

As incredible as it may sound, under the archaic law, miners can stake a "claim" and mine any federal land, except within National Parks and Wilderness Areas, simply by declaring that they think the land holds valuable minerals. They pay the government no rent on the land and no royalties on the minerals extracted, but they may lease the land to others and charge them rent and royalties. And if "valuable minerals" are found, the claimholder may "patent" or purchase title to the land for $5 or less an acre. Referred to as "the last of the great land giveaways," the law has sold off 3.2 million acres of public land—an area the size of Connecticut.

But the worst aspect of the law may be that it allows the managing agencies—the Bureau of Land Management (BLM) and the Forest Service—no recourse for environmental protection. Mining can take place in sensitive areas, and there is no provision for reclamation of damaged land. The result: 52 mines on the Environmental Protection Agency's (EPA) Superfund National Priorities List of the worst toxic waste sites in the country; 424,000 acres of unreclaimed federal land left by hardrock mining; and 10,000 miles of rivers polluted from mining operations, as estimated by the U.S. Bureau of Mines. Says Jim Lyon of the Mineral Policy Center in Washington, DC, "no one really knows" the whole scope of the damage. "Probably the worst problem from hardrock mining is degraded water quality from heavy metals and mine drainage. Once it starts, it's almost impossible to stop, it can severely impact a large area, and it will take generations to clean up."

Of the thousands of abandoned sites, only a handful are being restored. It's much more common for a company to abandon the mine site, file for bankruptcy to avoid future liability costs, and leave the public with the cleanup bill. The Atlas Asbestos Mine Superfund site in Fresno County, California, for example, was listed on the EPA's National Priorities List in 1984. Located on 435 acres of BLM land, it has three open pit mines and numerous piles of asbestos waste. After abandoning the site in 1979, Wheeler Properties filed for bankruptcy, leaving the public with the cleanup bill. Soil erosion on the steep slopes has prevented revegetation, and wind erosion and water runoff has caused the levels of asbestos in nearby surface water to exceed EPA stan-

***Reprinted with permission from E-The Environmental Magazine, September/October, 1993 subscriptions $20/yr., P.O. Box 6667, Syracuse, N.Y. 13217***

dards for human health. It now threatens the drinking water in the California Aqueduct and nearby communities. In 1991, the EPA approved a $4.29 million effort to contain asbestos at the site to prevent further contamination, but the agency is still seeking responsible parties to help pay the costs.

Apart from all the destruction, lost government revenues from the law has opened some eyes. According to the Mineral Policy Center, mining operations on federal lands produce $3 to $4 billion worth of minerals each year—and the public doesn't get a cent. Royalties, which nearly every other developed country charges, would add hundreds of millions of dollars to the federal treasury.

The 5,000-acre Stillwater Mine outside Yellowstone National Park holds platinum and palladium deposits worth $30 billion. The Johns Manville and Chevron companies have jointly applied to buy the land for $20,000—less than $5 an acre. "Clearly, the American people, who own the $30 billion worth of minerals, ought to get a fair return on that," Lyon says. "And the land should not be sold," he added. "That's in violation of the Federal Land Management Act, which says that the land policy should encourage the retention of public land, not create inholdings."

In recent years, legislators have introduced bills to reform the Mining Law, only to see them blocked by industry lobbying. But the current Congress offers better hope for change. The environmental community supports a bill introduced by Representative Nick Rahall (D-WV). It calls for an eight percent royalty on gross production, abolishes patenting (meaning that all land will be retained by the federal government), and charges a $100-a-year fee per claim for companies to hold the mineral rights. It also includes strong environmental standards, such as: requiring a company to demonstrate up front that its mining will not harm the environment; giving regulators authority to stop damaging operations; establishing strong inspection and enforcement provisions; allowing citizen lawsuits to give the public the opportunity to compel enforcement and to stop bad operations; and establishing an abandoned-mine cleanup program to be funded by the royalties, permit fees and penalties.

But the Rahall bill faces stiff competition. In the Senate, Dale Bumpers (D-AR) has introduced an identical bill, which can't get by the Energy Committee. Instead, Senator Larry Craig (R-ID) has proposed a bill that the Mineral Policy Center calls "sham reform." It asks for a two percent net royalty. Lyon calls this a "full employment program for corporate accountants" that won't bring the federal treasury any money. It also doesn't provide environmental constraints or a cleanup program.

The Clinton Administration, under the aegis of deficit reduction, last spring asked for a 12.5 percent royalty on minerals extracted from federal lands, but Western Democrat Senators convinced the President to back down, and Senator Bennett Johnston (D-LA), chair of the Energy Committee, agreed to move a reform bill forward this year only if it were taken out of the budget process. He plans to bring supporters of both Senate bills into conference with the House reformers to write a final compromise version this fall.

Industry has countless arguments against real reform. They claim that royalties and expensive environmental measures will close down mining operations everywhere. They present scenarios of thriving communities turned into ghost towns, skyrocketing consumer prices and general economic disaster. But Lyon replies: "Mining has never been a sustainable industry. It's always been boom or bust. We think that the idea of reform putting the industry out of business is a red herring." The 1872 Mining Law only covers one-third of hardrock mining in the U.S. The rest is on private, state or tribal lands, where the industry has no trouble paying royalties, and companies often lease lands to each other, charging royalties ranging from five to 20 percent. Even the family of Senator Max Baucus (D-MT) gets royalties for mining on its land. Yet Baucus has told Clinton that collecting royalties from federal lands will break the industry. "If royalties are good enough for the Baucus family ranch," Lyon asked, "why aren't they good enough for the American people?"

Contact: Mineral Policy Center, 1325 Massachusetts Avenue NW, Washington DC 20005/ (202)737-1872. ❑

**Questions**:

1. How many miles of rivers have been polluted by mining in the U.S.? How many acres of unreclaimed land are there?

2. How many abandoned mine sites are being restored? Why?

3. What percent of hardrock mining in the U.S. is covered by the 1872 Mining Law? On what land is the remaining mining done? Are royalties paid there?

Answers on page 301

# 33

*Caverns are much more extensive and unexplored than many people realize. They are also an increasingly threatened natural resource in the U.S. Caves contain vast natural wonders that can be found nowhere else, from fossil remains of giant Ice Age mammals to unique crystal and rock formations. They are also important in the hydrologic cycle of many areas, where they form part of the aquifer system and help purify ground waters. But caves are being increasingly despoiled in a variety of ways, including mining, logging, agriculture, grazing, urban sprawl, and dumping. Only recently have federal laws been enacted to protect caves that underlie public lands. But such laws, especially the 1988 Federal Cave Resources Protection Act, are not fully enforced and are full of loopholes, according to some critics.*

## Subterranean Blues

Caverns are an uncharted, and increasingly threatened, wilderness

by Betsy Carpenter

One would think that New Mexico's newly discovered Lechuguilla cave would be fairly safe from harm. It is located in a national part, first of all, and vandals would have a hard time even finding the entrance, tucked away as it is in a remote desert canyon. Furthermore, the only way into the cave proper is to rappel down a daunting 155-foot rock face.

Yet Lechuguilla, one of the wonders of the subterranean world, faces a far more insidious threat than the vandals who would plunder its dazzling geological formations. Though the cave falls within the Carlsbad Caverns National Park, some passages are a mere 900 feet from lands leased by the Bureau of Land Management for oil and gas exploration. Cavers fear that escaping gases could seep into Lechuguilla or nearby Carlsbad Caverns, contaminating the fragile underground realm and threatening tourists and cavers.

Lechuguilla is by many measures a unique cave, but its plight is not unusual. All over America, caves are being damaged and destroyed by mining, logging, grazing, agriculture, urban sprawl and dumping. Ironically, this environmental assault is coming at a time when caves are finally being recognized as something more than commercial tourist attractions. Scientists have begun studying this vast unseen world for insights into everything from mineralogical processes to the migrations of prehistoric humans. America's unappreciated underworld, they argue, not only is a repository for a host of valuable and uninventoried species but provides essential conduits for ground water. "Caves are part of aquifers," says hydrologist Tom Aley, president of the American Cave Conservation Association. "Asking why caves are important is really asking why pure ground water is important."

**Out of Sight**

Yet until recently, caves had few champions. Only in 1988 did Congress pass the Federal Cave Resources Protection Act, which for the first time directed federal agencies to manage public lands in ways that would protect the caverns below. But the feds have taken their time developing regulations to implement the legislation. The Interior Department finally produced its regulations last month, while the Department of Agriculture promises its rules soon.

By many accounts, the cave protection act is full of loopholes. For instance, it states that cave conservation should not interfere with the development of oil and gas and mineral resources and that only "significant" caves must be protected in any case. The problem with the latter provision, say critics, is that many caves are still undiscovered and unexplored. Indeed, until 1986, when a team of cavers dug through a short, rubble-filled passage, Lechuguilla looked pretty insignificant.

Today, however, with almost 70 miles of passageways mapped, there's no question that Lechuguilla is a world-class find. At 1,590 feet, it is the deepest cave in the country. It is fantastically "decorated," with an abundance of crystal formations that have taken millenniums to form. One of its most spectacular chambers is a ballroom-size space adorned with monstrous gypsum "chandeliers," some more than 18 feet long. Explorers have also found many previously unknown formations, including helictites—twisting, spaghettilike strands of calcite—that grow underwater.

The cavern seems to be populated primarily by an intrepid

band of microorganisms that derive their energy from sulfur and sulfide minerals. Scientists are searching this strange community for compounds that might have disease-fighting potential. Says Ohio State University's Calvin Welbourn, "Lechuguilla is such a weird environment, who knows what chemicals [the microorganisms] are secreting?"

**Human Presence**

Yet as scientists and surveyors probe this virgin wonderland, they are, despite themselves, changing it forever. Even when great care is taken to minimize physical damage, cavers trample formations. What's more, explorers undoubtedly introduce foreign insects and microorganisms as well as nutrients in the form of hair, tiny flakes of skin, droplets of sweat and crumbs of food. At the National Cave Management Symposium last month in Carlsbad, N.M., one researcher made a nannyish plea for "clean clothes, clean cavers and careful eating."

The cave's future now hinges on a long-awaited environmental impact statement. The BLM has nearly completed a document with a "preferred alternative" that would prohibit drilling in a newly created "cave protection zone" along the park's northern border. Further, the BLM would buy back oil and gas leases or portions of leases that fall within the zone. Outside the zone, drilling would be permitted, but only with special protective strategies for minimizing gas leaks should drilling problems arise.

The almost completed EIS raises the hackles of the oil and gas and mining industries, which historically have had their way with the BLM. "There are a lot of nervous people. The BLM has never taken back an oil and gas lease," says BLM geologist Joe Incardine. Fred Yates of Yates Energy Corp. contends that the impending EIS overstates the risk to Lechuguilla. "This is not a virgin drilling area," he says, pointing out that there are at least 25 gas-producing wells along the northern border of Carlsbad Caverns National Park. Yates Energy Corp. is part owner of an oil and gas lease adjacent to the park, part of which now falls inside the proposed protection zone. Yates argues that it may not make economic sense to develop only a portion of the lease as recommended, and he is seeking compensation. He estimates the lease is worth about $18.7 million.

The likelihood of gas's leaking into and contaminating Lechuguilla remains undetermined. One study, which the BLM conducted with the help of local cavers, suggests that any gas from the proposed wells would not make it all the way into Lechuguilla. But skeptical cavers and even a couple of renegade BLM officials point to another BLM study that mapped all known subsurface fractures in the area. While none were found to directly link Lechuguilla's passages with any proposed wells, several came very close. "Risking a world-class resource to make one company richer for a little while? It just doesn't make sense," argues one New Mexico caver. Either BLM chief James Baca or Secretary of the Interior Bruce Babbitt will make the final ruling on Lechuguilla after the EIS is completed in January.

While the controversy over Lechuguilla has grabbed the spotlight, the stiffer challenge for both the caving community and federal agencies may be the task of conserving the host of lesser known, less dramatic and still undiscovered caves on federal lands. Indeed, America's caves, like outer space and the ocean depths, are truly an unexplored wilderness. New Mexico's Guadeloupe Mountains, for instance, are a caving mecca, attracting aficionados from all over the world. Yet local cavers estimate that they have found only two thirds of the region's caverns; just last month, cavers discovered three new ones. Even the popular Carlsbad is still being mapped at a rate of 5 miles a year, and scientists don't really know how far it reaches.

Now that the cave protection act is being implemented, federal agencies are charged with compiling a master list of "significant" caves. For both cavers and the feds, however, the biggest hurdle to coming up with this roster may be their deep mutual mistrust. Cavers accuse the feds of having closed their eyes to the marvelous wilderness beneath their feet. "They are afraid to know too much. They figure the less they know, the less they have to manage," asserts Jeff Lory, Southwest region chairman of the National Speleological Society.

Many federal officials concede that their stewardship of caves has been lacking and that, as a result, the resource has suffered. But they say their job has been hindered by a culture of secrecy among cavers. In fact, cavers are notoriously taciturn about the location of "wild" caves, in part to protect them from vandals but also to avoid having federal officials prohibit cavers' access to their discoveries.

By all accounts, this adversarial approach is backfiring now that poor land-use practices pose the greater threat to caves. Says NSS President Jeanne Gurnee, "We have a choice now—either be on the outside looking in or cooperate and make this effort work." For her part, Gurnee is urging full cooperation. ❑

**Questions**:

1. Give two reasons why "cavers" are reluctant to reveal the location of "wild" caves.
2. Where is the deepest cave in the U.S.? How do cavers harm caves?
3. What is an EIS? Are these performed on caves?

Answers on page 301

## 34

*The Florida Everglades compose one of the largest, most unique natural habitats in the United States. It is not only critical for survival of many endangered species, it also plays a key role in the water cycle. The many plants in the 'Glades function to filter out pollutants in the water, helping to purify the water before it moves into the groundwater and ocean. Unfortunately, the 'Glades have been overwhelmed by human activities. Hundreds of tons of fertilizer, pesticides and other pollutants are pumped into the Everglades each year. These originate from agriculture and development to the north. For the last few years, the federal government has been trying to force the state of Florida and farmers to help restore the 'Glades. The Army Corps of Engineers is restoring river channels to their former state; pollution reduction measures are under way and surrounding tracts of land are being bought up and restored. Many argue that the burden of cost of this restoration should not fall on the average taxpayer, but those who created the problems. Developers, ranchers, loggers, farmers, and others should be required to contribute substantial amounts of money.*

# New Life for the 'Glades

Bruce Babbitt's fix-it compromise foretells the future of ecowars

by Michael Satchell

The evening of the second Sunday in January was blustery when some 700 people filed into the high school stadium bleachers in Clewiston, Fla., for a prayer vigil. It was an unprecedented gathering for folks in the small agricultural community on the south shore of Lake Okeechobee, but here—as elsewhere—these are worrisome times on the land. Out west, people in timber, mining and livestock towns had held similar gatherings to seek deliverance from the force of change that loomed over South Florida.

The threat to their livelihoods was tangible. Local real-estate listings had suddenly fallen flat. Banks began requiring extra collateral on loans. At the Sunday gathering, the Rev. Ken Reaves, pastor of the First Baptist Church, turned for inspiration to the Old Testament's Nehemiah, who triumphed in the face of adversity: "If God can move the heart of a pagan king, then God can move the hearts of those inside and outside the government." A political sermon was delivered by United States Sugar Corp. Senior Vice President Robert Buker. "Our opponents," he declared, "are uncaring, ruthless and dishonest."

**"Death sentence."** The principal object of all this fear was Interior Secretary Bruce Babbitt and his ambitious strategy to resurrect the polluted, desiccated and dying Everglades. Last week, Florida Gov. Lawton Chiles signed legislation to launch the first step in a decades-long, multibillion-dollar effort to restore the world-renowned wetlands biosphere. The nearly 2 million-acre ecosystem includes the Everglades National Park, the Big Cypress National Preserve and the Loxahatchee National Wildlife Refuge. Some 200 tons of phosphorus in farm effluent is pumped annually into the Everglades. Scientists say the nutrient promotes the growth of cattails that choke out saw grass and other native vegetation. This in turn affects the food supply and the habitat for fish, birds and other wildlife on dry land, in the marshes and in Florida Bay.

The initial $700 million cleanup, to be paid for by the sugar industry and Florida taxpayers, will create giant marshes to filter polluted farm runoff. Environmentalists immediately branded the plan a sellout to the sugar industry. They believe the farmers should pay the entire cost of any cleanup and are convinced this first restoration step doesn't go far enough. Says Tom Weiss of Clean Water Action: "I feel comfortable describing it as a death sentence for the Everglades."

For Babbitt, it was something quite different. The agreement marked the first major success in his consensus strategy to bring ecofoes together, halt expensive and time-consuming legal trench warfare, get both sides to give ground and avoid environmental "train wrecks" like the war over the spotted owl, which pitted environmentalists against timber interests. But in the ongoing, nationwide jobs-versus-environment conflict, neither side in this Florida battle is happy. Their discontent reflects future difficulties Babbitt will face in brokering agreements in other thorny natural resource controversies, such as mining, grazing, logging and saving the threatened Pacific Northwest salmon.

Babbitt's ambition in Florida is to restore an entire ecosystem, rather than opt for the traditional and easier piecemeal approach. For decades, farmers who grow sugar and vegetables on some 700,000 acres to the north of the great swamp have had first and free use of Okeechobee water to irrigate their fields and have been allowed to pump the fertilizer-polluted runoff into the wetlands and water conservation areas that feed Everglades National Park.

**Two decades.** After a fractious year of negotiations between a Babbitt Interior team and the sugar growers, a water cleanup bill was produced in April by the Florida Legislature. The farmers will pay up to $320 million, slightly less than half of the cost of the initial cleanup, which will take 20 years to complete. Florida taxpayers will be billed for the balance. Ending the contamination, however, is only the first phase of an ambitious restoration that could eventually cost $2 billion.

A government wish list calls for replumbing much of South Florida by purchasing farm acreage to create vast new wetlands and mile-wide flow ways to succor the parched and polluted ecosystem. The dramatic increase in clean water will revive not only the nation's most threatened crown jewel park by the barren, algae-choked and dying Florida Bay and the seriously threatened coral reefs that rim the Florida Keys.

Babbitt had sought a quick peace in the government's ongoing battle with the politically powerful sugar industry and the smaller citrus and vegetable farmers. Tired of their stonewalling tactics, he wanted a phaseout of their polluting practices, and he wanted them to finance a big chunk of the cleanup. Babbitt came close to a deal last July, but as negotiators bickered over the final details, the talks broke down in December amid a flurry of recriminations and farmers' mistrust.

In early April, as both sides prepared to resume the court battle—dashing Babbitt's hopes of a negotiated peace—the Florida Legislature revived the agreement and crafted a new Everglades cleanup bill that both sides' supporters in the Legislature backed. It leaves Babbitt happy, the sugar industry sullen but agreeable and most environmental groups furious.

**No new taxes.** The farmers are to drastically reduce their use of phosphorus fertilizer and pay their share of cleanup costs over 20 years, depending on how fast the water quality improves. This will buy a decade of peace in which Uncle Sam will impose no new environmental taxes or cleanup regulations for the first half of the 20-year agreement.

The money will be used to purchase some 40,000 acres of land to create six artificial filtration marshes that will act like giant kidneys to cleanse the farm runoff before it is pumped south. No one knows if the experimental restoration strategy will succeed, but unhappy environmentalists are threatening to try to block the deal. This fall, they plan a statewide ballot referendum that would impose a penny tax on each pound of sugar produced in the state. If successful, the tax could raise $875 million over 20 years—with the money to be used for restoration. Growers claim such a tax would put them out of business—a result some environmentalists would be pleased to see.

The greens question why the sugar growers shouldn't pay the entire cost of the phosphorus cleanup. The farmers are heavily subsidized with federal price supports and import quotas that keep prices artificially high for consumers, and critics charge they have basically been allowed to violate federal and state clean-water laws for the past 20 years. "It's a sellout," complains Joe Podgor, who heads Friends of the Everglades. "The farmers are the polluters, but they've been let off the hook. Babbitt wants to be the peacemaker: instead he's starting another war."

The farmers aren't particularly happy, either. They don't like the size of their cleanup fees, but they are tired of the costly legal battle with the government. It began in 1988 and had no end in sight until Babbitt wooed them to the bargaining table early last year. "We don't think the artificial marshes will work, we don't agree that phosphorus promotes cattail growth and we don't think we should pay so much," says Buker, one of the industry negotiators. As for Babbitt, he is delighted with the end of the long war of attrition and start of the restoration. "The river of grass has been given a new lease on life," he exults. "This is a model of cooperation."

While the farmers have bought 10 years of legal peace with the government, they are girding for what is certain to be a longer and more expensive battle. The government plan for the latter phases of Everglades restoration calls for recreating part of the natural sheet water flow that nourished South Florida before intensive development began in the 1940s.

Under the most ambitious scenario, the government would acquire and flood up to 200,000 acres of prime farmland—close to one third of the private holdings. The plan calls for creating mile-wide freshwater flow ways along sections of the 1,400-mile canal system that will inundate even more farmland. And there could be a ban on all new wetlands development throughout the region.

**Farm "annihilation."** The long-term plan, revealed last December, hit the farmer with the force of a bank foreclosure statement. Buker calls it "a blueprint for the annihilation of farming in our region" and raises the specter of flooded towns and thousands of lost jobs. Environmentalists scoff that growers are using the report as a "gloom-and-doom red herring" and argue landowners will be well compensated for their losses.

The families of Clewiston and other nearby communities face an uncertain future, but one fact is incontrovertible: After years of political drift and indecision, coupled with a recalcitrant sugar industry hamstringing the Everglades restoration at every chance, a plan has been inexorably set in motion. On April 23, for example, the Army Corps of Engineers launched a $372 million project to restore the 56-mile-long, arrow-straight Kissimmee River to its natural meandering state as it flows into Lake Okeechobee. This will improve water quality in the lake and in the overflow that trickles into farming areas and then south into the Everglades.

Fixing the final cost of restoration in years or dollars is speculative, although the $2 billion sum is often bandied about. The only certainty is that South Florida sugar and citrus growers—like developers, farmers, loggers, fishermen, miners, ranchers and others across the country—will have to bear a heavier burden than the rest of the taxpayers to restore natural resources they have overexploited or ravaged. As the nation's environmental values and ethics shift, changes are inevitable. For some, they will be very painful. ❑

## Questions:

1. Give examples of wetlands. What percentage of wetlands in the coterminous U.S. has been lost in the past 200 years? What are the most destructive effects on wetlands?

2. What human activities on rivers and streams have contributed to environmental problems such as soil erosion, low or absent dry-season flows, flash floods, and decline in plant and animal populations?

3. Is the Kissimmee River and Everglades restoration project the largest in the U.S.? Is this project directed at the federal or state level? Describe the river's original channelization. How many acres of wetlands were lost when the Kissimmee River was channelized?

Answers on page 301

**35**

*Two decades after passage of the Clean Water Act, U.S. waterways continue to be plagued by problems. The more obvious problems, such as rivers catching fire and raw sewage disposal, have been largely eliminated. But many more subtle problems remain, as evidenced by massive fish kills and the destruction of other aquatic organisms, especially freshwater clams. Clams are among the most endangered groups of organisms in the U.S. because they are filter-feeders and are very sensitive to pollution. Sources of these new, subtle problems include pesticides, mine wastes, and many other pollutants that flow into the waterways as runoff. Such "non-point" pollution is probably the major challenge in reducing waterway pollution in the future because it is so difficult to monitor and find ways to reduce.*

# Troubled Waters

Two decades after passage of the Clean Water Act, American waterways no longer catch fire with chemical pollution, but they do carry invisible toxic waste that threatens fish and people. Congress tackles the issue this fall, but the answers may be as intricate as rivers themselves.

by Annette McGivney

For Norman Maclean, Montana's Big Blackfoot River was a pristine and spiritual place where any faithful fly fisherman could enjoy a near-religious experience, partaking in the best that nature had to offer. The trout-filled waters of the Blackfoot shaped Maclean's life and inspired him to write a book, *A River Runs Through It,* filled with romantic descriptions of the Blackfoot, which inspired Robert Redford to buy the screenplay rights and produce a movie about the beautiful river of Maclean's youth. Flyfishing, brotherhood, growing up in Montana—the Hollywood production had all the wholesome goodness of homemade bread, except for one thing. By the time Redford was ready to start filming two years ago, some 16 years after the book was published, the Big Blackfoot lacked the aesthetics necessary to serve as the setting for the movie.

Like many rivers in Montana, and in the rest of the U.S., the mighty Blackfoot has been broken by the resource-guzzling activities of humans and is no longer the proud water that Maclean described. The Montana locals weren't surprised that Redford and his crew had to simulate the Blackfoot, which hasn't supported a healthy fish population for at least 10 years. Its banks are scarred by clearcuts, and foul-looking mine waste slurps against road crossings. "The Blackfoot has been over-mined, over-cut, over-fished, over-recreated, and over-looked," says Becky Garland, president of the Big Blackfoot Chapter of Trout Unlimited, which is helping to revive the river.

Taken out of context, Maclean's words are prophetic—all of our land use practices and wastes have indeed merged into one, and our rivers run through it. Since Cleveland's Cuyahoga River caught fire in 1969, federal and state governments have worked hard to get the junk out of polluted waters. It's rare now to see a river topped with chemical foam, flowing bright red from industrial discharges, or reeking from raw sewage. But in many instances, these waters are not much healthier than they were two decades ago.

The Blackfoot is part of 2,279 miles of rivers and streams in Montana listed in the Environmental Protection Agency's (EPA) 1990 National Water Quality Inventory as having "elevated toxins." Nationwide, more than 28,000 river miles contained a hazardous level of toxins in 1990, and some 26 million fish were killed by pollution. "The public perception is that the rivers are cleaned up, that they're better than ever, because the goo is gone," says Kevin Coyle, president of American Rivers. "But if water quality is improving, then why are all the fish dying? Eventually, our rivers will look incredibly clean and they will be deadly." What worries him most is the invisible pollution that spills undetected from industry, and flows with rain water from every paved street and farmed field into rivers and streams. "Back in the 70s, people thought of water quality in terms of gooey stuff washing up on the docks," he says. "Today, non-point source pollution is the major problem for rivers, and it's not usually something people can see. Poison runoff can even be aesthetic because the water looks so clear."

When the Clean Water Act (officially called the Federal Water Pollution Control Act) was passed in 1972, the mandate was broad: "to restore and maintain the chemical, physical and biological integrity of the nation's waters." But the immediate task was to stop

***Reprinted with permission from E-The Environmental Magazine, September/October 1993, subscriptions $20/yr., P.O. Box 6667, Syracuse, N.Y. 13217***

municipalities and industries from dumping untreated or poorly treated wastes into public waters. And urban rivers like the Delaware, Potomac and Hudson, which once stunk from blocks away, are now popular spots for fishing, canoeing and picnicking. Much of the progress has come from improved wastewater treatment. The federal government has spent $56 billion on new treatment plants since the early 70s.

But controlling point-source discharges is about as far as the nation has gotten in meeting the goals of the Clean Water Act. Under the deadlines set out in 1972, by 1983 all of the nation's waters were to be fishable and swimmable. And by 1985, the discharge of any pollutant into U.S. waters was to have been eliminated. According to EPA's 1990 Water Quality Inventory, 37 percent of rivers and streams assessed did not meet fishing and swimming standards. The figure could be higher, since only 36 percent of the nation's 1.8 million river miles were monitored. And industrial releases of toxic substances into American waters still average 360 million pounds a year.

"Rivers are the catch basins for all our land use practices—farmers spraying their fields, homeowners spraying their lawns, logging, grazing, everything," Coyle says. And our land use practices are causing aquatic species to go extinct at a rapid rate. The invertebrates are shouting out most loudly, telling us some rivers are dying."

A recent study by The Nature Conservancy found that 65 to 70 percent of bottom-of-the-food-chain aquatic species like crayfish and freshwater mussels are classified as rare to extinct, while only 11 to 14 percent of North American terrestrial species (birds, mammals and reptiles) are considered rare to extinct. Since freshwater mussels are filter-feeders and have little tolerance for water-borne pollutants, they have been seen as the proverbial canaries in the coal mine. "The unprecedented decline of mussels is nothing less than a red flag, a distress signal from our rivers, streams and creeks," says Nature Conservancy president John Sawhill. "We cannot ignore this warning and the degradation of water quality that it implies."

The Clean Water Act is up for reauthorization this year, and the need for the federal government to come to the rescue of rivers is just as urgent as it was in 1969. A preliminary bill was introduced in the Senate last June by Max Baucus (D-MT), head of the Committee on Environment and Public Works, and John Chafee (R-RI), the committee's ranking minority leader. Following summer hearings, the final bill is expected to be submitted to the Senate for approval early this fall.

"The House isn't as far along as the Senate with a reauthorization bill, but getting Clean Water Act legislation through this session is a top priority for Congress," says Robyn Roberts of the Clean Water Network, a coalition of 450 environmental and community organizations that want a "strengthened" Clean Water Act. Supported by national players like the Sierra Club, American Rivers and the National Wildlife Federation, as well as local grassroots groups, the Network wants a reauthorized Clean Water Act that will "eliminate the use of the nation's waters—and other parts of the environment—as dumping grounds for society's wastes." The Network also wants Congress to federalize water quality programs and make them less permissive, to focus on non-point source pollution, preserve wetlands, expand regulations of toxic discharges, and prevent clean waters from getting polluted. Many of these components already appear in the broad language and numerous amendments to the Clean Water Act, but federal and state agencies are failing to carry them out. "The law seems fine to me as it is written," says Nina Bell, executive director of Northwest Environmental Advocates (NEA) in Portland, Oregon. "It's just the implementation that stinks."

**Keeping Tabs**

Over the past decade, local groups like NEA have led the fight against river degradation by spending a great deal of time policing the polluters to stay in compliance with water quality standards. The Willamette River no longer has rafts of sewage floating atop its waters or slime lining its banks. Municipal wastewater treatment plants have been built, industries must obtain discharge permits, and some 200 miles of greenway have been acquired for a large riverfront park system. In 1989, the U.S. Army Corps of Engineers described the Willamette as "one of the cleanest streams of comparable size in the nation," and the EPA often refers to it as an example of a river restoration success story.

The Willamette, however, is far from restored and requires constant management. Water levels are regulated by a series of 13 dams, most of the fish are hatchery stock, and waste discharges are diluted with impounded water. A 148-mile strip of the Upper Willamette has unsafe levels of dioxin, a carcinogenic byproduct of pulp mills and sewage treatment plants that use chlorine. Other toxins violate water quality standards in scattered sections of the river, including DDT, PCBs, chlordane, arsenic and heavy metals. On the Lower Willamette, Portland uses 43 antiquated storm drains called combined sewer outfalls (CSOs) which dump raw sewage into the river when it rains. "I wouldn't take my child swimming in the Willamette," Bell says. "There's too much sewage. You can see toilet paper floating by after a rain."

Even more frustrating than the pollution is the lack of information about it. "The fundamental problem is we don't have enough monitoring. The states aren't required by EPA to collect certain water quality data, so, like with the Willamette, information is very sketchy," Bell says. The Oregon Department of Environmental Quality has admitted as much: "The river's current water quality is not completely known. Much of the existing information on the river describes conventional water quality measurements, such as levels of dissolved oxygen. Insufficient information exists on other topics, such as the impact of toxins on the overall health of the river."

The Clean Water Act requires states to designate uses for rivers—usually fishing and/or swimming—and then report to the EPA on whether or not those uses are met. The EPA offers guidelines, but the decision on which rivers to monitor and what testing methods to use is up to the states. Their information gets compiled by the EPA every two years for the National Water Quality Inventory, which typically only reports on about one third of the nation's river miles. And since every state has its own testing methods, the EPA can't combine all of the information into a national overview. "In times of limited resources, you have to monitor where there are suspected problems," says Elizabeth Jester Fellows, branch chief for monitoring in EPA's Office of Wetlands, Oceans and Watersheds. "Since only about one third of rivers are monitored in a two-year period, we are recommending that the states rotate their testing each period so more rivers are covered."

The EPA also encourages states to shift their focus from testing for the obvious—specific chemicals from point-source discharges —toward monitoring for non-point source pollution. But in 1991, the federal government gave less than $70 million to states for ambient water monitoring—assessing the quality of the aquatic environment rather than studying the effect of specific discharges. Federal monies for new wastewater treatment plants in 1991 totaled $6 billion. "Increased funding for monitoring is often viewed by Congress as a resource sink," says Fellows, "because it's hard to prove the money has been used effectively."

Richard Sparks, an aquatic ecologist with the State of Illinois, says, however, that river water quality won't improve unless the feds step up to the plate. "There needs to be more leadership from the federal government. States set up programs in response to federal requirements and grant opportunities. Most states are too strapped financially just to start up new water quality programs on their own," he says. "Our monitoring is supposed to provide a kind of 'state of the aquatic environment' report for the EPA, but I consider many of the measures the states use to gauge this as irrelevant." The EPA does recommend that states incorporate the use of biological criteria in their reporting and will require it by 1995. These methods, which involve counting the number of different aquatic species and the abundance of each, cost more than the chemical approach and require a staff of biologists—which most states don't have.

"The use of bio-criteria will result in the re-direction of money we currently spend on wastewater treatment plants to pay for the restoration of aquatic ecosystems," says James Karr, director of the Institute for Environmental Studies at the University of Washington. "Historically, states have concentrated on waters they suspected had toxic concentrations. Now, with bio-criteria, we're going to find problems in rivers the states thought were okay."

**Pointless Pollution**

When there is a blanket of dead fish covering a river just downstream from an industrial treatment plant, it's not hard for state and local authorities to find the source of the pollution and slap the guilty party with a fine. Non-point source pollution, however, is much more evasive—everyone is to blame, and no one is to blame. But there's no arguing with the fact that the agricultural industry is the primary culprit in polluting our nation's rivers. The EPA reports that 50 to 70 percent of impaired or threatened surface waters are affected by non-point source pollution from agricultural activities. According to the Conservation Foundation, nearly five tons of soil erode off of one acre of farmland each year in the U.S., carrying sediment (the number one pollutant), fertilizers, herbicides and insecticides into rivers and streams.

"The industrialization of agriculture over the last two decades has had a tremendous effect on rivers," says Kevin Coyle. "The U.S. decided it would maximize farm production and be the bread basket for the world. Farmers started spraying their crops with all kinds of chemicals, and spraying weeds with herbicides instead of tilling the land. There are more chemicals being used by farmers today than ever before. The EPA approves these substances, but they don't really monitor their potentially hazardous effects. And people wonder why the cancer rate is going up in Iowa?"

The 1990 Farm Bill encourages farmers to voluntarily participate in programs that minimize the use of chemicals and control agricultural runoff, but the financial incentives for these programs are not nearly as enticing as other U.S. Department of Agriculture (USDA) programs that encourage exploitation of the land. For example, more than two-thirds of U.S. cropland is enrolled in a program where farmers receive price supports based on the average crop yield over a five-year period. This discourages the farmer from rotating crops or letting the land lay dormant, which would reduce runoff into rivers as well as the need for pesticides.

An obvious step would be to place stricter land use regulations on the agricultural industry. But the notion of making farmers comply with federal requirements is politically taboo, and downright un-American in much of the country. "Voluntary action through education is more effective than regulation in addressing our environmental issues," says James Moseley, USDA assistant secretary for agriculture for natural resources and environment. "Prohibiting the use of certain chemicals and policing and fining polluters is not the best way to deal with water quality concerns, particularly in a diversified industry such as agriculture. Regulations undermine agriculture's flexibility in determining production options."

The other major cause of non-point source pollution is urban runoff. Poisonous substances wash down city streets and storm drains from everything and everyone—from residential landscaping to road construction to motor oil. But finding ways to control the conglomeration of wastes that merge into urban runoff is a new and evolving science. "There is a lot of emphasis being placed on figuring out how to deal with non-point source pollution," says Curtis Dalpra of the Interstate Commission on the Potomac River Basin. "A lot of methods for abating non-point in urban areas are not turning out to be as effective as people first thought they would be."

**Accidentally Toxic**

For all of the talk about non-point source pollution, toxic spills are still a serious threat to both water quality and public health. In 1990, 31 states reported concentrations of toxic contaminants in fish tissue that exceeded public health standards. Deadly substances like DDT, chlordane, PCBs and dioxin live for decades in a river's aquatic food chain. "Toxic spills and the legacy of toxic substances along riverbanks are a real problem," says Richard Sparks. "Industries have to meet requirements in order to get discharge permits, but accidents happen—and unless there's a fish kill, small spills often go undetected."

In fact, much to the surprise of unsuspecting river recreationists accidents happen quite often. A U.S. Public Interest Research Group (U.S. PIRG) study has found that 21 percent of the nation's 7,185 major industrial and public waste treatment facilities are in serious or chronic violation of their discharge permits, and another 19 percent of the facilities report at least occasional violations. What's more, the study says, many of these facilities violate their permits on purpose because the cost of paying minimal fines to the EPA is less than that of investing in the equipment needed to comply with water quality standards. An occasional toxic spill may seem like a negligible crime, but the long-term cost to public health can be high. Sparks says that it takes many of these toxic substances decades to degrade, and as time passes the toxin "bio-magnifies" as it moves up the food chain. "A fish can tolerate a certain amount of mercury," he says, "but a human that eats several of those fish could get too big a dose." EPA studies show that dioxin bioaccumulates at very high rates, and that even immeasurably small quantities cause a range of health problems in humans from immunological disorders to cancer.

Since states want to promote clean looking rivers for recreational users, it's not uncommon to find waterways where the fish are known to contain toxins but there are no signs warning people of the danger. On a stretch of Virginia's scenic Shenandoah River there's an advisory against eating fish tainted with mercury from an industrial spill 17 years ago, but there are no signs. "People tore the signs down a long time ago and we haven't put them back up," says a Virginia Department of Health official.

"Citizens have a right to know when significant threats to their health or environment are present in their communities," says U.S. PIRG staff attorney Carolyn Hartman. "We want the Clean Water Act to require public postings on waterways that don't meet standards, or have a fishing or shellfish ban." U.S. PIRG and the Clean Water Network also support legislation to require dischargers to post a sign on the entrance of their facility that details what is going into the river and who people can contact for more information.

Congress added a section to the Clean Water Act in 1987 to accelerate efforts to control and eventually eliminate toxic point-source discharges into rivers. It requires states to identify and put on three EPA lists all sources of "priority" pollutants, and then stringently regulate those sites. A 1991 report by the U.S. General Accounting Office found, however, that the process of who gets listed is often politicized at the state and local levels. "Nationwide, EPA deleted a total of 309 facilities from the lists," says the report, primarily because "the fear of the negative image associated with being listed as a toxic pollutant discharger prompted certain industries to pressure states to make their water quality standards less stringent. In Alabama, this action resulted in nine out of 10 paper mills being deleted from the discharger list because they were no longer in violation of the new, less stringent dioxin standard."

"I'm in favor of more federal control when it comes to toxic discharges," says Nina Bell of NEA. "States are undercutting human and wildlife health in order to appeal to industry. There needs to be a provision in the Clean Water Act that recognizes some pollutants are so dangerous that there is no safe level for discharging them into our rivers. Why are we haggling over how much substances like dioxin should be diluted? Congress had to ban DDT, but it should have been eliminated through the Clean Water Act."

But focusing only on the worst pollution may create problems of its own. "Our national fixation on cleaning up the nation's most polluted water resources to minimally acceptable standards has blinded us to the imminent decline of existing pristine water bodies," says a 1992 report from the National Wildlife Federation. "If the current course remains unchanged, the consequence of continued neglect of outstanding water resources in the United States will be equally mediocre water quality everywhere."

Fulfilling the goal of the Clean Water Act, "to restore and maintain the chemical, physical and biological integrity of the nation's water," remains a daunting task, but many environmental leaders say it is still possible. It will require federal leadership and plenty of funding, and most important, a renewed commitment to the original idea of the Clean Water Act rather than to the bureaucratic quagmire that it now represents. It will also require compromise on the part of business, government and environmental groups, but the one thing that can't be compromised is the river itself. "I hate the word 'mitigate,'" says Neal Emerald, grassroots coordinator for Trout Unlimited. "You can't make up for pollution that's already gone into a river. It's much better not to mess it up in the first place. What it comes down to is having a commitment to protecting the environment instead of raping and pillaging our natural resources." And if we don't? Beautiful, healthy rivers will go the way of the Blackfoot, and become the stuff of fiction rather than real life treasures.

*For information on what's being done to improve rivers and streams and how you can get involved, contact:*

- *American Rivers,* 801 Pennsylvania Avenue SE, Washington, DC 20003/(202)547-6900.
- *Clean Water Network,* 1350 New York Avenue NW, Washington, DC 20005/(202)624-9357.
- *EPA Office of Wetlands, Oceans, and Watersheds,* 401 M Street SW WH-556F, Washington, DC 20460/(202)260-7166.
- *National Wildlife Federation,* "Keeping Clean Waters Clean" Program, 1400 16th Street NW, Washington, DC 20036/(202)797-6800.
- *Northwest Environmental Advocates,* 133 SW Second Avenue, Portland, OR 97204/(503)295-0490.
- *U.S. Public Interest Research Group,* 215 Pennsylvania Avenue SE, Washington, DC 20003/(202)546-9707. ❑

**Questions**:

1. The EPA reports that what percent of impaired or threatened surface waters are affected by agricultural non-point sources? What bill encourages farmers to voluntarily minimize the use of chemicals and control runoff?

2. Historically, states have tested waters for toxic concentrations of chemicals. Will it cost more to test with biological criteria? What are some methods of using these bio-criteria? Does using bio-criteria provide a more complete test of the water's health?

3. Why are freshwater mussels regarded as similar to canaries in a coal mine?

Answers on page 301

# 36

*The Clean Water Act of 1972 is widely acknowledged to be one of the most successful pieces of environmental legislation ever passed by the U.S. Congress. While it has been costly, with over $541 billion spent on water pollution control since 1972, the act has also led to substantial improvements in the water quality in many lakes and rivers. The act must be reauthorized periodically, typically every 5 years, and the recent reauthorization has raised many questions about how to improve the act. For example, gathering water quality information has proven to be extremely costly and it is very difficult to measure the effectiveness of various pollution control projects in many places because the data are so fragmentary. Also, serious water problems are not covered under the existing act. Nonpoint water pollution, such as runoff from farms and urban areas, wetlands, and groundwater are three main groups of problems that are currently neglected.*

## 20 Years of the Clean Water Act

Has U.S. water quality improved?

by Debra S. Knopman and Richard A. Smith

Once again, it is time to reauthorize the U.S. Clean Water Act—formally known as the Federal Water Pollution Control Act Amendments of 1972. Just as on every previous occasion that this major environmental law has been considered, Congress and the current administration are short on information about the true state of the nation's water quality and the factors affecting it. Since 1972, taxpayers and the private sector have spent more than $541 billion on water pollution control,'[1] nearly all of it on the "end-of-pipe" controls on municipal and industrial discharges mandated by the Clean Water Act. After all this time and money, it would be desirable to know whether the act has worked. Is the water cleaner than it would otherwise have been and have the environmental benefits, however they may be counted, exceeded the costs? The answer to both questions is that no one knows for sure. Only now is research beginning to move beyond anecdotes and to improve substantially the understanding of water quality throughout the nation.

More important than justifying past expenditures on water quality, however, is improving policymakers' ability to choose the most economically efficient pollution control strategies for the future. The recent U.S. presidential election debates demonstrated that there is increasing interest in the relationship between environmental protection and the nation's economy. Simply documenting changes in environmental conditions does not provide enough information for dealing with this issue. Policymakers need to understand cause-and-effect relationships and be able to predict the effect of various control strategies on environmental quality.

Several key issues will influence the upcoming debate on reauthorization of the Clean Water Act, a process that began in 1991 and will continue through 1993. During the debate, Congress and the new administration will rely on a patchwork of local, state, regional, and national studies about water quality. Unfortunately, the existing information base is fragmentary at best, but new federal programs that are now being implemented may improve the quality of data. However, an attempt to apply the data that are currently available is necessary in order to begin answering the fundamental question: Has the Clean Water Act worked?

**The Information Gap**

Some of the most contentious political issues in U.S. environmental legislation have been more symbolic than substantive. In the coming debate on the Clean Water Act, however, resolution of such major issues as effective municipal wastewater treatment, nonpoint-source pollution controls, control of toxic chemicals, and wetlands protection could have broad impacts on human and ecological health. Unfortunately, these issues are being debated largely in the absence of conclusive information about existing water quality as well as about the influence of current control measures and of various land-use and basin characteristics on water quality.

This lack of conclusive information is nothing new. In the early 1970s, when Congress was first debating the Clean Water Act, there was a widespread perception that water quality was bad and getting worse. Rachel Carson's 1962 best seller, *Silent Spring*, mobilized a generation of citizens to try to

*From Environment, vol. 35, no. 1, January/February 1993. Heldref Publications, 1319 Eighteenth St., NW, Washington, DC 20036-1802.*

stop the ecological damage wrought by unregulated use of chemicals in the environment.[2] Attitudes about water quality were based mainly on anecdotes and images of the Cuyahoga River on fire and raw sewage spewing into San Francisco Bay. Despite having little reliable information on national water quality, Congress responded to public demands to clean up the nation's surface waters by passing the Clean Water Act.[3] The act was passed despite a veto by then President Richard M. Nixon, who described the price tag on the bill—$24 billion over five years—as "extreme and needless overspending."[4]

According to the Senate report accompanying the passage of the Clean Water Act in 1972, "much of the information on which the present water quality program is based is inadequate and incomplete. The fact that a clearly defined relationship between effluent discharge and water quality has not been established is evidence of that information gap.... The fact that many industrial pollutants continue to be discharged in ignorance of their effect on the water environment is evidence of the information gap."[5]

Scientifically defensible assessments of water quality in the United States, or in any nation for that matter, are far more complicated than Congress realized in 1972. In fact, an information gap was first pointed out in 1971, when M. Gordon Wolman wrote in *Science* that "few observational programs combine the necessary hydrology with measurements of water quality, river characteristics, and biology. While some long-term observations exist, the lack of coordinate observations makes long-term comparisons [of river quality] virtually impossible."[6] This was still true in 1981, when the General Accounting Office released a report that concluded that the federal government had no reliable method of measuring the environmental effect of spending $30 billion to construct municipal sewage treatment plants, which were the cornerstone of the Clean Water Act.[7]

In 1993, scientists still cannot reliably answer the most basic questions about national water quality: How much of specific pollutants do point and nonpoint sources contribute to streams and rivers, and how does the point/nonpoint ratio vary geographically and with basin and climate characteristics? What fraction of the nation's surface water and groundwater fails to meet water-quality standards for toxic substances and conventional pollutants? What effect have the $70-billion construction grants program and the Department of Agriculture's $8-billion Conservation Reserve Program—implemented in 1986 and designed to reduce erosion of topsoil—had on water quality? Indeed, the Clean Water Act may have generated considerable benefits simply by keeping water quality constant in the face of increases in population and in the gross national product (GNP). To date, however, little progress has been made in reliably measuring the water-quality benefits of the act in any terms. The information gap also affects the assessment of water-quality controls directed at groundwater, lakes, estuaries, and wetlands—all of which can influence the quality of streams and rivers.

One reason for this lack of information is that the mechanism for program evaluation in the original 1972 act is not capable of tracking progress toward water-quality goals. The reporting requirements, section 305(b) of the 1972 act, sound simple enough: States must designate each of their water bodies for such specific uses as drinking-water supply, recreation, and cold water fisheries. Then, each state proceeds with its own monitoring and assessment methods to determine whether these uses are being met. Every two years, the Environmental Protection Agency (EPA) compiles the state results and reports on them to Congress in the National Water Quality Inventory.

In 1972, Congress assumed that the states and EPA would coordinate their data gathering, storage, and retrieval functions. Most importantly, Congress supposed that EPA would take the steps necessary to ensure methodological consistency, at least.[8] However, until very recently, this was not the case; EPA's implementation of the reporting requirements in section 305(b) has not produced the kind of baseline and comparative database needed to evaluate past water-quality trends and to develop future strategies. For example, according to the latest National Water Quality Inventory, in 1990, 51 U.S. states, territories, and jurisdictions assessed 647,066 river miles. This number reflects 36 percent of the total river miles in the United States, or 53 percent of the total river miles in the states that reported. Of the total river miles in these states, 33 percent were judged to support their designated beneficial uses; 20 percent were judged to be either partially supporting, not supporting, or in danger of not supporting; and the remaining 47 percent were not assessed. Interpreting these numbers is problematic at best. No one knows how representative the assessed surface waters were. It is likely that the 64 percent of all U.S. river miles that were unassessed have very different water quality than those that the states targeted for monitoring. In fact, EPA cautions that the numbers should not be compared either state to state or with previous reports because of "inconsistencies among states in how these data were generated." With these limitations, it is difficult to extract much useful national-level information from the data.

Some of the shortcomings of this process may be a consequence of funding. In 1990, EPA estimates that the states spent about $11 million in federal grant money and twice that amount of their own money for ambient monitoring.[10]' The magnitude of these expenditures is small when compared to the tens of billions of dollars spent annually by the public and private sectors for water-quality control measures."[11] A comparison of expenditures for information gathering with expenditures for water-quality controls suggests that the information gap is at least partially the result of not enough funds allocated for monitoring.

In addition, there are conceptual and operational problems, both with the 305(b) process itself and with EPA's compilation of the state reports. In general, the state assessments are based on a combination of ambient monitoring networks, local scale studies conducted for regulatory purposes, and professional judgment of state water-quality officials. Although the local studies are useful for local management of water quality, their results are difficult to aggregate nationally because the amount of emphasis on each of these information sources is highly variable from state to state. There is little consistency among state programs in the temporal and spatial scales of data collection, the procedures used to collect and analyze the data, and the types of data collected. As a result of these methodological inconsistencies, comparing results from one biennial report to the next to monitor change is impossible.

The only two long-term, nationally consistent surface-water-quality networks in the United States are operated by the U.S. Geological Survey (USGS). The National Stream-Quality Accounting Network (NASQAN) comprises 420 stations located on large rivers. The stations are located near the outlets of major drainage basins so as to measure collectively a large fraction of the total U.S. runoff.[12] The network also has been used to detect water-quality trends over time, to test for correlative relations to land use, and to measure mass transport from rivers to estuaries, the Great Lakes, and between major river basins. Station locations were not based on the presence or absence of pollution sources in the vicinity nor on specific dominant types of land use in the drainage basin. Also, NASQAN does not monitor saline coastal waters, lakes, or reservoirs. In contrast, the USGS Hydrologic Benchmark Network comprises 55 stations located in relatively pristine, mostly headwater, basins. This network was designed to define baseline water quality and has been used to examine the effects of atmospheric deposition on water quality.[13] Both networks measure stream discharge and concentrations of the major dissolved inorganic constituents, nutrients, suspended sediment, fecal bacteria, and some trace metals.[14]

By far the greatest limitation of the existing national networks is that they do not measure toxins and ecological indicators, which have emerged as major water-quality issues in the past several years. Another problem with existing national water-quality networks is that, although station selection was based on hydrologic factors rather than on proximity to pollution sources, the networks were not designed to provide a statistical sample of streams throughout the nation. As a result, the networks can only provide a general overview of national conditions. Moreover, because NASQAN stations represent relatively large drainage basins only, observed trends "are likely to reflect water quality effects resulting from large-scale processes, such as changes in land use, and atmospheric deposition, rather than localized effects such as changes in the amount or quality of point source discharges." [15]

These shortcomings in the national database must be kept in mind when considering existing information on the effects of controlling municipal waste; nonpoint sources of nutrients, pesticides, and heavy metals; and point sources of toxins—all of which are key issues facing Congress in the upcoming reauthorization of the Clean Water Act. Although understanding of water quality has improved over time, inadequate assessment at the national level continues to be a hindrance. Depending on the scale of assessment—from a local river reach or portion of an aquifer, an individual river basin or aquifer, multiple basins and regional aquifer systems, to a national statistical summary—conclusions about water-quality conditions could vary considerably, and, therefore, a variety of monitoring approaches should be applied.

**Reducing Municipal Sewage Discharge**

The cornerstone of the Clean Water Act was the "construction grant" program, funded predominantly by the federal government, to upgrade municipal wastewater treatment plants to a uniform standard of secondary treatment.[16]' In the 1987 amendments to the act, the grant program was converted into a state revolving loan fund financed by direct federal appropriations scheduled to end after fiscal year 1994. However, Congress is under considerable pressure by the states to extend the loan program beyond 1994.[17] Senate bill S. 1081, which must be reintroduced at the start of the new Congress, proposes using the loan program to encourage greater private-sector involvement in financing treatment facilities.

Given the magnitude of public investment, one important issue is the influence of municipal sewage treatment plants on water quality. Understanding the amount and nature of change in water quality, at both the local and regional scales, would provide Congress and the administration with an indicator of what type of and how much federal investment in additional treatment of municipal sewage is worthwhile.

The preponderance of evidence, based on case studies, subjective opinion surveys, and anecdotal information, seems to show that better municipal waste treatment has improved water quality in certain respects. However, how much water quality has changed and in what ways are less clear. In a review of the effects of treatment upgradings at 13 municipal treatment plants, dissolved oxygen levels clearly had improved downstream at 10 of the plants.[18] The Association of State and Interstate Water Pollution Control Administrators—whose members administer the federal monies for municipal wastewater treatment plant construction—has periodically polled its membership on how water quality has changed; the consensus is that the act has produced significant gains in water quality.[19]

There is some very good local evidence that improved sewage treatment results in improvements in water quality. For example, when the Blue Plains Treatment Plant in Washington, D.C., was operating with only secondary treatment, dense blue-green algae growth was a normal occurrence in the Potomac estuary, particularly during the summer. Following full implementation of tertiary treatment and denitrification in 1983, phosphorus levels decreased by one-third, ammonia concentrations decreased, and nitrate concentrations increased.[20] The improved wastewater treatment contributed to the return of underwater vegetation, an increase in water clarity, and increased and balanced biological activity. In a more recent case, statistical analysis of water quality in the White River in Indiana showed clear improvements in dissolved oxygen as a direct result of advanced wastewater treatment.[21] Similar improvements have occurred in the Delaware River, the Neches River in Texas, and the Flint River in Georgia, although heavy metals and organics continue to pose threats to human and ecological health.[22]

It is impossible, however, to take anecdotal information and results of individual case studies and extrapolate them to the nation as a whole in any quantitative way. Statistical analyses of national-level monitoring data can provide a more quantitative, albeit limited, picture of national water-quality changes attributable to sewage treatment plant construction and upgrading. Although not designed to provide a true random sample of surface-water conditions, available water-quality records for a few hundred widely distributed sampling locations provide a generally representative view of conditions in streams and rivers throughout the nation. Two important water-quality measurements collected at these stations—fecal bacteria counts and total phosphorus concentrations—have shown improvement in recent years, which appears to be partly the result of better treatment of municipal effluent. For example, between 1974 and 1981, the most intensive phase of treatment plant construction, downward (improving) trends were estimated at 23 percent of the 295 stations for which data were available.[23] Results of a recent analysis indicate that these trends, which began in the mid 1970s, continued and even strengthened through 1989.[24]

Perhaps the most noteworthy finding from national-level monitoring is that heavy investment in point-source pollution control has produced no statistically discernible pattern of increases in the water's dissolved oxygen content during the last 15 years.[25] In contrast to case studies showing dissolved oxygen increases in the most oxygen-depleted waters immediately downstream of many sewage outfalls, the data from a 350-station network provide a broader view of stream conditions nationally. The absence of a statistically discernible pattern of increases suggests that the extent of improvement in dissolved oxygen has been limited to a small percentage of the nation's total stream miles. This is notable because the major focus of pollution control expenditures under the act has been on more complete removal of oxygen-demanding wastes from plant effluents.

This lack of visible change may be attributed partly to the probability that much of the investment in point-source pollution controls has simply kept pace with population increases and economic development. Available data indicate that oxygen-demanding waste loads declined substantially in the 1970s.[26] However, loads were nearly stable through the 1980s—despite public and private point-source control expenditures of $194 billion-because the population increased by 10 percent and real GNP growth by 30 percent.[27] The nearly constant levels of dissolved oxygen in streams over this period of increasing pollution appear to represent a clear environmental benefit of pollution controls. However, there is still great difficulty in obtaining consistent and comparable pollution-load data on individual point-source discharges, as a recent USGS study of the Upper Illinois River basin clearly shows.[28] Such information might improve the ability to correlate recorded changes in water quality with changes in pollution loads. Here again, although EPA has recently placed a greater emphasis on compiling consistent and comparable point-source data, the lack of reliable historical data makes a meaningful retrospective view of water-quality changes difficult to obtain.

**Nonpoint-Source Pollution**

A major deficiency in the existing Clean Water Act is the lack of control of nonpoint sources of pollution, such as agricultural lands, feed lots, urban areas, suburban developments, and even silviculture and grazing. The enormous difficulties in identifying, measuring, and controlling nonpoint sources of contamination complicate the allocation of responsibility for its control. In fact, the costs of controlling certain nonpoint sources could greatly exceed the costs of point-source controls.[29] The current debate focuses on the choice between a "command-and-control" regulatory strategy like that used for point sources and voluntary or economic incentives.[30] (Also see "Dealing with Pollution: Market-Based Incentives for Environmental Protection," by Robert N. Stavins and Bradley W. Whitehead in the September 1992 *Environment*.)

Section 319 in the 1987 amendments to the act, which established a grant program to assist states in developing management plans for nonpoint-source controls, seems to be having little impact.'[31] The voluntary approach used in section 319 reflected Congress's unwillingness to encroach on state and local land-use decisions in light of the lack of information both about the actual magnitude of the nonpoint-source problem (although they suspected it was quite large) and about which nonpoint-source control strategies are most effective. In addition, Congress was reluctant to make the federal government vulnerable to a costly operational grant program at a time of mounting budget deficits.[32]

The current Senate majority staff draft revision of S. 1081 proposes to strengthen section 319 by mandating that the states implement a nonpoint-source management program within three years of the reauthorization. States would be required to follow national program guidelines published by EPA that would include, among other elements, "methods to estimate reductions in nonpoint-source pollution loads" and "any necessary monitoring techniques to assess over time the success of management measures."[33] Several options related to nonpoint-source pollution control were discussed within the Bush administration, including economic incentives and adoption of a program similar to one that has been added to the Coastal Zone Management Act, but no specific recommendations were made.

The relative proportion of nonpoint-to point-source loadings of nutrients, suspended sediments, trace metals, agricultural chemicals, and other toxic organic compounds at the national level is unknown. However, an increasing number of studies are beginning to illuminate the scope and magnitude of nonpoint-source pollution in selected areas of the nation. For example, in the Chesapeake Bay, approximately 90 percent of all nitrogen loading is estimated to come from nonpoint sources.[34] Moreover, it could cost from $43 to $325 per pound to reduce nonpoint loadings of phosphorus entering Chesapeake Bay, though the cost of removing phosphorus from point sources is approximately $39 to $71 per pound. One study estimated that between 70 and 94 percent of total annual nitrogen loadings measured at the NASQAN station furthest downstream in four major river basins came from nonpoint sources.[35] Others have cited more equal proportions of point- versus nonpoint-source contributions of nutrients, however.[36]

A study of herbicides in surface waters of the midwestern United States found that large concentrations of herbicides were flushed from cropland and transported through the surface-water system as a result of late spring and early summer rainfall.[37] Median concentrations of atrazine and other herbicides increased by an order of magnitude during this period and then decreased nearly to pre-planting levels by the time of the fall harvest. Water samples from more than 50 percent of the sites exceeded the promulgated maximum contaminant levels for atrazine in drinking water (3 micrograms per liter) during the spring. Some of the herbicides persist from year to year, and degradation products of these herbicides were found to be both persistent and mobile. In fact, the pesticide DDT and its breakdown products have been detected in the agricultural soils of the Yakima River basin in Washington State even though the use of DDT was banned almost 20 years ago.[38] In the San Joaquin Valley of California, irrigation drainage practices have been mobilizing selenium in the soils, thereby causing widespread damage to waterfowl and degradation of surface-water quality when the high-selenium waters drain into reservoirs and wetlands.[39] Knowledge of soils and other hydrologic and geologic conditions, however, has produced a sophisticated and integrated understanding of the causes of selenium contamination and has led to improved management practices.

Nonpoint-source pollution is not limited to agricultural areas, however. Urban areas can also be significant sources. According to data from NASQAN stations located in coastal areas, median loadings of phosphorus in predominantly urban basins are higher than median loadings in basins dominated by croplands or forests. Furthermore, it is becoming increasingly clear that atmospheric deposition has become a major source of nitrogen in surface water.[40]

Wetlands are also a significant part of the nonpoint-source control debate. Wetlands are not only good for wildlife; they play a critical role in local and regional water quality, as well. (For an excellent discussion of wetlands issues, see Jon Kusler's article in the March 1992 *Environment.*) Protection of wetlands brings the debate between the protection of ecological resources and the rights of property owners to the most local level. A clash between the definition of water resources as public goods and the exercise of private property rights, the wetlands debate touches on such difficult policy areas as federal involvement in state and local land-use decisions and the appropriate process for establishing a scientifically defensible definition of wetlands.

Section 404 is the *de facto* wetlands protection program of the Clean Water Act. The Army Corps of Engineers issues permits for any dredge-and-fill activities in the "navigable waters" of the United States only after administrative procedures are followed to allow for public comment on a draft environmental impact statement (EIS). The final EIS is subsequently issued by the corps with the concurrence of other federal agencies. Wetlands proposals, introduced in the House and Senate as separate bills to amend the Clean Water Act, range from broadening the jurisdiction of the section 404 permit process to limiting its applicability to strictly "navigable" waters. Senate bill S. 1081 includes no new provisions for regulation of wetlands. Complicating resolution of the issue is the difficulty in delineating wetland resources and their change over time. This difficulty makes the assessment of proposed policy changes a highly subjective process.[41]

The complexity of the nonpoint-source problem is hardly academic. What is the appropriate strategy and level of federal, state, local, and private investment in nonpoint-source controls under such varying conditions? To properly address this question for the nation as a whole, regional-scale hydrologic, chemical, and biological processes must be better understood. Simultaneously, consistent and comparable national-level monitoring data that are representative of the varying land uses contributing to nonpoint-source problems are needed to describe and compare conditions and trends among regions.

**Toxics in Surface Waters and Sediments**

The toxics issue presents EPA with large technical and financial difficulties. The number of potentially toxic pollutants is large, measurement is expensive, and analytical methods are not always sensitive enough to detect biologically significant concentrations. Furthermore, there is no nationally consistent database to rely on for guidance on where control efforts should be focused. EPA has regulated only about one-fifth of the industrial plants that dump toxic substances into rivers, lakes, and estuaries.[42] The existing regulatory structure of the Clean Water Act is almost entirely built around the control of point pollution sources through the pretreatment program (Section 307); nonpoint sources of toxic pollutants are presently unregulated.

Senate bill S. 1081 includes several sections that would increase control of toxic point sources. The bill sets strict deadlines for EPA to issue rules to control existing point sources of toxics. Also, each publicly owned treatment plant would be required to implement a "toxic reduction action program" directed toward significantly reducing specific sources of toxics flowing into the plant. And finally, EPA would be required to submit to Congress an annual "environmental toxics release assessment" that summarized the industrial sources of toxics, their geographic distribution, and approximate discharges.

Little reliable information at the national or even regional scale can be reported for either inorganic or organic toxic chemicals in natural waters. Because of the vast array of toxic organic chemicals in the environment, it is highly impractical to monitor any but a relatively small number of toxic chemicals on a regular basis. A far more efficient and meaningful approach would be to use bioindicators and chemical measures concurrently, yet the United States presently has no nationally consistent biological database that could be used for this purpose.

The country also lacks a national database on trace metals, some of which are toxic to aquatic life even at very low concentrations, such as less than one part per billion. Indeed, recent evidence suggests that very low ambient concentrations of trace metals are actually dissolved in most streams and rivers, but these concentrations are controlled more by stream chemistry than by anthropogenic sources.[43] Anthropogenic sources can significantly raise concentrations of toxics bound to sediment particles (in comparison with the concentrations dissolved in water), which then are ingested by aquatic biota. As with toxic organics, bioindicators may be the most practical means of monitoring the occurrence and trends of trace metals.

Although bioindicators are valuable, few known indicators are suitable for national-level monitoring, and few biological studies have been national in scope. One exception is the National Contaminant Biomonitoring Program of the U.S. Fish and Wildlife Service. Whole fish, collected periodically between 1970 and 1986 from a network of 113 stations located on major rivers and in the Great Lakes, were measured for trace metals and organochlorine contaminant concentrations.[44] Although there are difficulties in interpreting the significance of the results,[45] the trends reported are still of interest. Concentrations of cadmium, arsenic, and lead decreased by approximately 50 to 63 percent between 1976 and 1986, and the concentration of mercury remained nearly constant. Fish tissue concentrations of the pesticide Dieldrin, DDT-related compounds, and total polychlorinated biphenyls (PCBs) decreased by more than 60 percent between 1970 and 1986, a reflection of the fact that use of DDT was stopped in 1972 and use of PCBs stopped in 1976. Dieldrin, however, was not banned until 1988.[46]

**Closing the Information Gap**

EPA is fully aware of the problems with the section 305(b) reporting procedures and is working with the states to reduce methodological inconsistencies in the program.[47] Several other steps are being taken throughout the federal government to close the information gap.

To support future national, regional, and local decisions related to water-quality management, USGS has undertaken the National Water Quality Assessment (NAWQA) program. The program builds on an existing base of ongoing water-quality activities within USGS, as well as those of other federal, state, and local agencies. The full-scale NAWQA program was implemented starting in fiscal year 1991 following a five-year pilot effort and an extensive evaluation of the program's concept by the National Academy of Sciences.[48] The goals of the NAWQA program are to describe the status of and trends in the water quality of large representative parts of the nation's streams and groundwater and to provide a sound scientific understanding of the natural and human factors that affect the quality of these resources. To meet these goals, the program integrates information about water quality at local, study-unit, regional, and national scales and focuses on water-quality conditions that affect large areas or that recur frequently on the local scale. Because of its regional focus, the NAWQA program is ideally suited to investigate nonpoint-source contamination and to define on a regional basis the relative contributions of major contamination sources.

The NAWQA program consists of comprehensive assessments in 60 areas of the nation—each of which cover about 20,000 square miles—that account for 60 to 70 percent of the overall water use and population served by public water supply. These 60 study units cover about 50 percent of the coterminous United States and include at least a part of every state.[49] Biological studies in surface water conducted as part of the NAWQA program include analysis of trace elements and trace organic compounds in fish tissues, measurements of sanitary quality, and analysis of ecosystem structure in relation to habitat and water-quality conditions. The chemistry of suspended sediment and bed sediment samples is also determined as part of the program.

This information base will form a national set of data and interpretations that will contribute to a national assessment of critical water-quality issues; aid the states in focusing their monitoring activities on the most important water-quality issues in the most sensitive settings; help the nation and the states define the status of and trends in regional water quality; and provide sound scientific information that can be used by policymakers and resource managers.

Another new initiative is the Environmental Monitoring and Assessment Program (EMAP), which is now beginning to be implemented. Planning for EMAP began in 1989 in response to a recommendation of EPA's Science Advisory Board to establish a statistically sound national survey for monitoring ecological resources. The Science Advisory Board saw this approach as a means for EPA to detect emerging environmental problems and characterize their geographic extent, and persistence over time. EMAP is intended to estimate "the current status, extent, and geographic distribution of ecological resources; the proportion of these resources that is degraded; trends in the conditions of these resources; and the probable causes of adverse effects."[50]

Biological response indicators will be EMAP's primary measure of ecological conditions. Other chemical and physical indicators will be used to measure stress and chemical exposure of ecological resources. The EMAP sampling framework consists of a systematic hexagonal grid of approximately 12,500 points for the coterminous United States. As currently envisioned, a subset of roughly 3,200 of these points will be randomly selected for surface-water field sampling. Each year, 800 of the 3,200 lake and stream sites will be sampled, giving a four-year resampling cycle. EMAP's results will represent statistical summaries of regional conditions, not descriptions of conditions at individual locations. The surface-water component of EMAP is in the process of development, and thus details are not available at this time. Groundwater is unlikely to be included in EMAP in the foreseeable future.[51] However, work is proceeding on other resources in the EMAP design, such as forests, inland wetlands, agroecosystems, near coastal lands, deserts, and grasslands. Intensive high-altitude aerial photography will precede actual field sampling to characterize as much of the land cover, land use, and extent of resources as possible within each of the sample hexagons.

The 1987 amendments to the Clean Water Act established the National Estuary Program to "identify nationally significant estuaries, protect and improve their water quality, and enhance their living resources."[52] The program uses Comprehensive Conservation and Management Plans (CCMPs) developed through the collaborative efforts of management conferences, whose purposes include assessing trends in water quality, identifying environmental stresses, and planning restoration activities. Congress named 12 estuary projects to receive priority consideration, and EPA may select others in the future. Quantitative comparisons of estuarine water quality may be possible if consistent methods are used among the studies; however, this is not an element of the program.[53]

All of the program initiatives will contribute to the improvement of the U.S. information base on water quality. The recently formed Intergovernmental Task Force on Monitoring Water Quality,[54] under the joint leadership of EPA and USGS, now provides an institutional mechanism to identify systematically gaps in monitoring efforts and particularly focuses on the need to evaluate the effectiveness of federally mandated pollution control strategies. The task force is charged with identifying what kinds and amounts of data are needed, determining what can be obtained by improving the integration and consistency of existing federal and state programs, and recommending extensions of existing programs or new programs to fill gaps in knowledge. In addition, the task force may recommend modifications to existing networks to provide a more statistically representative view of national conditions, but this is not under consideration at this time.

**Science and Policymaking**

Reauthorization of the Clean Water Act raises several important questions about the value of scientific information in policymaking. In theory, the five-year reauthorization cycle affords both Congress and the administration the opportunity to engage in serious program evaluation. In practice, decision makers usually deal with issues that arise from anecdotes rather than from a formal evaluation of monitoring data collected specifically to identify and quantify the causes and effects of existing and emerging water-quality problems.

The language of Section 305(b) indicates that Congress intended the National Water Quality Inventory to provide the data to evaluate the success of the act. For a variety of reasons, the inventory has not met expectations of national consistency in monitoring and analysis. Also, although the USGS has operated the NASQAN monitoring program since the early 1970s, a significant USGS program to describe regional and national water quality has only recently emerged with the initiation of NAWQA in the mid 1980s.

Existing U.S. databases present a fragmented picture of national water quality. Almost no information about potentially toxic organic compounds and trace metals is available. Consistent and comparable ancillary information on land use and point-source discharges at the national scale also is lacking. Use of bioindicators is desirable but currently out of reach on a national scale.

Critical steps are now being taken to produce policy-relevant information, but more could be done. The USGS is proceeding with the full implementation of the NAWQA program to provide the scientific basis for understanding cause-and-effect relationships in water quality. And throughout government, academia, and industry, research and development on bioindicators are increasing, which promises to provide meaningful integrative measures of the effects of both toxic and conventional contaminants on ecological resources. These integrative measures are needed in addition to the physical and chemical measures currently used. EMAP could provide the impetus, framework, and funding for more rapid developments in this area.

In addition, EPA, in close consultation and cooperation with other federal and state agencies, is proceeding with plans to improve the 305(b) reporting program by specifying comparable reporting procedures. States also need to report water-quality measures according to some nationally agreed-on ambient standards and designations of uses. The Intergovernmental Task Force on Monitoring could become the institutional vehicle for implementing these changes, the means of integrating the various federal monitoring and assessment programs, and the instrument to encourage more rigorous consistency from the states. Furthermore, modifications to existing networks could provide for a more statistically based description of water-quality conditions and trends on the national scale.

If the steps outlined above are carried through, then many questions about national water-quality conditions could be answered within the next five years. However, it will probably take more than a decade to amass the kind of data needed to document environmentally and statistically significant trends in water quality. In any case, a prerequisite to bringing accurate and meaningful water quality information to the forefront of the policy debate is recognition by decision makers that they do not now have the information they need to make wise decisions for the future. Only then will the effectiveness of the Clean Water Act be known. ❑

1. A Carlin and the Environmental Law Institute, *Environmental Investments: The Cost of a Clean Environment* EPA-230-11-90 083 (Washington, D.C: U.S. Environmental Protection Agency, November 1990).
2. R. Carson, *Silent Spring* (Greenwich, Conn.: Fawcett Publications, 1962).
3. Library of Congress Environmental Policy Division, *A Legislative History of the Water Pollution Control Act Amendments of 1972* (Washington, D.C.: U.S. Government Printing Office, 1973).
4. R. M. Nixon, "Message from the President of the United States Returning Without Approval the Bill (S. 2770) Entitled *The Federal Water Pollution Control Act Amendments of 1972*," in Library of Congress Environmental Policy Division, *A Legislative History of the Water Pollution Control Act Amendments of 1972* vol. 1 (Washington, D.C.: U.S. Government Printing Office, 1973), 137-39.
5. U.S. Senate Public Works Committee, page 1473, note 3 above.
6. M. G. Wolman, "The Nation's Rivers," *Science* 174 (1971):905-18.
7. Comptroller General of the United States, *Better Monitoring Techniques Are Needed to Assess the Quality of Rivers and Streams* (Washington, D.C.: U.S. General Accounting Office, 1981), 2 volumes; and P. E. Tyler, "After $30 Billion, No One Knows If Nation's Water Is Any Cleaner," *Washington Post* 14 May 1981, A2.
8. U.S, Senate Public Works Committee, page 797, note 3 above.
9. U S. Environmental Protection Agency, *National Water Quality Inventory: 1990* Report to Congress (Washington, D. C. U.S. EPA Office of Water, 1992).
10. U.S. Environmental Protection Agency, "Review Of Environmental Monitoring Costs Related to Environmental Indicators" (Washington, D.C.: U.S. EPA Environmental Results and Forecasting Branch, Office of Policy, Planning and Evaluation, April 1991).
11. Carlin, note 1 above.
12. J. C. Briggs, "Nationwide Surface Water Quality Monitoring Networks of the U.S. Geological Survey (Paper presented at the American Water Resources Association Symposium, San Francisco, 12-14 June 1978), 49-57.
13. R. A. Smith and R. B. Alexander, "Correlations Between Stream Sulphate and Regional $SO_2$ Emissions," *Nature* 322, no. 6081 (1986):722-24.
14. During the 1980s, inflation greatly eroded the fixed annual funding of $5 million. As a result, several variables were dropped, including organic carbon and the measurement of trace elements and nutrients in whole water samples. The ordinal design of the networks called for fixed interval sampling on a monthly basis. Now, because of inflation, the frequency of sampling is quarterly or bimonthly in both networks. As a result, total samples collected in NASQAN decreased from more than 6,000 in 1979 to about 2,000 in 1990.
15. D. P. Lettenmaier, E. R. Hooper, C. Wagoner, and K. B. Faris, "Trends in Stream Quality in the Continental United States, 1978-1987," *Water Resources Research* 27, no. 3 (1991):327-39.
16. A. M. Freeman III, "Water Pollution Policy," in P. R. Portney, ed., *Public Policies for Environmental Protection* (Washington D.C.: Resources for the Future, 1990).
17. Environmental and Energy Study Institute, "1992 Briefing Book on Environmental Legislation" (Washington, D.C., 1992), 13.
18. W. M. Leo, R. V. Thomann, and T. W. Gallagher, *Before and After Case Studies: Comparisons of Water Quality Following Municipal Treatment Plant Improvements* EPA Report 430/9-007 (Washington, D.C.: U.S. Environmental Protection Agency, 1984).
19. Association of State and Interstate Water Pollution Control Administrators, *America's Clean Water: The States' Evaluation of Progress:* 1972-1982 (Washington, D.C.: ASIWPCA, 1984).
20. V. Carter and N. Rybicki, "Resurgence of Submersed Aquatic Macrophytes in the Tidal Potomac River, Maryland, Virginia, and the District of Columbia," *Estuaries* 9, no. 4B (1986):368-75.
21. C. G. Crawford and D. J. Wangsness, "Effects of Advanced Wastewater Treatment on the Quality of White River, Indiana," *Water Resources Bulletin 27*, no. 5 (1991):769-79.
22. R. Patrick, *Surface Water Quality: Have the Laws Been Successful?* (Princeton, N.J.: Princeton University Press, 1992).
23. R. A. Smith, R. B. Alexander, and M. G. Wolman, "Water-Quality Trends in the Nation's Rivers," *Science* 235 (1987): 1605-15; additional results presented in R. A. Smith, R. B. Alexander, and M. G. Wolman, *Analysis and Interpretation of Water-Quality Trends in Major U.S. Rivers, 1974-81*, U.S. Geological Survey Water-Supply Paper 2307 (Reston, Va.: U.S. Geological Survey, 1987).
24. R. A. Smith, R. B. Alexander, and K. J. Lanfear, A Graphical Summary of National Water-Quality Conditions and Trends, Open-File Report 92-70 (Reston, Va.: U.S. Geological Survey 1992).
25. Smith, Alexander, and Wolman, note 23 above; Smith, Alexander, and Lanfear, note 24 above.
26. Smith, Alexander, and Wolman, note 23 above.
27. Carlin, note 1 above.
28. J. S. Zogorski, S. F. Blanchard, R. D. Romack, and F. A. Fitzpatrick, *Availability and Suitability of Municipal Wastewater Information for Use in a National Water-Quality Assessment: A Case Study of the Upper Illinois River Basin, Illinois, Indiana, and Wisconsin*, Open-File Report 90 375 (Urbana, Ill.: U.S. Geological Survey, 1990).
29. S. A. Freudberg and J. P. Lugbill, "Controlling Point and Nonpoint Nutrient/Organic Inputs: A Technical Perspective" (Paper presented at the conference "Cleaning Up Our Coastal Waters: An Unfinished Agenda," U.S. Environmental Protection Agency and Manhattan College, Riverdale, N.Y., 13 March 1990).
30. For some time, economists have queslioncd the economic efficiency and effectiveness of technology-based regulatory controls on point sources. For background on this issue, see P R. Portney, ed., Public Policies for Environmental Protection (Washington, D.C.: Resources for the Future, 1990); and J. J Boland, "Enticing the Colossus," Johns Hopkins Magazine, June 1992, 43-45.
31. R. W. Adler, "Returning to the Goals of the Clean Water Act," *Update Water Resources*, no. 88, (1992):23-30.
32. Senate debate on overriding veto of H.R. 1, 4 February 1987, in *A Legislative History of the Water Quality Act of 1987* (Public Law 100-4), vol. 1, November

1988, 322-23; Hearings of the House Committee on Public Works and Transportation, Subcommittee on Water Resources, "Reauthorization of the Federal Water Pollution Control Act (Nonpoint Source Pollution)," 24 April 1991, 701-911.
33. U.S. Senate Environment and Public Works Committee, *Majority Staff Draft: Amendment in the Nature of a Substitute to S.1081 Water Pollution and Prevention and Control Act of 1991* (Washington, D.C., 31 December 1991).
34. L. P. Gianessi and H. M. Peskin, *An Overview of RFF Environmental Data Inventory, Methods, Sources, and Preliminary Results* (Washington, D.C.: Resources for the Future, 1984), 111. Estimates of nonpoint-source loadings were derived using a regression procedure developed by T. A. Cohn of the U.S. Geological Survey in 1991; Freudberg and Lugbill, note 29 above.
35. W. G. Wilbur, U.S. Geological Survey, personal communication with the authors, 1992.
36. E. L. Tyler, "Reauthorizing the Federal Water Pollution Control Act," *Update Water Resources* no. 88 (1992):7-16.
37. E. M. Thurman, D. A. Goolsby, M. T. Meyer, and D. W. Kolpin, "Herbicides in Surface Waters of the Midwestern United States: The Effect of Spring Flush," *Environmental Science and Technology* 25, no. 10(1991):1794-96.
38. J. F. Rinella, P. A. Hamilton, and S. W. McKenzie, *Persistence of the DDT Pesticide in the Yakima River Basin*, USGS Circular 1090 (Washington, D.C.: U.S. Geological Survey, in press).
39. R. J. Gilliom et at., *Preliminary Assessment of Sources, Distribution, and Mobility of Selenium in the San Joaquin Valley, California*, Water-Resources Investigation Report 88-4186 (Sacramento, Calif.: U.S. Geological Survey, 1989).
40. D. C. Fisher and M. Oppenheimer, "Atmospheric Nitrogen Deposition and the Chesapeake Bay Estuary," *Ambio* 20, no. 3/4 (1991):102-08.
41. U.S. Fish and Wildlife Service, *Wetlands Status and Trends* (Washington, D.C.: U.S. Department of the Interior, Fish and Wildlife Service, 1991).
42. U.S. Environmental Protection Agency, *Report to Congress: Water Quality Improvement Study* (Washington, D.C.: U.S. EPA, 1988); and M. Weisskopf, "Default on Industrial Effluent," Washington Post, 30 December 1991, A11.
43. H. L. Windom, J. T. Byrd, R. G. Smith, Jr., and F. Huan, "Inadequacy of NASQAN Data for Assessing Metal Trends in the Nation's Rivers, " *Environmental Science and Technology* 25, no. 6 (1991): 1137-42. For additional information on the difficulty of measuring dissolved concentrations of trace metals, see A. M. Shiller and E. A. Boyle, "Variability of Dissolved Trace Metals in the Mississippi River," *Geochimica et Cosmochimica Acta 51* (1987):3273-77; and A. R. Flegal and K. Coale, "Discussion: Trends in Lead Concentration in Major U.S. Rivers and Their Relation to Historical Changes in Gasoline-Lead Consumption by R. B. Alexander and R. A. Smith," *Water Resources Bulletin 25* (1989):1275-77.
44. C. J. Schmitt, U.S. Fish and Wildlife Service, National Contaminant Biomonitoring Program, personal communication with the authors, 1992; and C. J. Schmitt, J. L. Zajicek, and M. A. Ribick, "National Pesticide Monitoring Program: Residues of Organochlorine Chemicals in Freshwater Fish, 1980-81," *Archives of Environmental Contamination and Toxicology* 14 (1985):225-60.
45. Smith, Alexander, and Lanfear, note 24 above.
46. G. W. Ware, *The Pestitide Book* (Fresno, Calif.: Thompson Publications, 1989).
47. A. E. Mayio and G. H. Grubbs, "Nationwide Water-Quality Reporting to the Congress as Required under Section 305(b) of the Clean Water Act," in *National Water Summary 1990-91-Stream Water Quality* (Reston, Va.: U.S. Geological Survey, in press).
48. R. M. Hirsch, W. M. Alley, and W. G. Wilber, *Concepts for a National Water-Quality Assessment Program*, USGS Circular 1021 (Reston, Va.: U.S. Geological Survey, 1988); and National Research Council, *A Review of the U.S.G.S. National Water Quality Assessment Pilot Program* (Washington, D.C.: National Academy Press, 1990), 153.
49. P. P. Leahy, J. S. Rosenshein, and D. S. Knopman, *Implementation Plan for the National Water Quality Assessment Program*, Open-file Report 90-174 (Reston Va.: U.S. Geological Survey, 1990).
50. U.S. Environmental Protection Agency, *Design Report lor EMAP*. EPA/600/3 -91/053 (Washington, D.C.: U.S. EPA Office of Research and Development, 1990); and idem, *Reducing Risk: Setting Priorities and Strategies for Environmental Protection*, SAB-EC-90-021 (Washington, D.C.: U.S. EPA Science Advisory Board, 1990).
51. D. A. Rickert, U.S. Geological Survey, personal communication with the authors, 1992.
52. U.S. Environmental Protection Agency, *National Estuary Program after Four Years: Report to Congress*, 5039-92-007 (Washington, D C: U.S. EPA April 1992).
53. T. Armitage, U.S. Environmental Protection Agency, personal communication with the authors, 1991
54. U.S. Office of Management and Budget, "Coordination of Water Resources Information," Memo randum No. M-92-01 (Washington, D.C., 10 December 1991).

## Questions:

1. What is the source of large concentrations of herbicides in surface waters of the midwestern U.S.?

2. What is the greatest limitation of the existing national networks that measure surface water quality?

3. What book mobilized a generation of citizens to try to stop the ecological damage of unregulated chemical use?

Answers on page 301

37

*Bacteria are a rapidly growing method of cleaning up the environmental wastes that occur in so much of our world: oil spills, landfill leakage, and toxic spills to name but a few. Such so-called bioremediation is often vastly cheaper than "high-tech" chemical cleaners and physical removal. Bacteria are introduced to the waste, where they not only detoxify it, but biochemically break it down. This recycles the component elements into the local environment. Bioremediation has been used to clean many types of wastes in many kinds of environments. However, the method is in its infancy, and many wastes cannot be digested by known bacteria. One method of solving this problem is the growing biotechnology field, which researches ways of taking genes from one bacterium species and inserting them into another species. This greatly expands the potential for bioremediation. For instance, an organism that can live in an oil spill, but can not digest it, can acquire genes from another bacterium that produces enzymes to digest oil.*

# Off-the-Shelf Bugs Hungrily Gobble Our Nastiest Pollutants

## The cleaners of environmental messes are now turning from brute force to a stealth strategy: enlisting nature's invisible army

by James D. Snyder

Oily water, contaminated soils, toxic residues leaching from landfills, nuclear wastes—these days we seem to be lost and befuddled in terrain filled with disagreeable and dangerous stuff. How, we ask ourselves, are we ever going to clean up this awful mess we have made?

Well, something is quietly eating away at that problem even now: bugs. Bacteria, to give them their proper name—legions of them. From sewage plants and oil spills to contaminated streams, muscular microbes are at work, digesting the nastiest pollutants and producing mainly water and carbon dioxide.

If anything good emerged in the wake of the *Exxon Valdez* oil spill in March 1989, it was the highlighting of the potential of bioremediation, a term that means employing microorganisms to restore natural environmental conditions. But it didn't happen instantly. In their first desperate efforts to save the 900 miles of affected shoreline, cleanup crews scrubbed individual rocks and stones, knowing that they couldn't really hope to reach the oil that had seeped between cracks and under gravel.

Next, they tried hosing the oil off beaches with hot pressurized seawater. While the high-powered blasts cleansed boulders and helped provide breathing room for a few large animals, those closer to the bottom of the food chain—snails, clams and barnacles—were cooked by the high heat or smothered when the beach was churned. By early summer, crews noticed that the few untreated beaches were beginning to show signs of returning to the picture-postcard environment that existed before the spill turned Prince William Sound into a black lagoon.

In Washington, D.C., meanwhile, the Environmental Protection Agency (EPA) was fielding calls and letters from members of a fledgling industry. Manufacturers and packagers of specialized bacteria—then numbering no more than perhaps 50—were asking for a chance to unleash their troops.

Unlike most of the "native" microbial strains that live in the ocean, the bacteria they offered used oil as a food source. Extra nutrients—usually nitrogen and/or phosphorous—are added at the work site to stimulate the microbes' metabolism and make them eat more voraciously. Once they do so, their only "wastes," besides water and carbon dioxide, are fatty acids that are in turn eaten by plankton and other organisms.

Bioremediation wasn't on the EPA's list of officially approved remedies for oil spills, however. After two months of wrangling, a specially convened scientific panel agreed to a limited test on one secluded beach. The ground rules: nutrients only; no "outside" bugs and no private manufacturers. Instead, an EPA-Exxon team sprayed ordinary liquid farm fertilizer (containing nitrogen) on a cobblestone beach plot.

**A Clean Rectangle on an Oiled Beach**

The test began that June. As the EPA later reported: "Approximately 2 weeks after the [liquid] fertilizer was applied to the cobblestone beach plot, scientists observed visible reductions in the amount of oil on rock surfaces. To the scientists who surveyed the test plot by helicopter, it looked as if a clean rectangle has been etched on the beach's surface."

Not so for an untreated area south of the test plot. "It remained considerably contaminated," said the EPA, "suggesting that nature alone could not account for the dramatic reduction of oil observed in the test area."

The bug makers were heartened. If ordinary ocean bacteria and farm fertilizer could do all that, then surely the EPA would recognize the potential of specially chosen microbes and more-potent nutrients. But the EPA cautiously spent the rest of the summer on more fertilizer tests. The following summer, when commercial microbes were finally admitted for field tests, the sun had pretty much set on the *Valdez* cleanup and the chance for any widespread bioremediation. Yet, the big Alaskan spill was a watershed event both for the EPA and for bioremediation. The agency has since assigned the National Environmental Technology Applications Corporation (NETAC)—a research group spawned by the EPA and the University of Pittsburgh—to develop and standardize protocols for treating specific types of spills, such as open sea and shoreline.

"Until the *Valdez*," says NETAC president Edgar Berkey, "there was knee-jerk fear of manufactured [industrially grown] microbes, probably brought about by too many movies like *The Blob That Ate Detroit*."

The EPA also changed a key policy. Previously, it had insisted that any technology used at one of the 1,200 waste-contaminated Superfund sites on its "nation's worst" list be certified at "Best Demonstrated Available Technology" (BDAT). Because incineration can destroy 99.99 percent of most organic wastes, it was invariably chosen over all other forms of treatment—even when others promised to cost much less. Biotreatment obviously takes more time and usually isn't more than 95 percent effective. But it's so much cheaper that the EPA is now urging staffers to stress "innovative technologies." Today, more than 200 EPA-policed cleanup sites use biotreatment.

All that progress would delight and surprise George Jefferies. Jerry Bower, too. In the early 1950s, Jefferies, an inventor from Salem, Virginia, was intrigued by the fact that bacteria retreat into a spore form (a sort of suspended animation) whenever they face hostile conditions such as extreme heat or cold. Live spores have been found in the chambers of Egyptian pyramids.

Jefferies' contribution was to patent a way to make microbes go in and out of this "suspended" state almost on cue. Thus, he reasoned, they could be grown and then packaged and inventoried until needed. But commercial uses were limited to such routine chores as treating septic tanks and clogged drains.

Jerry Bower might have fared the same if it weren't for luck and pluck. By 1958 Power was already a successful inventor thanks to his patent on the first plastic bread wrapper. A friend told him about a feed-additive salesman with an idea and no money. Ed Noeker, who sold animal antibiotics, often heard farmers complain that the antibiotics in his feed pills also tended to kill the naturally occurring microbes that help a cow digest her food. Why not develop another additive made of cultures taken from a cow's rumen? Once back in their warm, moist natural home, the bugs would snap back to life and rekindle the digestive process.

Bower had no sooner agreed to invest in the scheme when he suffered two setbacks. First, Noeker, the partnership's technical expert, died unexpectedly. Then university researchers determined that the cultures wouldn't work as a feed additive.

"Rather than give up, Jerry immersed himself in the business," says the man who bought him out in 1980. R. B. (Jones) Grubbs, the president of what is now Solmar/ InterBio in Orange, California, recalls that Bower tried variations of the formula for everything from cleaning restaurant grease traps to retarding the deterioration of corn stored in silos. What kept him going was a vision that large-scale sewage-treatment plants would one day accept bugs to help break down sludge. Today they do. "But at the time, sewage plants were getting so many calls from guys with 'miracle' chemicals that he was dismissed as another snake oil salesman," says Grubbs.

Then came Lady Luck, in the form of the *Queen Mary*. The venerable ocean liner had just been towed into nearby Long Beach harbor to be transformed into a floating hotel, only to be rebuffed by the grossest of indignities. "The local fire department said her bilges were so full of oil and tar they were a firetrap," Grubbs recalls. "One of the city councilmen knew Jerry and asked about the chances of his formula working on petroleum. Well, it was awfully primitive compared to what we use today, but he went aboard and dumped in a few boxes. In three days the volatile gases were gone, and within six weeks they were able to discharge the contents of the bilge right into the harbor with no toxic risk whatever."

Today, Solmar sells 26 formulations, and it's not unusual for some manufacturers to offer 40. Off-the-shelf formulas resembling dried herbs—usually sold in bags or drums—are designed for such uses as eating animal waste, keeping ponds free of algae and cleaning up gasoline leaked from service station storage tanks. Moreover, bug makers are getting into brand-name advertising and packaging with names like Oil Spill Eater, Alpha Biosea, Marine-D and Bioblend.

Some companies guard their formulas zealously. Carl Oppenheimer, a former University of Texas microbiologist turned manufacturer, talks with the fervor of a big-game hunter as he describes junketing from Russia to Antarctica to Italy collecting microbes renowned for their oil appetites. All are blended, he says, with a special secret nutrient. (It has been rumored that the additive is simply ground from dried cow pies, which, of course, are treasure troves of nitrogen and phosphorous.)

Jones Grubbs says he uncovered a "terrific" product for treating animal wastes because he "just happened to be visiting a pig farm one day and saw this lovely pile out behind the barn. It looked like peat moss and was absolutely odorless," he says. "But the farmer told me it had started out as plain hog manure. Now if you know what hog manure smells like, you'll know the instant respect I had for these particular bugs. I felt as if I'd just discovered some superstar athletes."

But just how do you get one bug to work on oil and another on animal wastes? "It's sort of like drafting a bunch of guys for a football team," shrugs Grubbs. "You try 'em out on the field until you've seen which ones play what positions the best. Take any sample out of the ground and you'll find a consortium of different bacteria at work. In that hog manure, for example, we isolated maybe 25 different types in the lab, then used some standard tests to identify the one type that was the real workhorse."

Finding the Right Stuff is crucial, of course, but there's much more to drafting microbes for the war on wastes. Some bacteria exist only in aerobic conditions (where oxygen is present) and need constant churning, or aeration. Others work only in anaerobic (oxygenless) surroundings, such as the kind that feed on sulfur compounds and give off that infamous rotten-egg smell (hydrogen sulfide). "One of the main reasons people seek out firms like ours," says Grubbs, "is for bugs that don't produce odors."

With other firms, what's salable may be a special nutrient or something else added to the basic formula. Last year, for example, Lockheed Missiles & Space Company announced a "breakthrough" oil slick remedy. The microbe strain involved is actually found in natural seawater, but an additive of powdered clay helps coagulate the spreading oil and keep it from sinking, and a nitrogen fertilizer excites the microbe's appetite.

As bug makers refine and expand their formulas, the range of new applications is growing as well. For instance, the type of bacteria that tamed the farmer's manure pile does the same for the human version in a flushless toilet. Sold by Planit Services, Inc. of Kelowna, British Columbia, for use in rural vacation homes and worksites, the ToileTronic resembles a conventional toilet except that its only link to the outside is a pipe leading to a ceiling vent. Below the seat, a stainless-steel chamber contains peat moss, which can hold twice its own weight in liquid. When the toilet is in use, a fan vents odors up through the ceiling pipe while an automatic rotator mixes the peat moss and human waste. Thanks to a built-in aerator (ever the key to keeping aerobic bugs robust), the result is what Planit's Raymond Perry calls "a harmless, virtually dry residue." A typical family of four, he says, can collect a month's worth of residue in a shoebox-size plastic bag and "toss it in the garbage or sprinkle it on the yard."

The same basic concept works on a grander scale for the palm Beach County Solid Waste Authority in Florida. Faced with dwindling landfill space and a state law that mandates recycling 30 percent of its waste by 1994, the county recently opened a $1.3 million facility to compost some 30 tons a day of sewage sludge and yard clippings. One of perhaps 20 like it nationally, the Palm Beach facility features four 250-foot-long troughlike bays under one roof. There, automation helps microbes munch out in minimum time. An agitator mixes and moves the waste materials down the long bays while blowers force air through them to keep the bugs frisky. The result, 21 days later, is an almost odorless, fluffy soil conditioner that the county uses on its own parklands and sells to garden shops.

Small businesses are beginning to discover microbe muscle as well. Until passage of the tough new Clean Air Act (CAA) amendments in 1990, the only toxic industrial emissions policed closely by federal regulators were those from potentially big troublemakers like paint and pharmaceutical manufacturers. Today, the CAA is forcing thousands of bakeries, printers, dry cleaners, auto body shops and others to filter out fumes deemed to be volatile organic compounds (VOC).

Some VOC control systems can cost $1 million-plus to install for a company that might not gross that much in a year. One cheap alternative, already flourishing in Germany and the Netherlands, is ideal for facilities that face moderate—but not severe—VOC emissions or odor problems. The gases are routed outside the building, humidified and channeled through a sort of horse trough that might run 100 feet in length. Positioned along the bottom are aerators. Piled on top are peat moss, wood waste or compost—loaded with bacteria. As the vapors seep up along the filter bed, target pollutants become a wet film covering the filter particles. Once trapped like that, they're attacked by the bugs and converted to the usual $CO_2$ and water.

Still, the big show for bioremediation is the hazardous waste cleanup site, usually one of the 1,200 falling under the EPA's $10 billion Superfund program. It may belong to an active petrochemical complex; it may be an abandoned "midnight" dump site or a former hazardous waste landfill teeming with everything from drums of solvents to discarded electrical transformers. Before the *Exxon Valdez* spill, the closest thing to large-scale bioremediation work was "land farming," in which a petroleum refinery would dump piles of oil wastes out on the south forty and churn them every so often to provide aeration. Time simply wasn't a factor in completing the job. Today, it's very much on the mind of every Superfund site manager who has to worry about pressure from local environmental action groups or how long he can afford to keep his leased equipment. His reaction to bioremediation is usually, "Thanks for the bugs. Now how can we make them work even faster?"

The upshot has been perhaps a hundred ingenious innovations. To wit, one way to make bacteria work more efficiently is to control their conditions. A cold snap, for example, can make bugs left in groundwater as sluggish as the rest of us. But pump the contaminated water up into a large heated tank and you can make it as hospitable as a cow's tummy. This bioreactor also allows you to monitor the rate of consumption and add nutrients or new bacterial battalions if needed. BioTrol Inc. of Chaska, Minnesota, has installed its patented bioreactor at several sites to remove everything from solvents and gasoline to the nasty wood preservative pentachlorophenol. BioTrol's reactor is filled with a highly porous packing material to which the microbes adhere. When air is supplied in the form of fine bubbles, the bugs are positioned like spiders to trap and dine on anything that floats by.

Typical EPA nightmare: an underground gas plume of volatile organic compounds threatens to waft into an aquifer that supplies drinking water. Conventional treatment calls for sinking small shafts, perhaps in the hundreds, to trap the VOCs and route them to expensive extraction equipment. But it's time-consuming and still leaves much of the VOC mass underground. Another way: dig a few narrow wells and force air through them. Underground, the VOCs are forced to rise to the surface, where most are captured via extraction wells. Meanwhile, the hollow spaces beneath the surface, now filled with fresh air, become dining rooms for bacteria. Bugs and nutrients, pumped in through the same air wells, serve as a sort of janitorial crew by filtering out the remaining VOCs.

Suppose the pollutants aren't in groundwater but imbedded deep in the soil. In remediation's medieval period—say pre-1990—conventional wisdom called for excavating the toxic soil and trucking it off for incineration, often at $400 a ton. Today it's possible to treat the soil in place for $80 to $100 per ton with technologies like the one used by In-Situ Fixation of Chandler, Arizona. Its five-foot-diameter twin-auger system consists of a pair of oversize drill bits fixed to a tractor for easy access to job sites. As the augers twist to depths of more than 100 feet, they inject and homogeneously mix in a potion of microbes and nutrients that have been poured down their hollow shafts.

As these and other innovative bioremediators establish their speed and cost effectiveness, the more they're spurred on to tackle more chemically complex forms of pollution. "Over 90 percent of the chemical substances classified as hazardous today can be biodegraded," contends Tom Zitrides of Bioscience Incorporated in Bethlehem, Pennsylvania. At an Air Force base near Fairbanks, Alaska, a remediation team is bioventing jet-fuel-saturated soil. The ECOVA Corporation in Redmond, Washington, employs bugs in a bioreactor to attack creosote (the tarlike wood preservative) found throughout 100 acres that had been home to a lumber-treating plant. General Electric and others are making headway in harnessing bugs to destroy polychlorinated biphenyls (PCBs), used in liquid form to insulate pipes and electric transformers until they were banned as carcinogenic in 1977. Lab tests have even shown promise in destroying such public enemies as zinc oxide, a common byproduct of electroplating, and TNT, which is strewn all over military proving grounds.

Just how big can bug cleanups get? Perhaps the largest project to date is the 40-acre site laden with Bunker C marine fuel that ECOVA took on near Los Angeles. By churning and aerating some 300,000 cubic yards of the thick, gooey stuff in stages that took ten weeks each, the contaminant levels gradually dropped 85-90 percent.

**The Midnight Dumpers' Toxic Cocktail**

More typically, a site may contain a smorgasbord of contaminants, challenging bug makers to brew teams of different microbes. When Allegheny County, Pennsylvania, earmarked 1,200 acres for its newly opened airport terminal, they included 4 acres that had been capped after being strip-mined in the 1890s and then used as a municipal garbage dump for another 50 years. When Waste Stream Technology (WST) and Sevenson Environmental Services won a contract for the cleanup job, they soon discovered that the alleged household-waste landfill had also been visited by anonymous industrial dumpers in the wee hours. "We found 15 compounds," says WST president Mike Barnhart. Typical of their initial levels were a horrific 21,300 parts per million (ppm) for oil and grease; 8,200 ppm for naphthalene (the main ingredient in mothballs until banned); 1,900 ppm for xylene (a paint and gasoline ingredient); and 480 ppm for the solvent toluene.

The remediators brought out a trailer-mounted bioreactor that, says Barnhart, "allowed us to ferment different formulations as if we'd been back home in the laboratory." After the soil was dug up and tilled by mixing in bugs, air and nutrients, Barnhart watched it change "from a black, refuse-filled soil to a brown, highly cultivatable soil with a much-reduced odor."

Within two months, naphthalene, xylene and toluene had all been reduced by more than 98 percent, and oil and grease by more than 80 percent. Why the difference? The first three groups all have simple molecular structures like small bubbles—bite-size snacks for bugs. Oil and grease have long molecular chains that stretch a meal to several courses. Even so, the remediated levels are considered low enough to keep the oil and grease from migrating to streams or deeper soil levels.

But the possibilities don't end there. Researchers are already exploring the outer edge of the microcosmos and finding awesome prospects for further restoring the natural order of things. Here are three areas already showing progress:

•Microbes can make some compounds more "user-friendly." Although coal may be our most plentiful fossil fuel, most of the nation's vast Eastern reserve can't meet clean air standards because it emits too much sulfur when burned. The aromatic compound dibenzothio-thene (DBT) clings tenaciously to hydrocarbon molecules and becomes sulfur dioxide when ignited. But just recently the Chicago-based Institute of Gas Technology patented a bacterial strain that eats the DBT (90 percent of it so far in lab tests) while leaving the hydrocarbons intact.

•Microbes can be genetically engineered, reducing the need for toxic chemicals. In the fall of 1991, the EPA approved the first genetically engineered pesticide. Called Cellcap, it incorporates a gene from one microbe that produces a toxin deadly to potato beetles and corn borers into a thick-skinned microbe that is hardier. Even then, the engineered bacteria are dead when applied to crops.

•Microbes—in roundabout ways—can help attack stubborn metals and radioactive wastes. Microbes have been used for decades to concentrate copper and nickel in low-grade ores. Now scientists are exploiting the fact that if certain bacteria are given special foods, they excrete enzymes that in turn break down metals and minerals. Researchers at the U.S. Geological Survey, for example, found that two types of bacteria turn uranium from its usual form, one which easily dissolves in water, to another, one which turns to a solid that can be easily removed from water. They may be able to do the same for other radioactive inhabitants of the global waste pile.

Ironically, the benefits of biotreatment seem to be bypassing that king of all trash heaps, the municipal landfill. The reason is that the warm and wet environment that bugs like isn't compatible with conventional landfill wisdom. We line trash pits with layers of polyethylene to prevent leakage into aquifers, then cover them each day to keep neighbors from griping about odors. The result, of course, is that you can still read newspapers dumped in 1951 (SMITHSONIAN, JULY 1992).

That, too, may change. A few innovators are designing wetter landfills so the bugs can do what they do best. In the process, the microbes produce methane, the chief ingredient of natural gas, which can be captured for use as a cheap, clean energy source.

Bioremediation is still in its infancy, and there is no telling how far it can go. Genetic engineering, by which the best features of several bugs would be combined in one, could multiply the potential benefits. Whatever happens, it appears a safe bet that microbes will be the Marines of future environmental cleanups. ❑

**Questions**:

1. What is the name of the first genetically engineered pesticide? What does it kill?

2. Why is contaminaged ground water sometimes heated up?

3. How can microbes help clean up air pollution such as VOCs?

Answers on page 301

38

*The costs of cleaning up toxic waste sites in the U.S. are truly staggering. The estimated 9,000 Superfund sites may cost up to $500 billion. When added to the cleanups planned by all federal agencies, the total federal cost is about $1 trillion, which is over four times the annual U.S. defense budget. Nor does this cost include the many state and locally funded cleanups. Unfortunately, much of the money designated for cleanup is not spent on cleanup but is wasted in legal fees and unneeded studies. Only about 12% of Superfund monies is actually spent on cleaning up the site. A main reason for this is the inefficient way that the thriving cleanup business is set up. Many businesses have no incentive to save money. A solution suggested by many is to allow more public input into the decision-making process. This would help break up the current bureaucratic network now monopolized by a few regulators and large corporations.*

# Can Superfund Get on Track?

## While poisons still seep and drift from the nation's waste sites, lawmakers decide the fate of our most ambitious clean-up legislation

by Karen Schmidt

Fourteen years ago, the future of 140 miles of Montana's Clark Fork River looked as bleak as any science fiction writer could have imagined. Forming the largest polluted area in the nation, some 50,000 acres of the river corridor were littered with arsenic, copper, zinc, cadmium and lead. Heavy rains swept these toxic metals into the river and periodically killed massive numbers of fish. Toxic dust denuded a forest where otters, deer, elk and osprey once lived, leaving it a barren moonscape. The cause? More than a century of extensive and poorly regulated mining and smelting activity.

Then came Superfund, formally known as the federal Comprehensive Environmental Response, Compensation and Liability Act (CER-CLA)—an ambitious plan to finance the cleanup of the nation's most hazardous abandoned dumps. Now, the Clark Fork region comprises four separate Superfund sites. Companies testing waste treatment strategies have sprouted around Butte, Montana, and more than $100 million worth of cleanup is underway. Tailings are being scooped from stream banks in Butte, tainted soil is being removed from the town of Anaconda, and a clean new drinking-water system is operating in Milltown.

But for all that, surprisingly little has changed. The hazardous material still abounds, and people and wildlife are still suffering from the pollution. As residents who live near Clark Fork can attest, cleaning up is hard to do. Here, on a grand scale, can be found the pitfalls of Superfund—variations of themes playing at most of the country's Superfund sites, including a history of lawsuits upon lawsuits, cases of poor management and confusion about how best to proceed. As for finances, most of the millions spent in the Clark Fork area, as at other sites, have gone until recently to solving legal disputes and conducting preliminary studies, while the pollution persists.

This year, as Superfund makes its way through reauthorization in Congress, just about everyone involved is voicing dissatisfaction—government officials, industry representatives and environmental organizations alike. Even President Clinton has said flatly, "Superfund has been a disaster."

It was not supposed to be this way. Recalls Congressman Al Swift (Democrat, Washington), chairman of the subcommittee on Transportation and Hazardous Materials, "Superfund was passed in 1980 to address what many believed to be a serious, but relatively limited, problem. The EPA was instructed to find the 400 worst hazardous sites, known as the National Priority List (NPL). Most believed that cleaning up a site was relatively inexpensive and involved removing containers or scraping a few inches of soil off the ground."

Now, legislators envision the Superfund program as lasting for decades. So far, 221 long-term cleanups have been completed and 55 have been removed from the list. But the National Priority List now names more than 1,192 sites and is expected to continue expanding. To clean up the sites presently on the list will cost more than $28 billion, EPA estimates.

This year, Congress will try to identify Superfund's successes, diagnose its ills and come up with a corrective prescription: amendments to the law, changes in EPA administration or the much more unlikely creation of a new kind of public works program.

At the heart of the legislation is the protection of both human health and natural resources. At Superfund sites, EPA evaluates public health threats. One example is the exposure of Milltown residents to metals in their drinking water. However, the law assigns the duty of defending natural resources to other federal agencies, states and Native American groups. For example, the state of Montana is assessing how toxic pollution around Clark Fork has harmed wildlife and water resources. The state is suing ARCO, the multinational giant responsible for the site's cleanup, and hopes to one day coordinate a plan for revitalizing the ecosystem.

Congress must also decide how much money to authorize EPA for funding enforcement of the law. The agency orders companies that contributed to the pollution to pay for a site's restoration. When EPA can find no responsible party, it draws money out of the "Superfund," a trust fund of taxes collected for the most part from the chemical and petroleum industries.

The question of the law's fairness is also generating considerable debate. Industry says it forces big companies to pay too much—in the form of taxes, defensive litigation and clean-up costs. "We're not fighting the polluter-pays concept," says Morton Mullins, Vice President of Regulatory Affairs for the Chemical Manufacturers Association (CMA). "But we're against the polluter paying and paying and paying."

CMA, the American Insurance Association (AIA) and other industry groups are urging Congress to convert Superfund to a public work program funded by taxpayers—or, at the very least, to overhaul the rules for determining who pays. Industries blame the current rules for gumming up the works with too many lawsuits (which they themselves often initiate). As the AIA explains in its Superfund briefing book, companies "challenge EPA clean-up decisions throughout the investigation and remediation process in order to protect themselves."

Environmentalists believe the law's mandate to collect funds whenever possible from polluters, called "joint and several liability," is the best solution—and the least likely to further burden taxpayers. But that still leaves the way EPA administers the program. "This program has become bureaucratized to the nth degree," says Erik Olsen of the Natural Resources Defense Council.

Despite the criticism and the dire statistics, so far the nation has no better solutions for cleaning toxic dumps. Fortunately, Superfund's very existence has lit a fire under companies to reduce waste. From 1987 to 1991, the chemical industry cut releases of toxic chemicals to the environment by 35 percent, largely in response to Superfund. Says CMA's Mullins, The threat of liability strengthens the commitment to pollution prevention."

Referring to the tragic toxic dump in a residential area that first provided the impetus for Superfund, EPA Administrator Carol Browner has credited the program with saving "thousands of communities all over the United States from become Love Canals." Under Superfund, EPA has responded to 3,300 emergencies that involved safely removing chemical dangers such as leaky drums of hazardous waste and radioactive medical materials improperly stored in urban warehouses.

Until the act passed in 1980, the poisoned Montana region and hundreds of other sites had little or no hope of reclamation. In a sense, the first Superfund cleanups have been pioneer projects. Site managers at Clark Fork had to develop their own safety standards for lead in soil, for example, because no general federal guidelines were in place. Says Robert Fox, EPA's Clark Fork Superfund coordinator, "We've had to break new ground in many areas." Indeed, the future of Superfund may well depend on whether the program can learn from the lessons of these pioneer sites. Take the case of New Bedford, Massachusetts. From the mid-1940s to 1978, factories that manufactured electrical capacitors spewed tons of wastes laden with polychlorinated biphenyls, or PCBs, into the town's picturesque harbor. The cancer-causing chemical has settled in sediments and entered the food chain. As a result, fiddler crabs have declined in number and soft shell clams are not reproducing well. One study found that the PCB levels were highly toxic to some members of all major groups of organisms. In 1979, the Massachusetts Department of Health banned fishing and seafood harvests from the harbor, but that hasn't stopped the tides from transporting PCBs even farther into the ecosystem.

Today, New Bedford residents are inhaling small, but not insignificant amounts of PCBs, says Gayle Garman, EPA's sixth and current project manager for the site. The chemical evaporates from the harbor and is carried by the ocean breeze. EPA posted danger signs a decade ago near the most heavily contaminated area, a 5-acre zone called the "hot spot," which contains an estimated 120 tons of PCBs. But people still catch fish there and teenagers drink Cokes on the banks.

After New Bedford harbor joined the Superfund list in 1982, EPA took eight years and $25 million to identify a relatively safe way to remove the "hot spot" sediment without dispersing more PCBs. In 1990, the agency made its recommendation: Carefully dredge the sediment, pump it directly onshore and incinerate it in New Bedford.

As engineers prepare to run a trial burn this spring, they face bitter local resistance to the plan. Though the goal is a permanent reduction in risk, during the four-month burn there will be a slight increase in risk of exposure to emission of toxics such as dioxin. "This is a Catch-22 situation," laments Jackie Duckworth, a young mother who lives near the harbor. "We can't let the PCBs sit there, but it's dangerous to take them out." Angry local citizens and members of Congress have demanded that an allegedly safer (but untested) technology for burning hazardous waste, called EcoLogic, be tried. According to EPA, the risks associated with the burn are very small, and the EcoLogic option would further delay the cleanup by several years.

As for the harbor's wildlife, the national Marine Fisheries Service (NMFS) emerged victorious from one of the first resource damage cases ever filed. The landmark case, which took 10 tortuous years, should now pave the way for other resource damage claims, says Thomas Bigford, chief of NMFS's Habitat and Resource Protection Division in Gloucester, Mass. To restore New Bedford's wetlands and marine environment, however, could take another 13 to 20 years, he estimates.

In a very different setting, downtown Pensacola, Florida, is marred by two Superfund sites and a third one waiting to join the list. At that last site, the Escambia Treating Company, the low-income minority community has been critical of the emergency action taken there. A local group called Citizens Against Toxic Exposure complains that the opinions of residents have not been respected by federal officials. President Margaret Williams quips, "When EPA talked to us, they acted like they were doing us a favor by cleaning up a site." Tillman McAdams, EPA North Florida site assessment manager, points out that the agency has removed the immediate threat to the town's drinking water. "We're working as hard as we can to address the problem," he says. "That's our job."

Across town, several thousand people live within a mile of one of the two listed Superfund sites, American Creosote Works, where a plant once treated telephone poles with wood preservatives. After years of dumping creosote, which contains cancer-causing hydrocarbons, and a compound called pentachlorophenol (PCP), which degrades to toxic dioxin, the company declared bankruptcy. The money for cleanup now comes from the federal Superfund pot and the state of Florida.

In 1983, EPA took emergency action there to reduce the community's potential exposure to the chemicals. Workers drained contaminated lagoons, dried the toxic sludge and then capped it with a layer of impermeable clay. Now, ten years and five project managers later, the "temporarily" capped piles still sit on a lot bigger than a city block. According to Mark Fite, EPA's project manager for the site, cleanup has been slow because federal and state agencies have had to reach agreement every step of the way. "It's not technical problems that are slowing things down here," he says. So far, a plan has been approved for treating the site's groundwater contamination. Plans are still on the drawing board for treating the 100,000 cubic yards of toxic sludge and soil.

Suburbs too have their share of Superfund sites. In Southern California, a 145-acre unlined gravel pit in Monterey Park began in 1948 to accept all kinds of industrial and residential wastes. When the Operating Industries, Inc., (OII) landfill closed in 1984, houses had been built next to the resulting pile. In places it was 300 feet deep, garbage rotting along with a toxic brew of more than 300 million gallons of hazardous liquids. When the site joined the Superfund list in 1986, toxic vapors and liquids were migrating toward neighborhoods, and the steep slopes of this poisonous mountain—which now had the Pomona Freeway cutting through it—threatened to give way.

Since EPA took emergency action to stabilize the slopes, 180 companies that contributed to the dump have settled with the agency and are now leading the cleanup. They're working toward controlling the spread of the hazardous liquids, the landfill gas and the groundwater contamination. Engineers have nearly finished their design for a huge multilayer cap that will cover the site. Although cleanup has moved along at a good pace, the OII landfill will require decades of monitoring and maintenance, according to Roy Herzig, an EPA project manager.

Cleanup often requires that level of attention. Along with that difficult lesson has come a greater understanding of the magnitude of the tasks at hand. One hundred kinds of chemical pollutants exist in various combinations and mixtures at Superfund sites. The substances are poisoning settings as diverse as mountain streambeds, coastal harbors, urban blocks and suburban neighborhoods. And yet only a limited number of technologies can be counted as ready tools for cleaning up the contamination.

For destroying toxic materials, only incineration has a proven record of effectiveness and high standards for safety. But even though it may be the only option available for treating highly concentrated and hazardous substances, incineration has become unpopular, as EPA's Garman has found in New Bedford with residents who fear the slight risk of exposure to toxic emissions. "I'm concerned that because of politics we may lose this technology," says Garman, an environmental engineer. "We need it in our bag of tricks to clean up the environmental messes out there."

One promising new alternative for treating some contaminated soils is bioremediation, an inexpensive strategy in which microorganisms break down certain toxic chemicals into harmless substances. However, bioremediation has yet to be critically evaluated and refined for use in the field. So far, the technique does not appear to work on metals, highly concentrated chemical wastes or complex mixtures of toxic substances. At the American Creosote site in Pensacola, bioremediation failed to detoxify the sludge. Although the organisms readily digested the simpler organic chemicals in the mixture, EPA found that the microbes did not degrade more than 30 percent of the carcinogenic compounds.

Another lesson comes from the program's sometimes poor administration, which EPA has been working to improve. After studies found that three of every five African Americans and Hispanic Americans live near toxic dumps and that Superfund sites in minority areas took as much as 42 percent longer to clean up, the agency opened an Office of Environmental Equity in November 1992. In June 1993, EPA also outlined a model for accelerating Superfund cleanups. The plan calls for a single comprehensive assessment of a site and "cookbook" remedies for certain common and well-studied contamination problems, such as wood-preserving chemicals in soil.

The agency is also rounding up more polluters these days. According to Bruce Diamond, director of EPA's Office of Waste Programs Enforcement, EPA now gets companies to pay for more than two-thirds of Superfund cleanups—double the rate five years ago, when the majority of projects drew from the trust fund.

Fixing Superfund, however, may take more than vigorous enforcement and other administrative improvements at EPA. "We've acknowledged that administrative changes are not enough; the law will probably be changed," says Timothy Fields, director of EPA's Superfund Revitalization Office.

EPA Administrator Carol Browner has said she is committed to the law's polluter-pays principle but has expressed interest in changing other parts of the legislation. She may, for example, propose that Congress require EPA to actively involve communities early during cleanups and set national clean-up standards.

Though Congress hopes to set the federal program on a new and accelerated course, streamlining Superfund, like reclaiming toxic dumps, won't be easy. Adding and removing clauses to the already dense and complex law could even slow cleanups at first as site managers adjust to the changes.

Even so, cleanups are finally gaining momentum at many sites, including Clark Fork. There, ARCO and the community of Anaconda are preparing to build a world-class golf course where a mountain of toxic tailings once stood. And negotiations between ARCO and the state of Montana on restoration of the ecosystem are wrapping up. Some day, maybe everyone involved with Superfund will be able to agree with D. Henry Elsen, EPA's Clark Fork legal coordinator, who says. "The Superfund horror stories are mostly from years past."

**Questions**:

1. What is the name of a cheap way to treat soils that uses microbes to break down toxic chemicals?

2. How many sites are on the National Priority List? How many long-term cleanups have been completed?

3. What town is coping with tons of PCBs in its harbor?

Answers on page 301

*The U.S., along with many other industrialized nations, is running out of landfill space. Increasing safety and environmental regulations and increasing land prices have led to skyrocketing prices for new landfills. A new landfill costs $500,000 per acre. As a result, only 364 landfills opened up between 1986 and 1991, while 2,200 landfills were closed, mainly because they had filled up. One solution to this problem is incineration of waste, which is what currently happens to about 10% of the waste produced in the U.S. This not only reduces waste volume to 1/10 its original size, but the heat of combution is often used to heat buildings or generate electricity. Unfortunately, incineration produces toxic air pollution as well as toxic ash that must be disposed of. Nearly all environmentalists therefore agree that recycling and especially source reduction are the best ways of decreasing waste in the "waste stream." Source reduction means buying products with less packaging and finding other ways to reduce the volume of the things we consume. Recycling is when we reuse cans, bags, and any kind of waste in some useful way instead of discarding it. The average American produces 6 pounds of waste per day, more than the average for any other country. Furthermore, this amount is climbing, so there is a strong need for emphasis on source reduction.*

# Waste Not, Want Not

## As landfills fill up, a new breed of waste experts is finding ways to reduce the flow of trash

by Thomas A. Lewis

The nightmare came true for Richard Weber on May 29, 1992. As director of the Department of Environmental Resources in Loudoun County, Virginia, Weber had been put in charge of the county landfill early in the year. The landfill, serving an extended, 60,000-population suburb of Washington, D.C., was due to be full in about 14 months, "Trash," Weber says ruefully, "is a pretty unforgiving deal. You never have as much time as you think.

"I can live on the edge as well as the next guy," says Weber, "but I like to know exactly where I am. I'm a soil scientist by trade, and I live by maps. So I had a complete survey done of where we were." On May 29, the survey results reached his desk. The landfill would be full not in 14 months, but in 28 days.

By wangling some dispensations from the state and money from the county board of supervisors, Weber was able to extend the landfill by one acre. But he could not negotiate more time. At one point, he recalls, he and some fellow office workers found themselves "exhuming part of the landfill with shovels," in order to splice a new liner—plastic that prevents groundwater contamination—to the one in use. Their labor completed, the landfill extension opened for business with just two days to spare. And Weber has now arranged to keep his county's trash buried until, perhaps, 1996.

When asked to predict just when the next crisis will occur, Weber uses words new to the vocabulary of waste managers. "That depends," he says, "on how well we do with waste avoidance." Waste avoidance. Otherwise known as source reduction. Neither term has much of a ring for the rest of us. But they have become music to the ears of a new breed of waste experts. And more and more government officials who are either imagining or experiencing Rick Weber's nightmare are paying attention. Even the U.S. Environmental Protection Agency has made source reduction its top priority for solid-waste management.

The oldest source-reduction program in the country, Seattle's, is less than 12 years old. Minnesota's program, one of the best, is just over 3 years old. When Ken Brown, a former U.S. Forest Service geologist, applied for the job of source-reduction specialist with the Minnesota Office of Waste Management, he recalls, "I didn't have much experience. But neither did anybody else." The field he is helping invent is critical, says Brown, "because source reduction can actually move us forward, to a better standard of living, toward sustainable economic development."

All over the country, landfill managers like Weber are finding that space is running out for the endless tidal wave of trash they endure. According to the U.S. Environmental Protection Agency (EPA), Americans discard 4.3 pounds of trash per day for every man, woman and child alive. That's without taking into account debris from construction sites or sludge from sewage treatment plants. When *BioCycle* magazine, respected industry-wide for its exhaustive annual survey of state solid-waste officials, added in both of those additional waste sources, the figure jumped to a whopping 6 pounds per person—200 tons per day in Loudoun County alone.

The average person is likely more familiar with the idea of waste reduction than source reduction. In the semantics of the world of trash, waste reduction has for years meant burning, recycling or composting. Source reduction, on the other hand, means drying up the stream. In the words of Diane Wesman, director of Minnesota's Office of Waste Management, "The best way to manage waste is to not create it in the first place."

She ought to know. With aggressive policies, Minnesota has achieved one of the lowest landfill rates in the country. The state's strategy includes grants to local governments and businesses for finding ways to generate less trash. Elsewhere, less successful experiments have run the gamut from vague pamphlets urging consumers to buy less packaging to specific laws such as the Suffolk County, New York, ban on polystyrene food containers, quickly invalidated by a challenge from the plastics industry.

Yet a consensus is emerging, and one elegantly simple tactic—charging for bags of garbage, called "pay as you throw"—is yielding early dramatic results. But first, to fully appreciate the new discipline, one might consider the sobering lessons of the American dump (rechristened the sanitary landfill after managers learned how to get rid of the most noxious odors and obnoxious rodents). The most basic lesson is that landfills, of course, eventually fill. According to the National Solid Wastes Management Association, more than half of U.S. landfills will run out of room in fewer than ten years—and the deadline in eight states (including Virginia) is in no more than five years.

Establishing new landfills, with today's strict safety regulations and high land prices, costs an average of $500,000 per acre. And these days, proposals to locate a new landfill just about anywhere draw fervent opposition. Between 1986 and 1991, 2,200 landfills closed, while only 364 opened.

The EPA recommends that local governments practice both waste reduction (by incinerating 20 percent and recycling 25 percent of the materials already discarded) and source reduction (decreasing by 25 percent the amount of trash being discarded). The goal: sending only about 30 percent of the waste stream to landfills. By *BioCycle*'s count, 32 states currently bury 80 percent or more. One of the major tools for waste reduction is the incinerator, able to quickly reduce trash to one-tenth its volume. Incineration is much reviled by the public and distrusted by environmentalists—but admired by waste managers.

America currently sends 10 percent of its waste to 171 incinerators, most of which generate electricity or steam. But combustion emits pollutants into the air, including such persistent toxins as dioxin and furans, and the remaining ash can be highly toxic. Small wonder that, despite EPA recommendations that the country double the amount of waste "managed by combustion," proposed incinerators tend to stall in the early planning stages after they inflame opposition.

In stark contrast with burning and burying, recycling is increasingly popular with the general public. A 1992 poll conducted for the Grocery Manufacturers of America estimated that 64 percent of Americans are regularly recycling aluminum, 56 percent are recycling newspapers and more than 40 percent are recycling plastic and glass.

While environmentalists are enthused, waste managers are ambivalent about the steady increase in recycling. In 1991, according to *BioCycle*, though recycling reclaimed 14 percent of the waste stream, the surge in the supply of recovered resources overwhelmed the ability of industries to process and reuse them. Boston, for example, instead of profiting from its newsprint recycling, *paid* $26 per ton or more to have the stuff hauled away.

Still, recycling is here to stay. It is reducing the waste flow, saving energy and resources and creating jobs. According to ALCOA (Aluminum Company of America), in 1991 aluminum recycling employed 30,000 people—twice the number working in aluminum manufacturing. Though this particular model may not widely apply to other forms of waste, it gives hope to managers. "The supply revolution is well under way," Phil Bailey of the Buy Recycled Business Alliance has said, "but the demand revolution is just beginning."

Recycling can work for organic as well as manufactured materials. Increasingly, municipalities are banning yard waste from landfills, and 2,000 composting facilities in 46 states now turn leaves and grass clippings into soil conditioner. The city of Houston, Texas, hopes to compost as much as 30 percent of its 600,000 tons of trash per year at a location not far from the Astrodome.

But even if every community met EPA guidelines for alternatives to landfills, fully one-quarter of today's trash stream would remain. And the only way to staunch it is to alter a ubiquitous and mundane feature of American life: packaging, which makes up one-third of our waste. In the United States, according to *State of the World 1992* by the nonprofit Worldwatch Institute, "as much as half of all paper production and nearly a quarter of all plastics sold go into packaging." Reducing packaging will involve changing habits and challenging assumptions at every step. The stuff protects products during shipping, presents merchandise attractively in the retail store and provides a uniform shape for stacking.

The EPA is quick to point out that not all the news about corporate America's waste habits is grim. EPA source-reduction coordinator Carol Wiesner is collecting such success stories as the Michigan furniture manufacturer (Herman Miller, Inc.) that is saving $1.4 million a year using reusable and recycled shipping containers; and the switch by the McDonald's fast-food restaurant chain from its polystyrene clamshell hamburger container to paper wrap, with a consequent 70 percent reduction in its waste stream.

As for the EPA's ability to measure the success of its initiatives, ironically Wiesner blames the Paperwork Reduction Act for thwarting EPA's ability to survey local governments. That's one reason the *BioCycle* magazine survey is widely used for assessing trends in waste disposal.

The survey reveals great disparity among the states. Many have recently taken a more active role in waste management, motivated perhaps by the knowledge that if the localities are overwhelmed, the states cannot escape responsibility. Some have set mandatory targets for waste reduction, primarily by recycling, and nine have passed "bottle bills" that impose a refundable surcharge on beverage containers. Individual states have banned specific forms of packaging, such as nonrecyclable juice boxes and plastic six-pack carriers.

Minnesota, as a counterpart to its low landfill rate (39 percent), has one of the highest recycling rates (31 percent). At the other end of the qualitative scale, Wyoming landfills 97 percent of its trash and recycles a mere 3 percent. The other 48 states are somewhere between these two.

Measuring source reduction is another matter. You can weigh, sort and analyze trash, but how do you measure the absence of trash? Perhaps the school curricula on source reduction in California, Ohio and several other states and the pamphlets and demonstration projects in Minnesota, Vermont, Pennsylvania and elsewhere, are working. How would we know? "All we can do," says Gene L. Mossing, solid-waste manager in Olmstead County, Minnesota, "is put it out there and hope."

So far, the "pay as you throw" strategy is producing perhaps the most dramatic, immediate, measurable results. It replaces the usual flat rate for garbage disposal with a charge for each bag or can of refuse. Seattle pioneered the idea in 1981, replacing its $18-per-month fee with a charge for each can of garbage. Now, the average Seattle household puts out one can per week, down from three and a half cans in 1981. When authorities in Perkasie, Pennsylvania, began charging residents about 4 cents a pound to dispose of trash, the town saw its waste stream drop by 60 percent. Overall in the past several years, the number of localities using the technique has grown from 20 to more than 200.

Despite the isolated success stories, however, most Americans continue to discard a dismaying amount of waste. "We've been making decisions as if our resources were unlimited," says Minnesota's Ken Brown. "It's time to come to terms with the consequences of waste." ❑

**Questions**:

1. The EPA recommends that local governments send only about what percent of the wast streat to landfills? How much do they recommend for source reduction of the waste stream?

2. What are some advantages of recycling? According to ALCOA, does aluminum recycling employ more people than aluminum manufacturing? If so, how much more?

3. What is the "pay as you throw" strategy? How much was wast reduced in Perkasie, Pennsylvania, when this tactic was used?

Answers on page 301

# 40

*Asbestos is a term applied to certain fibrous minerals that have been used primarily for their heat-resistant properties. Asbestos has been implicated as a cancer-causing agent and is therefore a potential environmental concern. But not all of the fibrous minerals included under the term asbestos, including the most commonly used mineral, have been shown to cause cancer. Because of the high cost of removing all asbestos, which may be as much as $100 billion, knowledge of the particular mineral comprising the asbestos is necessary in dealing with this environmental problem.*

## Asbestos

Abatement of asbestos may cost upwards of 100 billion dollars cumulatively. Over 90 percent of all asbestos used in the U.S. consists of a mineral which current research indicates poses minimal health risk, except under conditions of heavy industrial exposure.

by The American Institute of Professional Geologists

**The Substance and Its Uses**

Asbestos occurs primarily in metamorphic rocks. "**Asbestos**" was once a loosely used commercial term for fibrous minerals used in heat-resistant fabric. "Asbestos" has a newer meaning as a group of a few fibrous minerals that are specified as health hazards by federal statutes. Asbestos includes two very different mineral groups—sheet minerals and amphiboles.

The sheet mineral group includes **chrysotile** (a hydrated magnesium silicate) which comprises about 95% of asbestos used in commercial and industrial applications. Chrysotile appears under the light microscope as thin needles, but a scanning electron microscope reveals that these fibers are actually made of tightly coiled sheets—much like a roll of gift-wrapping paper. Chrysotile occurs in metamorphic rocks which were produced when hot subsurface (hydrothermal) waters reacted with magnesium-rich igneous rocks. Some areas of the world where chrysotile has been commercially mined include Quebec, South Africa, Russia, Italy, Cyprus, Rhodesia and, in the United States, the west flank of the Sierra Nevada Mountains, Arizona, New York and New Jersey.

The asbestos minerals of the amphibole group occur as true needles, not coiled sheets. Those specifically regulated by federal law include **amosite** (also called *brown asbestos*—an iron-magnesium silicate), crocidolite (also called blue asbestos—a sodium-iron silicate), anthophyllite (a magnesium iron silicate), **tremolite** (calcium magnesium silicate), **actinolite** (calcium magnesium iron silicate) and **ferroactinolite** (calcium iron silicate). Together these make up only about 5% of commercial asbestos.

All of these asbestos minerals have a naturally fibrous mode of occurrence and tend to separate easily into minute fibers. The fibers themselves display flexibility, high tensile strength, heat and corrosion resistance, and physical durability. These qualities have made asbestos minerals important to a broad spectrum of industrial and home products.

**The Dangers of Asbestos**

Exposure to specific kinds of asbestos and asbestos-containing material has been found to cause severe and often fatal lung diseases. To date, asbestos is documented as a health hazard only if it becomes friable (easily torn apart and dispersed) and airborne so that it may be inhaled. Health problems arise because amphibole asbestos fibers cannot be broken down in the lung. Fibers retained in the lung constitute a constant irritant, and the body's reaction over 10 to 40 years eventually produces cancerous cells at the site. Once these cells overcome the body's immune system, rapid mortality usually follows.

Melvin Benarde in his book, *Asbestos, The Hazardous Fiber,* notes that exposure to cigarette smoke has been found to increase the incidence of lung cancers by a factor of 10 to 18 times that of nonsmokers with occupational exposure, and 5 to 9 times that of smokers without occupational exposure. Current (1993) research shows that the best step anyone can take to reduce risk from lung disease is simply to stop smoking. The risk of dying of diseases caused from smoking is about 1 in 5; the risk of dying from asbestos without smoking is about 1 in 100,000—about 1/3 the risk of being killed by lightning.

Demographic studies, such as those carried out in the Thetford area of Quebec, Canada, support the hypothesis that asbestos which is ingested orally does not pose a significant health problem. Even diets that contained 1 percent asbestos (amosite), an astronomically huge dose, produced no enhanced mortality in laboratory rats.

**The Controversy about Asbestos**

Most of the proof of toxicity to humans centers around the amphibole types of asbestos, particularly crocidolite. Statistics are indeed grim for that mineral. One study conducted at the Dana-Farber Cancer Institute in Boston, Massachusetts, showed that, of 33 men from one factory who were involved in 1953 in cigarette filter manufacture that utilized crocidolite, 19 had died of asbestos-associated disease by 1990. Crocidolite is considered the most dangerous form of asbestos. In contrast, chrysotile (95% of all asbestos used commercially) has not been clearly implicated as a hazard except in unusually heavy industrial exposures over long periods of time, but it is nevertheless legally defined as a hazardous substance.

It is difficult to gather data from general workers because there is usually no control on what kinds of asbestos individuals have been exposed to. Working groups such as carpenters, laboratory chemists, plumbers, builders, air conditioning and heating duct installers, or residents of asbestos shingle homes—all of whom have lived daily for most of their lives with chrysotile—do not report heightened mortality attributable to asbestos. Those few studies which have been able to discover and evaluate a population (one example is from Thetford, Quebec) exposed to high concentrations of chrysotile (and no other form of asbestos) show no significant enhancement of mortality. The preponderance of current data indicates that working or attending school in a building that contains the common asbestos, chrysotile, poses little risk. This data is in serious conflict with federal policy which defines chrysotile asbestos as a hazardous substance and mandates its abatement in public buildings. If chrysotile is not actually a serious health hazard, then an economic drain on our economy that may extend beyond 100 billion dollars is taking place to abate a material which is not a proven threat to health or longevity.

Misidentification of asbestos minerals is also a chronic problem. Lack of standards that result in misidentification of samples places huge, unwarranted financial burdens on some clients and sometimes leaves the public at risk. Analysts often consist of graduates from a brief (4 days to one week) short-course, and misidentification becomes more common as poorly trained microscopists enter the profession of asbestos identification. A current (1992) program exists to accredit *laboratories* through test-round samples (through the National Voluntary Laboratory Accreditation Program—NVLAP) but not the *analysts* who perform the identifications. Therefore it is possible to have analysts of both high and low proficiency working in a NVLAP-accredited laboratory. NVLAP test-round samples can receive special attention not received by samples that are routinely processed in large numbers, so the passing of periodic test-rounds by participating laboratories does not guarantee reliability. Because reliability in asbestos identification depends so heavily upon the personal skills of individual microscopists, periodic testing for certification of individual analysts, rather than for an entire corporation's or agency's laboratory, seems reasonable. There is currently (1993) no required qualification, testing or review process to certify every asbestos analyst who actually performs the work.

**Regulation and Asbestos**

Asbestos detection, management, abatement, and disposal are highly regulated. At the federal level, the United States Environmental Protection Agency (EPA) is the organization concerned with protection of the environment and occupants of school buildings. The health and safety of workers in the private sector is the concern of the Occupational Safety and Health Administration (OSHA). Further regulation and control occurs at the state level.

The Asbestos Hazard Emergency Response Act (AHERA) became effective December 14, 1987. This first phase of legislation mandated specific asbestos management procedures for all schools. The second phase, the Asbestos School Hazard Abatement Reauthorization Act (ASHARA), became effective November 28, 1992, and extends to other public buildings. These acts have produced a sudden demand for an entirely new group of asbestos specialists including asbestos analysts and asbestos abatement workers. AHERA also spurred a growing cognizance of legal liabilities (and resulting lawsuits) within the asbestos mining and manufacturing industry. This impacted manufacturers of asbestos-containing materials (ACMs), building contractors, construction companies or other employers who may unknowingly (or knowingly) expose their employees to airborne asbestos fibers.

Current (1993) policy by EPA mandates that both the amphibole and the chrysotile forms of asbestos be treated as identical hazards. This means that liability extends to owners of buildings containing asbestos, to real estate agents selling such buildings, to banks and lending institutions issuing loans for purchasing those buildings, and to purchasers of land upon which asbestos (or other hazardous materials) may be present in building materials. Homeowners who have asbestos inside their homes will, in many states, have difficulty selling their home without first abating the asbestos. Those involved in ownership or transfer of land where some asbestos may have been buried or dumped are similarly vulnerable. ***At present (1993) these legal liabilities exist regardless of whether the asbestos present is benign chrysotile or one of the dangerous amphibole types.*** Potential liabilities have spurred the development of an asbestos abatement industry which includes building inspection and sampling in order to minimize the cost of doing unnecessary building renovation, avert future liabilities, and prevent investments in property that may require major clean-up.

**Roles of the Geologist in Asbestos Abatement**

**In identification.** Because of their unique education in mineralogy and laboratory techniques, geologists should assume the lead role in the identification of asbestos. True asbestos consists of minerals, and only professionals who have substantial formal education in mineralogy and experience in determinative methods for minerals should perform identification work. It is common for unskilled analysts to mistake cellulose wood fiber for chrysotile, or the benign calcium silicate mineral, wollastonite ($CaSiO_3$), for the regulated amphiboles, actinolite and tremolite.

A few other fibrous minerals may yet be added to the list of materials implicated in lung disease. One zeolite (a group of silicate minerals that contain water in their crystal structures), erionite, has been recently implicated in lung disease in Turkey. However, the worry that all fibrous minerals may be serious health hazards appears groundless. Only a few minerals are hazardous, and it is important that they be clearly distinguished and not confused with the much more abundant and diverse group of benign substances.

**In education.** Laymen are unfamiliar with the risks of asbestos and they are unable to distinguish asbestos from other fibers or between the various mineral types of asbestos. Secondary schools should provide information about asbestos as part of the "rock and mineral" section of earth science courses, and students should be familiar with what kinds of materials may contain asbestos. Geology programs at the university level should be aware of the need for skilled optical mineralogists in asbestos monitoring and abatement and be prepared to respond to the market needs. Colleges can prepare graduates for asbestos work by providing special course work in advanced microscopic techniques, asbestos analysis, and microscopy of man-made and earth-related materials for their students. These courses would also provide excellent electives for students who are entering the environmental, waste disposal or industrial hygiene fields.

**In developing public policy.** Government policies that mandate remediation and abatement should be based only upon the well-reviewed scientific evidence of the health risks and not on interests of those who may benefit financially or politically from growth of an abatement industry. Asbestos abatement has developed quickly into a multi-billion dollar business. Abatement is expensive, and businesses, homeowners, churches and school districts can be bankrupted by it. Large sums of money have been spent in abating chrysotile, which is probably a benign substance (except in unusual occupational settings that are beyond 20 fibers per cubic centimeter of air—about 10 times that of even present exposures experienced in modern asbestos mines). For perspective, it should also be added here that even quartz, the most common earth material and the main constituent of beach sand, is clearly implicated as the cause of a disease (silicosis) when inhaled as dust in unusual quantities over considerable time, such as might occur in heavy occupational exposures as the result of not wearing protective equipment.

The difficult decisions about what actually demands abatement must be based on information provided by experts with competence, ethics and no conflicts of interest. Geologists who are specialists in mineralogy can serve as team members with health professionals to ensure that studies that link a disease with a mineral make that link with a very specific, correctly identified mineral. In 1991, 2,900 people died simply from choking on food or ingested objects. The fact that a huge abatement industry, along with a massive regulatory component of government, has sprung up to control an asbestos "hazard" which probably causes less than 100 deaths each year requires serious examination of priorities and current public policy. All asbestos is not the same, and the most common variety seems to pose little actual threat to health or longevity. The asbestos "hazard" appears at time of this writing (1993) to have been overrated beyond reason. ❑

**Questions**:

1. What two groups of fibrous minerals are called asbestos?

2. How does asbestos cause cancer?

3. Which mineral(s) are toxic to humans, and what percentage of the total commercial asbestos in use does this mineral(s) constitute?

Answers on page 301

41

*Radon is a radioactive gas that is produced naturally from radioactive decay in many types of rock. Although this radon is the result of natural processes and is present in many areas, enclosed human structures over radon-rich bedrock often lead to dangerously high concentrations of the gas. This has become a serious concern in many regions in the last 10 years. As we have learned more about this problem, it has been found that preventing the buildup of radon and reducing the concentration in buildings already contaminated may not be that difficult. There is still much debate regarding the large-scale hazards posed by radon and how much should be invested in its remediation.*

# Radon and Other Hazardous Gasses

Natural gasses are geological hazards that have caused much recent concern. Radon is unique among gas hazards because it is radioactive, its toxic effects in amounts usually found indoors are obvious only after a long period of time, and its effects are not yet well assessed. Although radiation is doubtless damaging to the cells of living organisms, there is no such thing as a "radiation-free" environment on this planet. What actually constitutes a radon level we can live with as an acceptable health risk is one of the most heated controversies of all current environmental debates. At stake is the health risk for millions of people if low levels are serious hazards, or the needless drain of many billions of dollars from the national economy for remediation if risks are negligible.

by The American Institute of Professional Geologists

**Recognizing the Radon Problem**

The annual dose of radiation we receive comes from many sources. In addition to radioactivity from soil and rock, we receive radiation from diagnostic medical X-rays, from devices such as color televisions, computer screens and smoke detectors, from the sun and other cosmic sources, and from radiation inside our own bodies that enters with food nutrients (in the form of potassium-40 and carbon-14). It has been estimated that about 40% of the average American's dose of radiation comes from indoor radon.

1984 marked the beginning of the recognition of radon, an invisible, odorless, radioactive gas, as a major geologic hazard. In that year a worker at a nuclear power plant in Pennsylvania repeatedly set off the alarms used to detect employees' exposure to radiation at the plant. Intensive investigation finally revealed that the source of the worker's exposure was not his place of employment, but instead was his own home! The unusual accumulation in the home came from natural radon released from the soils and rock beneath the foundation, and this natural, radioactive gas provided constant, dangerous, radiation exposures for the employee and his family. The soils and rocks were not a uranium ore deposit; instead, the rocks were a **mylonite**, a rock type that is produced in fault zones where rock is sheared and ground together. Mylonites are less common than other rocks, but they are certainly not rare. An obvious question was, "*How many other homes might be similarly affected?*" Subsequent follow-up studies over several years now confirm that radon gas is a natural contaminant in virtually every state. This revelation may not have occurred for many years had it not been for the intense monitoring of employees by the nuclear industry.

Most homes and buildings do not have hazardous concentrations of radon. Yet owners should realize that awareness of potential radon hazards is growing, and future regulation may assign responsibility for disclosure of indoor radon levels to the seller. In some areas of the nation, a building is not marketable without an indoor radon survey.

**Radiation from Geological Materials**

**Radiation** is produced when unstable forms (radioactive isotopes) of a few chemical elements "decay." Decay is a process wherein radioactive atoms release both energy and mass. Release of energy occurs as emission of high energy radiation (gamma rays). Mass is lost when very small heavy and light particles (alpha particles and beta particles, respectively) are shot out of the atom at high velocities like tiny bullets.

Isotopes are forms of a chemical element that differ from

one another by weight. Some isotopes are radioactive; others are stable and never decay. Isotopes are distinguished in writing by presenting the name or symbol of the chemical element along with a number which denotes the atomic weight. Thus the radioactive isotope of carbon that weighs 14 units can be denoted as carbon-14 or as $^{14}C$, and the stable isotope of carbon, $^{12}C$ that weighs 12 units can also be written as carbon-12. $^{14}C$ forms naturally in the atmosphere.

Low-level radiation occurs in virtually all geological materials. In fact, natural low-level radiation has long been recognized, and its measurement has been used for decades as a useful means to distinguish different rock types in drilled wells. Potassium-40 is the most common geological radiation source and occurs in high levels in potassium-rich rocks such as granites, gneisses, slates, some schists, many shales and sandstones, in most types of glacial deposits, and in special potash salt deposits formed by evaporation of sea water. Potassium-40 decays in a single step that releases only gamma rays and produces harmless stable daughter products, calcium-40 and argon-40. Potassium-40 is not considered to be a health hazard.

Most of the remaining natural radiation come from three other isotopes: uranium-238, uranium-235 and thorium-232. In terms of radon, the important isotope is uranium-238. Uranium-238 decays in a series of steps that eventually changes uranium-238 into stable lead-206. **Radon** gas (radon-222 to be more precise) is a radioactive isotope produced during one of these intermediate steps of decay of uranium-238 on its way to becoming lead-206.

**Relationship between Indoor Radon and Geological Materials**

Mylonites are now known to be very concentrated sources for radon gas. Uranium is also enriched in granites, gneisses, schists, slates, some sandstones and some glacial deposits. Uranium is highly concentrated within black shales deposited in marine environments and in rocks used as commercial phosphate deposits.

Low content of uranium in the bedrock beneath a structure favors low radon in that structure, but the character of the bedrock is not always a foolproof indicator. Limestones and dolostones are normally low in uranium, but they can weather to leave behind insoluble clay-soils that are rich in radioactive minerals. As a result, homes situated in limestone-derived soils can be high in radon even though the unweathered bedrock is low in uranium. This is particularly true in humid or temperate areas, such as the northern and eastern U.S., where weathering builds thick soils. Caverns in limestones can also accumulate radon, primarily because they contain floors of clay muds derived from weathering of the surrounding limestones. Fracture fillings within limestones and dolostones can contain uranium mineralization brought into the fractures at an earlier time by groundwater. Coals are usually not high in radioactive elements when deposited, but the black shales that constitute roof and floor rocks of coal seams, particularly of coals that were deposited in brackish water near marine coasts, may be high in uranium. The numerous fractures in coal (called cleats) can accumulate this radon. Elsewhere, particularly in low-grade coals (lignites), ground water may add large amounts of uranium to these coals after they are deposited and make them sources of radon. In summary, even houses sited on "clean" rocks may exhibit high radon readings if soil-forming conditions are favorable to concentrating the uranium-bearing minerals in the soil.

Radon-222 has a short half-life (time for 50% of a given amount to decay—in this case 3.8 days), which allows it to remain a toxic, radioactive gas for less than a month. About 75% of radon produced at a given time decays in about one week after its formation. These constraints mean that radon must come either from uranium within a very immediate source, or there must be conduits available for very rapid migration from below. Such conduits might be open fractures, caves, old mines, or open well bores. Most radon contamination occurs as leakage from soil and bedrock into basement foundations. Sometimes, but rarely, construction materials made from certain kinds of rock may be a source of radon, and sometimes, but rarely, water supplied by private domestic wells may be a source where radon is carried into the home dissolved in the water. There everyday activities such as showering, flushing, heating or boiling of this water release the dissolved radon into the atmosphere of the building's interior. A factor which may increase indoor radon concentrations is the use of tighter, more energy-efficient construction that greatly reduces the draftiness associated with older buildings.

Remediation, except in cases of very high concentrations of radon, can be simple. Keeping radon gas out of basements is an exercise similar to keeping a basement dry. Caulking along joints of basement floors, sealing sumps with a tight-fitting plug, or coating porous walls with a sealer can be very effective in reducing radon content. Low amounts may also be brought down to acceptable levels simply by getting additional ventilation into the building. Opening basement windows and using the fan of the central heating and air system or installing a basement window fan to bring in fresh air are helpful measures, although in most climates they could be considered as only temporary seasonal remedies. In a few cases, a more complex system (sub-slab suction) may be required to flush soil gas from beneath the foundation floor.

**Why Radon Is a Hazard**

Radon is a hazard because it decays quickly to yield a radioactive solid (Polonium-218) which lodges permanently in the lungs. The solid then begins a rapid decay cycle that releases alpha and beta particles inside the lung within a matter of minutes. The result is damage to lung cells that may give rise to lung cancer. Ordinary concentrations of radon in the atmosphere are unavoidable and likely do not pose any danger. The gas becomes a serious threat when concentrations build up within poorly ventilated enclosures. It is particularly dangerous to smokers who are exposed to high concentrations.

**Radon Measurements**

Radon concentrations are reported in the United States in **picocuries** per liter of air (pCi/L). One picocurie-per-liter is about the level of radiation produced by the decay of two radon atoms within a one liter volume over the course of one minute. The average level of outdoor radon is about 0.2 pCi/L and the average indoor level is about 1.5 pCi/L or 5 times the outdoor levels. At present, the EPA calls concentrations above 4 pCi/L "action level" concentrations and recommends remedial action when radon content is above that level.

Homeowners and managers of public buildings can perform rapid short-term measurements with inexpensive test kits. These are exposed to the building's atmosphere and are then sealed and mailed to a laboratory for prompt analyses. Test kits commonly include charcoal cannister kits and alpha-track kits. These may be purchased at many hardware stores, and the cost of the kit should include the analysis.

The inexpensive charcoal cannister kits are used to monitor for only short-term periods (one to four days) and therefore they provide a convenient means for a quick evaluation. However, a four-day survey represents only a little over 1% of the year. Results from just a single short-term study should not serve as the sole basis for remedial action. A few short-term tests through the seasons will give a more realistic picture of the degree of hazard. An unusually high result should lead quickly to more precise evaluation.

Alpha track detectors are also inexpensive and can be deployed through an entire year. They require at least 3 months' exposure to obtain good data. These kits should be placed initially in those areas of the home that are likely to receive radon from external sources (such as in the bathroom where radon might enter through a domestic water source) and in living areas where radon may be concentrated by reduced circulation (such as a basement; crawl spaces don't qualify as living areas). Because radon is a heavy gas, it may be concentrated more at lower levels of the house. However, foundations constructed with hollow materials, such as some types of concrete block, can divert the radon to the first floor rather than to the basement.

Indoor radon concentrations vary markedly through the year in accord with weather changes. Maximum readings should be expected to occur during the time of year of minimum ventilation. This is usually winter, when the building is operated to conserve heat loss and when snow and ice prevent easy escape of radon from soils into the atmosphere.

**The Radon Controversy**

Some scientists hotly contest the need for widespread concern and expensive remediation. Disagreement comes because assessing the radon hazard is very difficult at this time. To assess the degree of hazard clearly, scientists must be able to attribute to the hazard the specific fatalities that result from it within a large population. The case for radon is especially troublesome because (1) radon produces lethal consequences only after many years of exposure and has only recently become recognized as a hazard and, (2) lung cancer, the terminal disease produced by radon, is produced by many other lung irritants as well, the greatest of which is smoking and not the least of which may be passive inhalation of smoke. To use the mortality rate for lung cancers directly, scientists need to distinguish specifically the lung cancers that are caused by radon. At time of this writing, that data is just not available for a sufficient population to directly assess the radon risk.

At present, evidence that low concentrations of radon in homes may pose significant risk must come from two indirect observations. The main indication of potential danger comes from known populations of miners exposed to radon in uranium mines. These workers, particularly *the workers who smoke as well as suffer occupational exposure* to radon, do have a significantly increased frequency of lung cancer. There is an undeniable risk for smokers at these levels of exposure. In contrast, another line of evidence which shows no significant risk lies in recent demographic studies which have tried to relate average life span over an area to regional concentrations of geological background radiation. These fail to support any link between regional radiation levels and expected longevity. In fact, North Dakota, which has high background radiation and some very high radon levels, is the state with the greatest longevity.

Problems arise when such indirect evidence becomes the basis for policy. First, the population of miners who received high doses of radiation from both radon gas *and* mineral dust is not representative of the populace that receives only low doses of radon gas at home. Radon concentrations such as those found in uranium mines and in the home of the nuclear plant worker in Pennsylvania are rare occurrences, and using these exceptions as the basis to develop policy for the general populace may lead to unwise mandates and needless terror. In contrast, the regional background radiation studies do not directly address the indoor radon concentrated in specific homes for specific individuals, and thus these regional studies may lead to an interpretation that induces false security.

In summary, concentrations of radon in a home can be inexpensively monitored and measured. Once a level has been established in pCi/L, the risks of the actual likelihood of contracting lung cancer, as estimated by the United States EPA, can be obtained from the risk table on radon levels (Ed. note: see page 33 of "The Citizens' Guide to Geologic Hazards"). Remember that the table is based only upon indirect evidence. If it is correct, then indoor radon should produce about 20,000 fatalities in the U.S. each year.

**Other Toxic Gasses**

Although radon has been a frequent topic of the media in the late 1980s and early 1990s, there are other natural gasses that sometimes pose deadly hazards under special conditions. About 1,000 people a year in the U.S. die through accidental poisoning by gas; many of these poisonings are due to **methane** used as "natural gas" in home heating and cooking. Residents who are not miners or drillers by profession, and who do not live with the special conditions listed below, will not likely be affected.

**Methane, carbon monoxide** and **carbon dioxide** are gasses that were particularly dreaded by coal miners prior to modern design, monitoring and ventilation of mines. The gasses were colloquially dubbed "fire-damp" or "choke-damp" in reference to the deaths of miners by fires, explosions and suffocation. These are not simply hazards of the distant past; 200 miners died in 1992 in Ankara, Turkey, in a methane explosion.

**Hydrogen sulfide**, a natural gas associated in lethal concentrations with some petroleum deposits, is even more toxic than the "cyanide" gas used in executions. The human body can sense hydrogen sulfide gas by its unpleasant "rotten egg odor" in quantities of less than a few parts per million in air. The gas is still dreaded by oil drillers, and whole drill crews have been killed instantly when this "sour gas" belched from a well onto a drill floor.

All gasses have the ability to migrate into shallow openings such as pore space in soils or rocks or into man-made openings such as wells, mines and tunnels, shallow sewers, septic tanks, utility conduits, drains or basements. Where seeps of these gasses occur naturally, their presence is evident by phenomena such as foul odors or "burning springs," and therefore the area is avoided for residence. Some danger occurs when development occurs over old mine workings or old leaking well casings. When gasses reach sufficient concentrations in developed areas, dangers arise from the potential for combustion (methane), explosion (methane), toxic inhalation (hydrogen sulfide) or suffocation (carbon dioxide and nitrogen). Methane gas from gas mains or sewers may leak and migrate through underground openings over a large area and explode with fatal consequences. Such a methane explosion occurred in Guadalajara, Mexico, on April 22, 1992, and left nearly 200 dead.

Gasses can also be released during earthquakes. Broken gas and sewer mains are obvious gas sources, but gasses can also be released with mud and sand boils and from ruptured cover over waste sites.

Volcanoes are well known for release of toxic and nauseous gasses that include sulfurous and chloride acid vapors. In Cameroon in 1986, over 1,700 people as well as wildlife and farm animals perished beneath a plume of carbon dioxide that belched forth from beneath a 200-meter-deep lake located within a volcanic cone.

Underground fires in coal seams, abandoned coal mines and spoil piles can perpetually release foul-smelling and health-damaging fumes on unfortunate residents nearby. These fumes consist of sulfurous gasses and hydrocarbons with an asphalt odor that are released and distilled from the coal and associated organic-rich roof and floor rocks.

Finally, man's disposal of wastes, particularly organic wastes and trash into the geological environment in "land fills," can often be a source of hazardous gasses. The same microorganisms that are responsible for converting natural organic matter into methane in swamps and marshes are also responsible for converting man-made organic wastes in landfills to that same gas. In instances where careless land use planning occurred, structures built over old land fills accumulated methane within basements, sewers and utility conduits with disastrous consequences of fire and explosion. Production of methane gas at waste disposal sites is so prolific, that it can at times serve as a useful local source of fuel.

**Roles of the Geologist in Reduction of Gas Hazards**

**In mapping.** Geologists can tell from simply looking at a geologic map where a number of areas of risk are likely to be located. The ability to make this interpretation comes from formal university training obtained in a number of courses and from field experience. This same training permits geologists to map deposits, both at the surface and at depth, in areas that have not been previously mapped, and to interpret older maps and reports in the light of newer, state-of-the-art knowledge.

**In public service.** As evidenced from this entire booklet, sites for new buildings should be evaluated by a qualified geologists before construction or design plans are begun. (Ed. note: reference is to "The Citizens' Guide to Geologic Hazards") Often the state geological survey can provide help of a general nature with maps that outline those areas that are underlain by bedrock of known or probable elevated concentrations of uranium. Aeroradioactivity maps, maps of surface radiation made from readings taken in low-flying airplanes, have been produced by the USGS for some areas. The state survey will likely have these maps available for review. The state surveys should also know about particular ground water sources that may be high in radon. As of 1993, the Branch of Sedimentary Processes at the USGS in Denver completed a map of the radon potential of the United States. The map is published with an accompanying booklet for each state.

**In research.** As a result of the recently recognized radon problem, geologists have learned surprising facts about the types of geologic conditions that lead to increased radon in structures, and about how radon becomes enriched in soil, ground water and rock. New work will reveal more, and the data provided by geologists will likely lead to a more precise statement about actual dangers of radon to homeowners.

**Summary**

Widespread education is the most significant step that can be taken against hazards from geological materials. However, some individuals teaching earth science, physical geography, physical geology and introductory geology have never themselves taken formal courses in mineralogy, petrology or soils. Property owners are the main victims of economic losses, where swelling soils or inappropriate construction materials can wreak havoc with structures. Safeguards there include procuring good information about the nature of the site from a professional geologist, and obtaining appropriate insurance to cover possibility of loss.

Radon and asbestos are two topics that are frequently in the national media. Massive amounts of money are already being spent to remedy these hazards, and some estimates to remediate these exceed the combined costs of all geological hazards. State-of-the-art knowledge about these hazards seems to be unknown or misunderstood by policy makers and journalists. The public remains ill-informed without realizing the substance behind the controversies that surround asbestos and radon. While some segments of the populace suffer needless fear and unwarranted financial loss, others are oblivious to real dangers. Massive regulatory actions that are not based upon solid science may be some of the most expensive blunders of this century. ❑

**Questions**:

1. What rock types are commonly sources of radon-222?

2. What is radon-222 a decay product of?

3. How does radon-222 cause lung cancer?

Answers on page 301

*Depletion of the earth's ozone layer is usually ranked one of the two most potentially harmful environmental challenges facing humanity today. (The other is global warming, discussed elsewhere.) Ozone is a form of oxygen that exists very high in the atmosphere and protects life from high-energy solar radiation. Ozone is constantly forming and being destroyed by sunlight-driven chemical reactions. An important theory held by many scientists is that manmade chemicals, especially CFCs, diffuse up into the ozone layer after being released and destroy ozone molecules through chemical reactions. Aside from the laboratory evidence for this is the evidence of the "ozone hole" that has developed over the South Pole in recent years. This led to the Montreal Protocol, an international agreement to phase out CFC use. However, there is much debate over whether CFCs are causing the ozone hole. For one thing, natural events such as volcanoes can affect the ozone layer. For another, some chemists argue that most CFCs do not reach the ozone layer because they react with some other gas first or that CFCs are not as destructive as we thought. This article is not only a good summary of the ozone issue, but is relatively neutral about what has become a heated topic.*

# Ozone

Scam or crisis? Somewhere between Camp Apocalypse and Camp Hogwash lives the not-so-simple truth. The real crisis here is our gullibility

by Patricia Poore and Bill O'Donnell

No journalistic "lead" for this story. No weary scientists peering through the frozen antarctic mist after a restless night spent dreaming of blind rabbits. No international chemical-industry conspiracies. We won't try to hook you. We won't promise an easy read, either. But we will try to remain neutral, even as we suggest that conventional wisdom may be deeply flawed.

We have spent the better part of two months deeply engrossed in the literature of ozone depletion. Our editorial interest had been aroused by the recent popular books[1] which present vastly different interpretations of science. We moved in further with the original scientific papers of Dobson, Rowland and Molina, Teramura, et al. We then read expert reviews of the literature by Elsaesser, Singer, et al.

We found compelling arguments on both sides, only to find credible contradictory information in the next day's reading. Let us warn you up front, anything resembling a final conclusion eludes us.

Although the science around ozone depletion is a continuum—an as-yet incomplete mosaic of postulates and findings—politics and media hype have transformed the argument into a battle between philosophical adversaries: Camp Apocalypse and Camp Hogwash. We may be accused of adding to the confusion by developing our arguments around this artificial divide, but there is really no choice. The escalating volley of counterpoints and rebuttals has given rise to overstatement, simplification, and a growing tendency to lie by omission, suppressing information that doesn't fit. It is to that situation that this article must be written.

Representing Camp Apocalypse are Sharon Roan, author of *Ozone Crisis*, and Vice President Al Gore, author of *Earth in the Balance*. They were preceded by, and continue to find support from, the environmental group Greenpeace and other environmentalist public-interest groups.

Representing Camp Hogwash, we found Rogelio (Roger) Maduro, primary author of *The Holes in the Ozone Scare*, and, less hysterically but just as dismissively, Dixy Lee Ray, primary author of *Trashing the Planet* and *Environmental Overkill.*

From our reading, we find that the truth may lie in the middle—in this arena, a radical place to be. Our views echo those of Ronald Bailey in his book *Eco-Scam*[2], and Dr. S. Fred Singer (*Global Climate Change*), a physicist and climatologist oft-quoted by both camps.

We are comfortable making these statements for the record: (1) Both camps are engaging in hyperbole and lies of omission. (2) The chemical industry is not "dragging its feet" in coming up with substitutes for CFCs; indeed they are responding deftly to market demand and will make huge profits from the phase-out. (3) There is currently no "crisis," nor any documented threat to human health, inasmuch as (a) the so-called ozone hole is an ephemeral disturbance over a mostly unpopulated area; (b) ozone thinning over populated latitudes, if it exists, is within the range of natural fluctuation and is seasonal; (c) no documentation exists for actual sustained increase in UV-B radiation at ground level.

**Reprinted with permission from the September/October 1993 issue of GARBAGE the Independent Environmental Quarterly, 2 Main Street, Gloucester, MA 01930.**

We believe, however, that there may be cause for concern. We are skeptical of the debunkers' claims. No, it is not chlorine from oceans and volcanoes which is responsible for creating the effect seen in the polar region; yes, chlorine from CFCs does apparently make it to the stratosphere and does destroy ozone.

We consider the lack of disclosure and debate regarding the real costs of a precipitous halocarbon phase-out to be near-criminal. The costs that will be involved (monetary, opportunity, and human) receive scant coverage, in spite of the fact that the "threat" of ozone depletion is demonstrably lesser than any number of real threats both worse in scope and more immediate.

**What Was the Question?**

Let's review briefly the points of debate, both to define the scope of our discussion and to inform those who have not been reading carefully the science sections of newspapers in the past four years.

A theory advanced in 1973 by F. Sherwood Rowland and Mario Molina (Univ. of California-Irvine) holds that chlorofluorocarbons (CFCs) and related molecules, owing to their tremendous stability, eventually reach the stratosphere and, photolyzed by intense ultraviolet radiation, split to release energized chlorine atoms that then destroy ozone ($O_3$) molecules. Obviously, $O_3$ is always being created and destroyed, but it is thought that the increased destruction from man-made chemicals has tipped the balance and caused a temporary net depletion in the $O_3$ concentration. This is potentially harmful because $O_3$ is one mechanism that controls the amount of UV-B radiation reaching the biosphere. (UV-B is an ultraviolet wavelength that affects the body, in ways that are both life-sustaining and, in excess, damaging.)

The "ozone hole" is the graphic name given to a phenomenon that so far exists only seasonally over the Antarctic region; namely, a thinning of the concentration of $O_3$ molecules that represents a depletion of up to two-thirds for a limited amount of time. Scientists on all sides of the debate agree that preexisting conditions unique to the Antarctic zone create the possibility of the "ozone hole."

Those scientists and politicians calling for an immediate CFC ban believe that very recent minor, unsustained, localized depletion of ozone in the stratosphere in some areas outside the polar regions is caused by CFCs, and is a harbinger of increasing future depletion. The issue was presented strongly enough that a phase-out was planned in 1987 in an international treaty called the Montreal Protocol; the dissemination of certain findings since that time resulted in a decision for a near-total ban by 1996.

In the meantime, a strong counterpoint to the ozone theory and the proposed ban has been presented, which many people refer to as a "backlash." Regardless of the personal political motivations of the skeptics, good reasons for this "backlash" include the discovery of information suppressed in the politicking for a ban, as well as a growing realization of the extraordinary costs inherent in phasing out the relatively safe, useful and ubiquitous chemicals.

**Points of Contention**

Each camp purports to interpret the original data. Only since the publication of the popularized books, or since about 1989, has the debate escalated into a war, with each side battling the other point by point. Let's look at Camp Apocalypse first, as it gained favor earlier and still holds sway. Here are their major points:

(1) CFCs, HCFCs, and halons are proven to be responsible for a sudden (in a 40-year period) depletion of stratospheric ozone, which created the "ozone hole" over Antarctica. The effect of CFCs on the stratosphere will lead to a similar hole over the Arctic, and will change the ozone layer over populated areas for at least 100 years, even with a phase-out and ban.

(2) The "ozone hole" gets worse every year, starting sooner or growing in area or breaking up later.

(3) Ozone loss causes increased UV-B at ground level, which has resulted in severe sunburns and will cause increases in skin cancer, cataracts, and immune deficiency.

(4) Increased UV-B at ground level will affect the food chain, from phytoplankton to soybeans, and may have apocalyptic results.

As you can see, their points are a mix of the known and the projected. Context is often missing from the arguments of both camps, as well. For example, saying "the ozone hole gets worse every year" sounds definitive (and terrifying), but the period referred to is only 14 years—an insufficient baseline from which to chart real deviation. Now, the salient points from Camp Hogwash, themselves a mixture of red herrings and truth:

(1) CFCs are heavier than air and therefore can't get up into the stratosphere.

(2) In any case, chlorine from CFCs pales in comparison to chlorine released by the oceans and volcanic eruptions.

(3) There has always been an ozone hole, we just didn't know how to look for it.

(4) There is no long-term "thinning" of the ozone layer.

(5) The relationship between stratospheric ozone concentration and UV-B at ground level is unknown, and no sustained increase in UV-B at ground level has been demonstrated.

Both camps bring up the same topics; only their conclusions differ. Let's look at those topics, one by one, with a more neutral perspective, based on a reading of both and reference to some of the original documents.

**The Ozone Hole**

The argument regarding whether or not the "ozone hole" existed before CFCs remains murky. The question, apparently, is what did ground-breaking researcher Gordon Dobson really find when he examined ozone concentration in the 1950s (i.e., before the proliferation of CFCs). Did he discover the ozone hole or not? Some in the hogwash camp have publicly asserted that Dobson found ozone levels as low as 150 Dobson Units over Halley Bay [on the Antarctic continent at approx. 75°S.].

We looked it up ourselves. Here is Gordon Dobson reviewing his findings of the late 1950s in a paper written for *Applied Optics* in March 1968—long before the controversy erupted.

*"One of the more interesting results on atmospheric ozone which came out of the IGY [International Geophysical Year] was the discovery of the peculiar annual variation of ozone at Halley Bay. The annual variation of ozone at Spitzbergen* [a Norwegian Island at approx. 80°N.] *was fairly well known at that time, so, assuming a six months difference, we knew what to expect. However, when the monthly telegrams from Halley Bay began to arrive and were plotted alongside the Spitzbergen curve, the values for September and October 1956 were about* 150 units lower than we expected. [our emphasis] *We naturally thought that Evans had made some large mistake or that, in spite of checking just before leaving England, the instrument had developed some fault. In November the ozone values suddenly jumped up to those expected from the Spitzbergen results. It was not until a year later, when the same type of annual variations was repeated, that we realized that the early results were indeed correct and that Halley Bay showed most interesting difference from other parts of the world. It was clear that the winter vortex over the South Pole was maintained late into the spring and that this kept the ozone values low. When it suddenly broke up in November both the ozone values and the stratosphere temperatures suddenly rose."*

So, while Dobson's group didn't find levels as low as those measured in the mid 'eighties, it's clear from his language that he was shocked at how low ozone concentrations were over Halley Bay, and at a loss to explain how such a phenomenon could exist. Whether or not the "hole" (that is, levels as low as 150 D.U.) is a recent occurrence, it is clear that the physical environment particular to Antarctica had depleted ozone in the austral spring before CFCs could be credibly implicated.

Dobson's group didn't have converted spy planes, high-tech satellite imagery, and countless researchers available to them. They had one instrument in one place. Today, we see the exact position of maximum ozone depletion shifting location from year to year. Could it be that Halley Bay was outside of the "hole" in '56 and '57? We can never know.

**Can CFCs Migrate to the Stratosphere?**

"CFCs are much heavier than air, and so could never reach the stratosphere." It is clear to us that this is a bogus argument. While it's true that CFCs weigh anywhere between four and eight times as much as air, and will sink to the floor if spilled in a laboratory, in the real world, they won't stay on the ground. Our atmosphere is a very turbulent place. Says Rowland: "The atmosphere is not a quiescent laboratory and its mixing processes are dominated to altitudes far above the stratosphere by the motions of large air masses which mix heavy and light gaseous molecules at equal rates. Throughout most of the atmosphere, all gaseous molecules go together in very large groups, independent of molecular weight.

"By 1975, stratospheric air samples ... had been shown regularly to have CFC-11 present in them. During the past 17 years, CFC-11 and more than a dozen other halocarbons have been measured in literally thousands of stratospheric air samples by dozens of research groups all over the world."

**What about Natural Sources of Chlorine?**

Say the skeptics: "The amount of chlorine hypothetically released by CFCs pales in comparison to that available from natural sources." They are talking about seawater evaporation and volcanoes. Dixy Lee Ray tells us in *Trashing the Planet:* "The eruption of Mount St. Augustine (Alaska) in 1976 injected 289 billion kilograms of hydrochloric acid directly into the stratosphere. That amount is 570 times the total world production of chlorine and fluorocarbon compounds in the year 1975."

The hogwash camp has said that one billion tons of chlorine are released into the atmosphere from natural sources, as compared to a theoretical 750,000 tons from man-made sources. Taken at face value, these seemingly scientifically arrived-at proportions would lead one to believe that man-made sources are insignificant.

Most unfortunately for the hogwash camp, Ray had made a terrific blunder. Her calculation came not from Alaska in 1976, but from a theoretical extrapolation of the total HCL released (not necessarily reaching the stratosphere) by a mammoth eruption 700,000 years ago. She may have made the same argument (which rested on a 1989 paper by Maduro) even with accurate numbers, but noise over the mistake has eclipsed the question.

Indeed, what about volcanoes, spewing chlorine compounds at high velocities? Again, the amount released by volcanoes is not the same as the amount reaching the stratosphere. Yet Maduro insists: "No matter what figure is used, the basic point remains that the amount of chlorine emitted by Mother Nature through volcanoes dwarfs the amount contained in man-made CFCs."

Ozone-depletion researchers counter that whatever the amount of chlorine compounds released through natural sources, all of it is washed out in the lower atmosphere through precipitation—before it has reached the stratosphere.

In summary, the hogwash camp is vastly overstating the importance of natural sources of chlorine. The apocalyptic camp entirely dismisses the importance of natural sources of chlorine because it is removed by rainfall, with negligible amounts reaching the stratosphere.

Whom to believe? Is it really true that only organic, water-insoluble compounds (e.g., CFCs) can deliver chlorine to the stratosphere? Are we to believe that there's enough precipitation in the antarctic night to wash out all the chlorine being emitted by Mount Erebus (a volcano, continuously active since 1972, six miles from the monitor station at McMurdo Sound)—before any of it can move up to the stratosphere in the great, turbulent polar vortex?

**Global Ozone Depletion? From What Baseline?**

For the record, no solid evidence exists to suggest ozone depletion over the northern latitudes poses any health hazard. Are you shocked? It's no wonder. Environmental groups and the popular press tell us the threat is now.

Case in point: On February 3, 1992, NASA "interrupted their research" to announce their prediction of a full-scale ozone hole over much of the U.S. and Europe: the infamous "hole over Kennebunkport" referred to by then-Senator Al Gore. It didn't happen. The October 1992 Greenpeace report entitled *Climbing Out of the Ozone Hole* claimed: "The formation of an ozone hole over the Northern Hemisphere in the near future, and possibly as early as 1993, now appears inevitable." Greenpeace's "inevitable hole" over the Northern Hemisphere didn't materialize, either.

That the alarms were false didn't stop them from becoming common knowledge. The July 1993 issue of the women's fashion magazine *Vogue* tells us that "thorough sun protection is the cornerstone of any summertime beauty strategy. As government scientists report ozone over the Northern Hemisphere is at its lowest level in fourteen years." The ominous warning appears in a feature article called "Beauty and the Beach," which shows page after page of bathing beauties soaking up the summer sun in the latest bikinis.

What the article fails to mention is that the 10 to 15% reduction government scientists reported occurred in March and April, when the amount of UV-B reaching the northern latitudes was but a small fraction of what the summertime sun delivers. We have to be careful when we interpret these diminished percentage-point results. A ten percent depletion over Kennebunkport in April (and its corresponding as-yet theoretical increase in UV-B) is still but a small fraction of that received in New York or Boston in June— when people really are out on the beach. We also must be careful to understand what baseline is being used to report these "depletions."

Of those who either discredit the degree of ozone thinning or differ on its range of effects, few carry greater weight or generate more controversy than S. Fred Singer, who holds a doctorate in physics from Princeton University and is now president of an Arlington, VA-based think tank called Science and Environmental Policy Project.

Dr. Singer is skeptical about claims by other scientists that, on average, global ozone levels are falling: "One cannot estimate whether there has been any long-term change from short-range observations because the natural fluctuations are so large." According to Singer, long-term analyses are compounded by daily ozone fluctuations that double naturally from one day to the next [without any cataclysmic outcome, by the way]. "Seasonal fluctuations, from winter to summer, are as much as 40% and the eleven-year solar cycle is three to five percent, on a global average. Extracting long-term variations from a few percentage points of change in a decade is like observing temperatures for one season and judging whether climate has changed over the long term. It can't be done.

"It is not possible to eliminate the chance that what we are seeing is a natural variation.

"The Antarctic hole is a genuine phenomenon," Singer concedes. "But it is nothing much to worry about because it lasts such a short while and has already stabilized. Besides, it is controlled more by climate than by CFCs."

**The UV-B Question**

The scary part of ozone depletion is, of course, the correlation to increased UV-B penetration. The most often-cited theoretical relationship is that for every 1% decrease in stratospheric ozone, we can expect a 2% increase in ground-level UV-B. It would seem a good check of diminishing ozone claims would be to quantify the penetration of UV-B. Problem is, the few who are looking can't find any increase at the Earth's surface.

Despite the analysis of TOMS (Total Ozone Mapping System) satellite data released by former EPA-administrator William Reilly indicating *springtime* average ozone levels over the United States have dropped 8% in the last decade, there are no data to suggest increased penetration of UV-B on the ground. In fact, a report published in the September 28, 1989, issue of *Nature* cites a study that found a 0.5% average *decrease* in UV-B between 1968 and 1982, despite an overall decrease in ozone column density of 1.5% over the same period.

Ozone doomsayers counter by arguing: 1) The monitors used are not capable of making distinctions between UV-A and UV-B radiation, and 2) UV-B is not reaching the surface because it's being absorbed in the troposphere by man-made pollutants. They reason that we shouldn't count on our fouling of the lower atmosphere to protect us from damage we're inflicting above.

If the monitors are antiquated, you'd think we'd be funding new ones, given our fear of the sky. The second argument is a red herring. The reported 8% depletion in stratospheric ozone (which should theoretically create a hard-to-miss 16% increase in UV-B) occurred during a decade where tropospheric pollution was decreasing over the U.S.—courtesy of the Clean Air Act.

**The Connection to Human Health**

All claims regarding human health risks associated are related not to ozone thinning *per se*, but to increased UV-B exposure. So far, researchers have not in fact tied increases in skin cancer and cataracts to increased UV-B exposure due to thinning ozone. There is no epidemiological evidence of suppressed immune function due to UV-B exposure caused by thinning ozone.

(No one questions that people get more UV exposure than in the past. Only a few generations ago, a tan was considered unhealthy. Only since the 1950s have so many people had the leisure and desire to be out in the sun, wearing scant clothing. And only with technological advances have so many white people been living in previously inhospitable "sunbelts.")

But is there more UV-B, overall, sustained, at ground level? What would it mean if we can find ozone depletion without a corresponding rise in UV-B penetration to ground level?

**Where We Are—As of August '93**

Public policy is driven by the public, not by scientists. A recent survey gave these results: 67% of Americans consider themselves "extremely concerned about the environment." But only one in five is aware that CFCs are used in refrigeration, and only one in 30 is aware that CFCs are used in air conditioning. Are a well-meaning public and the politicians who serve them not well enough informed to make global decisions that will cost hundreds of billions of dollars? Will future generations look back at the "ozone crisis" as the greatest waste of resources in human history? Or will they thank us for taking lifesaving action without delay? (The apocalyptics talk about political foot-dragging "for 14 years," but the Montreal Protocol is perhaps the fastest, largest non-military global response to a perceived threat in human history.)

The following observations are based not on our own scientific experiments, of course, but rather on a rational analysis of the facts following a great deal of reading. We have no vested interest in either camp.

1. Attributing the Antarctic "ozone hole" to CFCs is overstatement to the point of fallacy. Natural conditions have always existed which deplete the concentration of ozone in that region during a specific time of year. However, scientific data do support the theory that stable, man-made chlorinated molecules are implicated in a localized net ozone loss during the natural cycle.

2. Ozone depletion is not an epic crisis. Remember, even if ozone maintains 100% of its "normal level," skin cancers will still occur. On a day when ozone levels over Punta Arenas, Chile, are reduced by 50% because of the "hole," the theoretical maximum increase of UV-B levels would be equal to only 7% of what reaches the ground at the equator on the same day.

3. We must monitor UV-B at ground level to see if in fact there is any correlation with stratospheric ozone fluctuations.

4. An outright ban on CFCs and other useful halocarbons (before adequate substitutes are available) would cause more human suffering and economic mayhem than the theoretical increase in ozone depletion under a more managed phase-out. In the U.S. we have the financial means and perhaps the political will to accept the challenge—albeit at tremendous cost and lost opportunities. In other parts of the world, an already insufficient supply of affordable refrigeration would be exacerbated. The result will be more disease from food-borne bacteria, and greater hunger.

What do you think?

1. *The Holes in the Ozone Scare/ The Scientific Evidence That the Sky Isn't Falling* by Rogelio A. Maduro and Ralf Schauerhammer, 21st Century Science Associates, Washington, D.C., 1992. *Ozone Crisis/The 15-Year Evolution of a Sudden Global Emergency* by Sharon Roan. John Wiley & Sons, New York, 1989. *Trashing the Planet/How Science Can Help Us Deal with Acid Rain, Depletion of the Ozone, and Nuclear Waste (Among Other Things)* by Dixy Lee Ray with Lou Guzzo, Regnery Gateway, Washington, D.C., 1990. *Earth in the Balance/Ecology and the Human Spirit* by Al Gore. Houghton Mifflin Co., Boston, 1992.

2. *EcoScam/The False Prophets of Ecological Apocalypse* by Ronald Bailey. St. Martin's Press, New York, 1993. ❑

**Questions**:

1. What atom in the CFC molecule is said to destroy ozone? When was this theory first advanced by Rowland and Molina?

2. Is there evidence of low ozone concentrations in the South Pole before CFCs were widespread? What units are used to measure ozone?

3. Do the authors think an outright CFC ban is now called for? Why?

Answers on page 301

**43**

*In the 1990s, market principles are often promoted as a solution to many environmental problems. This contrasts with the 1970s when "command-and-control" legal solutions were often promoted. For example, the Clean Water Act and many other pieces of federal (and state) legislation were aimed at regulating pollution by setting strict limits. However, many economists point out that such legal restrictions are often very costly; they require constant monitoring and costs of enforcing such standards are high. In some cases, the "magic of the marketplace" can reduce costs, by providing incentives for people to reduce environmentally damaging behaviors. A well-known example is the "carbon" or "BTU" tax often proposed, to increase gasoline costs and promote conservation and alternative fuel development. Higher gas prices would promote such activities without contant government monitoring. However, some argue that this "free-market environmentalism" has gone too far. A valid argument that they make is that some, indeed many, things simply cannot be fixed with a price. What is the precise economic value of clean air?*

# The Price of Everything

## The market speaks on environmental protection. Put up or shut up, it says.

by Thomas Michael Power and Paul Rauber

Maybe it has something to do with the approaching millennium: the lion will lie down with the lamb, and toxic polluters will drink herbal tea with environmental activists. The wonderful new development is lauded in the press and preached from scores of think tanks. No longer, we are told, do we have to rely on threats of fines or jail time in order to get industry to do the right thing. The business leaders of today, working together with enlightened environmentalists, have discovered in the magic of the marketplace a cheaper, more effective, and less contentious remedy to just about any environmental ailment.

The debate over environmental protection in the 1990s fills the ideological vacuum left by the end of the Cold War. It is now fashionable, for instance, to compare government regulation to the "command-and-control" economic arrangements of the former Soviet Union. As the Soviet system failed, the analogy suggests, so too will a regulatory system based on the *diktat* of federal bureaucrats telling industry how much pollution to reduce and how to reduce it. "Command and control" is said to cost U.S. businesses $140 billion a year, handicapping the economy, hobbling the recovery, and unfairly vilifying many environmentally concerned Americans who just happen to own polluting industries.

The alternative to this clumsy, old-fashioned, and vaguely unpatriotic-sounding system is "free-market environmentalism" (a.k.a. "new resource economics"), which promises to harness the vigor and inventiveness of capitalism to heal the earth. To do so, it proposes to vastly expand our present notion of private property, to sell that property to the highest bidder, and then to let the logic of the market sort things out.

Already a new property right has been created: the right to pollute. One section of the 1990 Clean Air Act allows plants that pollute below certain levels to sell pollution "credits" to dirtier concerns; innovative, clean industries profit from their cleanliness, while the dirty industries pay for their sins until they can get around to cleaning up their acts. A market in these "pollution credits" has been established at the Chicago Board of Trade, where rights to emit tons of sulfur dioxide are bought and sold like pork bellies or soybean futures.

Having set prices on pollution, free marketeers are also trying to figure out what those who enjoy environmental quality should be made to pay for it. What will the market bear for the use of a regional park? Hopefully the public will pay more for Sunday hikes than the local developer will for condos, because if not, farewell forest. And if people want wolves in Yellowstone National Park, free marketeers argue, they should be willing to pay for them, cash on the barrelhead. It's just a question of settling on the price.

Not all proponents of free-market environmentalism subscribe to all of its logical but occasionally wacky conclusions. Every ideology has its ideologues; in this case, they are the libertarian-minded think tanks and academics who have provided the theoretical spadework for the new discipline. More common, however, are those who seek to pick and choose at the free-market table, ignoring dishes that don't coincide with their interests. Many businesses, for example, are enthusiastic about market solutions, but only when they result in a further giveaway of public resources. Contrarily, some environmentalists advocate market mechanisms in the name of efficiency, reasoning that making environmental responsibility cheaper will result in a corollary reduction of political opposition, the end result being the possibility of greater protection.

This, crudely put, is the position of the Environmental Defense Fund, the most market-oriented of the major environmental groups, as well as of some individuals within the Sierra Club. "We're finally getting past the debate about whose position is morally superior and moving on to a point where we will accomplish real reductions in pollution and resource use," says Dan Dudek, a senior economist at the EDF. His organization, which helped write the pollution-credit section in the Clean Air Act, looks forward to the establishment of national markets for nitrogen oxides, and perhaps even global markets for CFCs and carbon dioxide.

A big plus for free-market environmentalism has been its bipartisan support; neo-liberal Clintonian Democrats and anti-regulatory Bob Dole Republicans embrace with equal enthusiasm. *Mandate for Change*, candidate Clinton's policy blueprint, contained a chapter ("The Greening of the Market") calling for a harnessing of the "daily self-interest" of firms and individuals to replace "command-and-control" regulations. During the campaign, Clinton himself said that we must "recognize that Adam Smith's invisible hand can have a green thumb," and called for a "market-based environmental-protection strategy."

This is a bitter draught for many environmental activities, weaned on regulatory triumphs like the National Environmental Policy Act and practiced in lobbying the government to toughen environmental laws, not abandon them. Most environmentalists are innately suspicious of economists anyway. They are the ones, after all, who tend to portray environmental quality as an expensive frivolity; who tell us that pollution controls hamper productivity and threaten private property; that zero levels of toxic releases are a naively impossible goal; and that protecting endangered species without regard for the economic consequences is irrational—as perhaps, are many environmentalists.

(This suspicion of the dismal science is well warranted historically. From its beginning, the intellectual mission of Anglo-American economics has been to demonstrate the secret logic of allowing businesses to maximize profits, unfettered by social controls. That was, after all, Adam Smith's goal—to depict the selfish, even antisocial actions of private commerce as ultimately benefiting the public. No wonder the business community enthusiastically supported the intellectual venture that came to be known as economics.)

Yet these same wary environmentalists frequently endorse the use of economic instruments—perhaps without quite realizing it, and often to the profound distress of the affected industry. They insist, for example, that a price be put on empty beverage containers to create an economic incentive for recycling. (The deposit idea is now being considered for other, more dangerous solid wastes, such as automobile batteries, or refrigerators containing CFCs.) They argue that water "shortages" in arid regions result from the absence of incentives to conserve when the low price of government-subsidized irrigation reflects neither what the water costs to provide nor its value in alternate uses. They attack government subsidies for destructive programs such as the U.S. Forest Service's below-cost timber sales. Yet they remain queasy about extending this approach to all other environmental problems—with good reason, as it turns out.

This ambivalence reflects a healthy respect for the limitations of market "solutions." Economic instruments are tools, but using them does not require us to embrace a new ideology or to jettison all government regulation. It *does* require environmentalists to determine when such tools can be used productively, and which specific sort of tool is appropriate to a given situation or industry. It requires the adoption of an explicitly pragmatic approach to solving environmental problems. Most importantly, it requires that political problems be faced first.

Whenever environmental policy is made, three crucial issues must be involved:

- What level of environmental protection is desired in each particular location?
- Who is going to pay the direct costs of achieving the targeted level of protection?
- What policy tools will be used to achieve these levels and to impose the costs?

Since economic instruments are merely policy tools, it's no use talking about them until the first two far more contentious questions have been settled. Otherwise, market mechanisms will end up doing what they have always done, i.e., maximizing profits by ignoring pollution or shifting environmental costs elsewhere. Market measures, then, are appropriate in situations with firmly established pollution-control objectives, where conventional environmental regulation would result in pure economic waste. Say, for example, that we have decided to reduce the amount of solid waste going to a city's landfill. Instead of issuing a decree ordering such a reduction by every citizen and business, we change the way garbage fees are paid; instead of extracting them from property taxes, as is usually the case, we start charging by weight or volume, and institute curbside recycling at the same time. Recyclers get a break, and others pay in relation to the amount of garbage they produce. Here economic instruments have something to offer, but only *after* the basic political questions have been settled.

This was not entirely the case in the pollution-permit market created by the 1990 Clean Air Act and implemented earlier this year. This pet program of the free-market environmentalists was designed to ease the pain for industries required to halve their 1980 level of $SO_2$ emissions by 2000. While promising in theory, however, its actual implementation revealed a number of hidden problems.

Unaddressed, for instance, was the question of who had to live with continued high levels of pollution. When the geographic area over which pollution credits can be traded is very large—nationwide in the case of $SO_2$—the effect of the market can be to stick some people in dirty areas with the bill. In some parts of the country air quality dramatically improves; in others, serious pollution problems persist with the full blessings of the market and the law. (The geographic question is what got World Bank Chief Economist Lawrence Summers in such hot water last year when his memo about the "impeccable" logic of dumping toxic waste in the "underpolluted" Third World was leaked to the press.) At its worst, trading pollution rights can legitimize continuing pollution. In one of the very first acid-rain trades under the new program, the Wisconsin Electric Power Company sold a Pennsylvania utility 20,000 tons of pollution credits. Under Wisconsin law, however, the company would not have been entitled to emit the pollution in the first place—yet federal law still allowed it to be peddled to Pennsylvania.

Meanwhile, East Coast utilities have been selling their $SO_2$ credits to midwestern power plants, allowing them to continue burning high-sulfur midwestern coal. But what goes around comes around: the midwestern emissions ultimately drift back through the Atlantic and New England states, where they fall as acid rain.

Since it was concern over acid rain in the Northeast that led to the Clean Air Act's $SO_2$ caps in the first place, New York is now trying to prevent its utilities from selling $SO_2$ permits to upwind states. In the Midwest, on the other hand, ratepayers pay higher bills in order to finance their utilities' $SO_2$ purchases, but don't see any reduction in local $SO_2$. Had their utilities been forced by regulation to reduce emissions, at least the higher bills would have been offset by cleaner air; now the public pays for "pollution control," but gets none.

Supporters of emission trading argue that New York's fears are overstated, and that the benefits of local utilities cleaning up enough to sell credits far exceed the relatively small excess $SO_2$ blown in from upwind. Indeed, the Sierra Club itself has intervened with Ohio's public utilities commission in support of an acid-rain-reduction plan consisting of emission trading, energy efficiency, and use of low-sulphur eastern coal. "By forcing the marketplace to the lowest-cost solution that really works," says Sierra Club Ohio Chapter Energy Chair Ned Ford, "environmentalists gain credibility and enhance the opportunity for further reduction."

Of course, a simpler market mechanism could have been employed by taxing emissions above a certain level. While this would have had the same effect of rewarding the clean and punishing the dirty, it is anathema to free-market ideologues, whose interest is the creation of new private-property rights—in this case, a right to pollute.

Ironically, some of the businesses pollution-trading systems are supposed to assist don't want to play ball. The Ohio Power Company would prefer simply installing a scrubber. In Southern California, two dozen major businesses are opposing an emission-trading scheme proposed by the South Coast Air Quality Management District, claiming that it would "substantially raise the costs" of pollution control, and pleading to be allowed to continue with "command and control."

Many environmentalists also attack pollution-credit trading as fundamentally flawed. That is not necessarily true. The real problem is that the program was established before all the basic questions were answered—in this case, what level of environmental quality should be assured for all areas covered by the trade. Those answers can only be reached through a political process, not bought at the market.

There is a fundamental conflict here that goes far beyond the use of economic instruments. Environmental protection necessarily involves the transfer of control over very valuable resources from one group of people to another. These resources are the wealth of the natural world, extractable and otherwise, and the limited capacity of the air, water, and land to assimilate the wastes associated with economic activity. Historically (and to a considerable extent today), this natural wealth has been the province of industry. The Forest Service sells the national forests for a song to giant timber companies; the Bureau of Land Management allows gentlemen ranchers to denude the public range for a pittance; the 1872 Mining Law gives away the public's mineral wealth to multinational corporations. Until very recently, the cost of waste disposal was however much it took to build a smokestack, or a drainpipe to the nearest river.

Taken for granted, these hidden subsidies—economists call them "externalities"—are not reflected in the price of commodities. Timber is cheap because the Forest Service gives it away; driving is cheap because drivers don't pay for air pollution. The real cost, of course, is paid in sick children, eroded farmlands, vanished fisheries, and extinction of species, and is shunted to the public, preferably the public of future generations.

Over the past several decades, the environmental movement has attempted to transfer control of these natural resources—worth, literally, trillions of dollars—from the commercial sector to the public. A transfer of wealth of this magnitude cannot take place without considerable conflict. In the past, such power shifts have required revolutions.

The continuing struggle over environmental policy, therefore, is hardly surprising. Economic instruments can make a modest contribution toward resolving it, to the extent that they can reduce the cost of environmental protection. But the fundamental conflict over who controls the use of our air, water, and landscapes cannot be decided merely through a change in the instrument of enforcement.

Business sometimes argues that it doesn't really matter who pays the direct costs of pollution control, since the costs will ultimately be borne by the general citizenry in the form of higher prices anyway. But in this case environmentalists have Econ 1-A on their side: it is elementary economic theory that markets can change behavior only if the full costs of an activity—the externalities—are incorporated into the immediate prices paid. If the price of gasoline included the direct costs of maintaining a permanent fleet (let alone fighting a war) in the Persian Gulf, our transportation system would reform itself in a hurry.

If the free marketeers based their program on charging the true environmental costs for all resources used, the environmental movement would sign up en masse. But ideological free-market environmentalists often seem more concerned with the market than with the environment; they tend to feel that equity—the distribution of access to scarce resources, or the right to a clean and healthful environment—is less important than economic efficiency and property rights.

This is exactly what is being demanded in the current attempt to expand the legal concept of "takings" to include environmental regulations. (See "Look Who's Taking," September/October.) By this theory, any environmental regulation that results in lost profits requires government compensation. This assumes, of course, that people have a "property right" to pollute or damage the environment in any way they wish, and that the public has to pay them if that right to damage the environment is changed or revoked. Oddly, free marketeers somehow always assign property rights to those doing the polluting rather than to those being damaged by the pollution.

This is one example of the huge ideological gulf that separates the vision of a good society shared by most environmentalists from that of the free-market enthusiasts. Environmentalists act collectively to preserve certain qualities associated with the natural and, often, social environments. In this sense, they are fundamentally conservative: they wish things to remain the same, or even to return to a previous preferred condition. It is ironic, then, that their suspicion of market instruments is sometimes taken as proof of their "watermelon" character: green on the outside but red on the inside.

Free marketeers, on the other hand, are enthusiastic about the constant change that a market economy encourages, and are suspicious of any efforts to guide the direction of the economy or society collectively. They see such attempts as authoritarian, economically destructive, and tantamount to socialism. The economy for them is an adventure of unknown destination. Columnist George Will, for instance, writes fondly of the "billions of daily decisions that propel a free society into an exhilaratingly unknown future." We should learn to enjoy the excitement and change, and trust that the overall result, whatever it may be, will be much better than anything we could collectively arrange.

Ideological free marketeers insist that we should not use individual market tools without buying the whole package . Advocates such as John Baden of the Foundation for Research on Economics and the Environment (FREE) object strenuously to the use of market instruments "simply as tools for the efficient delivery of environmental goals . . . [while] the goals themselves remain collectively determined." Again, environmentalists are plainly the conservatives to the radical free marketeers, who are willing to trust everything to their faith in the inevitably positive outcome of market forces.

It is not necessary for the environmental movement to respond to one type of extremism and ideological wishful thinking by adopting another. Incentives do matter, and market instruments can help us, collectively, to protect the environment. Consider, for example, the following possibilities:

•In western rural areas, streams often run dry during peak summer irrigation periods. Because irrigation water is usually provided at very low cost to farmers, it is often used inefficiently (growing rice in California's Central Valley, for example). One solution is to allow government fish-and-wildlife agencies, water-quality agencies, or private-sector environmental groups to purchase water rights from farmers and use them to protect streams and their associated fisheries. These rights could be purchased on the basis of a willing buyer and a willing seller, a straightforward market transaction. Another approach is simply to raise the price of the water to the farmers to more closely approximate its real cost, thus discouraging ecologically foolish uses.

•Many of the most serious urban environmental problems—congestion, air pollution, noise—are associated with the automobile. Driving is rewarded in many ways, such as when businesses provide free parking for employees. But what if employers paid employees the cash value of parking privileges in higher wages, and then charged full cost for the parking? Those who choose to use mass transit or carpools would have higher net incomes, but no one would be worse off. Resources might well be saved and environmental costs reduced. Similarly, public agencies could charge commuters the full costs—including environmental costs—of using private automobiles. Increasing rush-hour tolls for lone drivers while forgiving them to car-poolers are steps in this direction.

•In the same vein, dramatically raising the price of gasoline to reflect its real costs—a ready military, poisoned ecosystems in Alaska, polluted low-income neighborhoods next to refineries—would shortly result in increased fuel efficiency and reduced automobile usage. Other auto-related costs, like collision insurance, could also be included in the gas price.

These examples are purposely speculative to give a feeling for the range of environmentally productive uses for economic instruments. Once environmentalists begin to think in this direction, they are likely to generate many more ideas. Call it "the magic of the marketplace."

Some free-market ideas, while undeniably creative, need careful scrutiny. A good example is the proposal to charge increased fees to recreational uses of public lands. (See "What Price a Walk in the Woods?" May/June.) The idea is to provide a positive incentive for bureaucrats whose revenues are closely tied to the amount of economic activity they generate. The more timber they harvest, the more mines they permit, the more land they lease for grazing, the larger are their budgets. (Since the government does not factor externalities into the equation, these activities show up as pluses on bureaucratic balance sheets, even when the activity results in a new loss to the public.)

Because recreationists pay few if any fees, the argument goes, no revenue is associated with their interests and the land managers ignore them. Hence the notion to charge hikers, campers, and skiers whatever the market will bear, thus producing a cash flow that will impress the bureaucrats enough to preserve and enhance recreational values. This, we are told, will automatically provide protection for public lands, because recreation fees would bring in far more than the timber, forage, and mineral charges that now largely finance these agencies.

Recreation-fee advocate Randal O'Toole explicitly suggests that if public lands were in private hands, the widespread environmental damage we observe in the West would be much reduced. This is hard to believe for anyone who has ever flown over the Pacific Northwest and seen the checkerboard of clearcut private lands next to still-intact bits of public forests, or peered beyond the beauty strips in Maine.

There is no doubt that when an agency develops a financial stake in serving a particular clientele, it becomes a strong advocate for that clientele's interests. Consider the many state fish-and-game agencies. Funded primarily by the sale of fishing and hunting licenses, they are single-minded defenders of fishing and hunting interests. In Montana and Wyoming, fish-and-game agencies have resisted wolf reintroduction because they fear that wolves will reduce the number of ungulates available to hunters. The Montana agency has opposed listing the grizzly bear as an endangered species because it would ban grizzly hunting, and has also refused to support the introduction of bighorn sheep in any area where they could not be hunted.

The moral is that when cash flow alone guides government agencies, some perversity always follows. In order to imagine how a recreational fee system might work, one need look no further than those national-forest areas that have been surrendered to intensive downhill-ski development. Nor does giving recreationists more influence in the management of wild areas necessarily guarantee a haven for backpackers and bird watchers. The Bureau of Land Management could well find that more money could be made sponsoring off-road-vehicle rallies than from either backpackers or cattle. Perhaps currently roadless wildlands would produce a larger cash flow were they open to motorized tours: snowmobile trails, Going-to-the-Sun-type roads through all of the spectacular mountain country. Already the solitude of the Grand Canyon is marred by the noise of sightseeing aircraft; in the Wasatch Mountains of Utah, backcountry skiing has been almost completely displaced by heli-skiing.

There is an important distinction to be made here. When we are talking about relatively common commodities such as timber or forage or minerals, it is perfectly reasonable to expect market approaches to work. After all, we already trust the production and use of those commodities to commercial markets. But to most of us, the management of our public lands is not (or should not be) primarily about adjusting slightly upward or downward the quantity of 2x4s or sheep pasture or phosphate rock that make it onto the market. The issues at stake in the management of public lands—biodiversity, wilderness, sustainability—go far beyond the world of commodities and the language of economics.

The problem with this method of influencing public-land management is that many of the things we want public lands managed for do not and cannot have dollar value attached to them. There is no way for cash-flow analysis to put an accurate price tag on a spotted owl or a grizzly bear, nor to indicate what wilderness is worth. Rather, we seek to protect wilderness and grizzlies not as playgrounds and playthings for tourists and fee-payers, but because we wish as least some small part of our heritage to continue to exist apart from us and our cash registers. Wilderness is "valuable" to us precisely to the extent that it is *not* used by humans; consequently, "use" is meaningless as a measure for accurately valuing it.

Markets are not neutral, technological devices. They are social institutions whose use has profound consequences. All societies purposely limit the extent of the market in order to protect their basic values. We, for instance, do not allow the buying and selling of votes and judicial decisions; we do not allow the selling of the sexual services of children; we do not allow human beings to be sold into slavery.

Free-market enthusiasts assume that market-oriented, calculating, self-regarding (i.e., "greedy") behavior is all that is needed for a good, responsible society. Such behavior, they assert, should be encouraged, not constrained. But what kind of decent society can depend upon this type of motivation alone? Selling certain things, in fact, degrades them: selling praise, spiritual favors, intimacy, the privileges of citizenship, or the outcomes of athletic events does not enhance their value, but reduces or destroys it.

Even the commercial market is built on basic morality that takes the larger society and its values into account. Well-functioning markets do not simply spring into being spontaneously. Rather, they are regulated by elaborate public and private social institutions like courts, contract law, industrial associations, and our stock, bond, and commodity exchanges. When the regulatory apparatus breaks down, so do the markets, as can be seen in the recent history of the savings-and-loan industry. Without social structures the pursuit of commercial gain degenerates into banditry, as is evident in the drug trade, in frontier societies (such as our own in the last century), or in countries like Somalia where those social institutions have collapsed. Unadorned market-oriented behavior leads to "gangster capitalism," not to the good society.

This raises a disturbing aspect of the use of economic instruments to solve environmental problems. A basic assumption of the free-market approach is that motives don't matter, only results. If bribing polluters out of polluting works, fine. If giving civil servants bonuses for obeying the law is effective, pay them. Assume the worst of all human beings and arrange incentives to harness the basest of human motives. This is the social logic of free-market environmentalism.

But to most of us, motives *do* matter. A lie is not the same as a mistake; murder is not the same as self-defense or manslaughter; prostitution is not the same as love. Oliver Wendell Holmes once said that even a dog distinguishes between being stumbled over and being kicked. Most of us have at least the sensibilities of that dog. We do care about the motives of our fellow citizens; it matters if someone seeks to protect the community and its land base because they actually care, as opposed to doing it only to protect their pocketbooks (or pick ours). Ethics and conscience matter, and markets can undermine both.

The basic operating principle behind a "free market society" is an antidemocratic one: that peoples' preferences, whatever they may be, should be accepted and given an importance in proportion to the dollars that back them up. But most of us—including those without a great deal of money—have moral and social values that lead us to be very critical of some preferences, the expression of which we seek to block regardless of their financial backing. Even when we do not support the use of legal restrictions to constrain their expression, most of us would be uncomfortable passively accepting all market outcomes as legitimate. Instead, we seek social and cultural means to discourage some and encourage others. That is what "manners" and "public opinion" and "community standards" are all about. One of the worrisome things about allowing the use of public lands to be determined entirely by the highest bidder is the implicit legitimization of those outcomes. Destructive behavior should not automatically become legitimate and acceptable simply because it is backed by the largest wad of cash.

Environmentalists should be concerned with the waste of all resources: natural, social, and political. Policies that are unnecessarily costly and that do not accomplish their objectives involve pure waste, and should be reexamined. Economic instruments for controlling pollution and managing public lands can offer more efficient way to reach our objectives; at the very least, by reducing the cost of environmental control, they open up the possibility of attaining higher standards of quality. Many market plans begin by setting a goal, and then proceed to attain it in the least costly way. That is a positive approach. Finally, economic instruments can solve certain types of environmental conflicts to the mutual satisfaction of all involved—a rare and attractive option in an otherwise contentious struggle.

But this mutual satisfaction can never be achieved by the commercial-market mentality alone. On the contrary, that mentality tends to gnaw away at the ethical underpinning of society, and can even undermine the foundations of makets themselves: witness the insider trading, market manipulation, and regulatory corruption scandals of the 1980s. The social and community values that environmentalists hold should not be abandoned to an ideological fad. They are crucial to building a healthy society, and should form the basis for the pragmatic decisions that will get us there.

**Questions**:

1. Are market solutions a panacea for environmental problems? Give an example.
2. Taxing "bads" such as hazardous wastes could generate how much money per year in the U.S.? What could this money be used for?
3. Let's say that you discard two bags of garbage and your neighbor discards ten, but you both pay the same collection fee. Is this situation a market solution for lowering waste production? If not, how would it become so?

Answers on page 301

# Part Five

# Global Climate Change—Past, Present, and Future

# 44

*Climatic conditions in the early history of Earth have important consequences for the origin and evolution of life. The earliest portion of Earth's history included a period of intense bombardment by asteroids preventing the continual existence of life. With the decrease in number of large impacts life could exist continually. During this same period the continents began to develop and grow. The following article shows how the interaction of impacts, the formation of continents, and microorganisms during Earth's early history laid the foundations for the development of climatic conditions suitable for the evolution of the more complex organisms which populate the more familiar world of the Phanerozoic.*

# Asteroid Impacts, Microbes, and the Cooling of the Atmosphere

by Verne R. Oberbeck and Rocco L. Mancinelli

Earth's surface temperature constrained microbial evolution, according to Schwartzman et al. (1993). Their hypothesis states that the maximal temperature that extant organisms of a given type tolerate is the surface temperature occurring when that type of organism arose. Schwartzman and his colleagues concluded that the temperature changed from 100°C to 50°C between 3.75 billion years ago (BYA) and 1 BYA. These temperatures are consistent with those derived from oxygen isotope ratios in ancient sediments (Karhu and Epstein 1986, Knauth and Lowe 1978). The 100°C surface temperature they derive for 3.75 BYA is also the same as Earth's surface temperature in 4.4 BYA (Kasting and Ackerman 1986).

In this article, we address the cause of the delay in surface cooling until 3.75 BYA, and we explore the implications for microbial evolution of a high temperature on early Earth. We propose that three effects of the early heavy bombardment of Earth by asteroids and comets, until 3.8 BYA, could have delayed onset of surface cooling.

**In the beginning . . .**

Soon after Earth accreted 4.4 BYA, there was a global ocean, and Earth's atmosphere had as much as 20 atmospheres of carbon dioxide, which caused a high level of greenhouse heating. Temperatures could not have declined until the carbon dioxide was removed through weathering of continental rocks, which liberated metal ions that combined with carbonic acid to form carbonate rocks.

A considerable period of time may have been required to form the first continents. Planetesimals, asteroids, and comets that hit Earth's surface during the first 700 million years of its history could have created the first land masses. However, impact basins formed during the early heavy bombardment would also have volatilized any carbonate rocks that had formed as a result of weathering of continents; this volatilization would have released carbon dioxide back to the atmosphere and would have prolonged the initial high levels of greenhouse heating even in the presence of continents.

Finally, until 3.8 BYA, large impact events could have prevented the habitation of land masses by microbes that would, according to Schwartzman and Volk (1989) and Schwartzman et al. (1993), have been able to accelerate chemical weathering of continental rocks, removal of carbon dioxide from the atmosphere, and cooling. We propose that the sequence of microbial evolution consistent with the early heavy bombardment and a high temperature of early Earth is: heterotrophs, chemoautotrophs, and finally photoautotrophs.

**The case for high temperature on early Earth**

Earth's surface temperatures, derived from the biota (Schwartzman et al. 1993), are actually compatible with the temperature estimates that had been derived previously from oxygen isotope ratios of sediments (Knauth and Epstein 1976). Nevertheless, Perry et al. (1978) rejected the high-temperature interpretation of isotopic data for early Earth because they interpreted the Precambrian tillites to be glacial deposits. Theoretical climatic models indicated that, if glaciation occurred, the maximal Precambrian temperature must have been on the order of 20°C (Kasting and Toon 1989). However, because an as-yet-unknown fraction of Earth's ancient tillites and diamictites (clastic

***Reprinted with permission of the authors from BioScience, March 1993, vol. 44, no. 3, pp. 173-177, copyright 1994, American Institute of Biological Sciences.***

deposits resembling glacial tills) are deposits of impact craters rather than deposits of glaciers (Marshall and Oberbeck 1992, Oberbeck and Aggarwal 1992a, Oberbeck and Marshall 1992, Oberbeck et al. 1993, Rampino 1992, Sears and Alt 1992), the existence of early glaciations and a 20°C upper temperature limit for the Precambrian climate has been questioned (Schwartzman et al. 1993).

Perry et al. (1978) proposed that the variation in oxygen isotope ratios in sediments was due to the temporal change in $^{18}O$ in seawater, or alteration of sediments during burial in a closed system and not climatic variation. However, Knauth and Lowe (1978) published a study of Precambrian cherts, including those formed at the sediment/water interface, that appears to exclude the possibility of diagenesis. The temperature history of cherts between 3.5 BYA and the present is quite similar to that which has now been inferred from the evolution of the biota by Schwartzman et al. (1993). Finally, Holmden and Muehlenbach (1993) concluded, from the study of oxygen isotope profiles of two-billion-year-old ophiolite mineral deposits, that the low oxygen isotope ratios of Proterozoic rocks and cherts do not reflect temporal depletions of $^{18}O$ in seawater. Thus, we adopt the view that oxygen isotopic data suggest a high-temperature Precambrian climate.

Additional arguments have been given against early glaciation and thus for a high-temperature early Earth. Schermerhorn (1974) challenged the prevailing view that Precambrian tillites and diamictites, especially those that occur along ancient continental breakup margins, are of glacial origin. He identified the textural criteria that had, until 1974, been used to identify tillites and diamictites as glacial deposits, and he argued that the deposits could have been produced by tectonism. Additionally, Salop (1983) pointed out that sediments bounding many of the ancient tillites and diamictites are those formed in warm-water environments. He postulated that the abrupt low-temperature excursions at the tillite and diamictite horizons can only be explained by glaciations if glaciations occurred suddenly in the midst of a hot climate. Knauth and Lowe (1978) also suggested that glacial periods do not preclude hot interglacial periods.

The ideas of Schermerhorn (1974), Salop (1983), and Oberbeck et al. (Oberbeck and Aggarwal 1992a,b, Oberbeck and Marshall 1992, 1993) strengthen the longstanding, and previously largely ignored, isotopic evidence for a high temperature of early Earth (Knauth and Lowe 1978). To this evidence we can add the inferences of Schwartzman et al. (1993). It is noteworthy that the temperature curves derived by them are quite similar to those for the same time periods derived from isotope data by Knauth and Lowe (1978).

Schwartzman et al. (1993) pointed out that their temperature history (between 3.75 and 2.5 BYA) was consistent with isotopic data by higher than other estimates based on equilibrium conditions for mineral precipitation. For example, Walker (1982) cited primary evaporite gypsum precipitation in 3.5-billion-year-old sediments to determine the upper temperature limit of 58°C for surface-water temperature rather than 73°C, which would be the upper limit if the sediments were precipitated in freshwater. The limit would be even lower if gypsum was precipitated in seawater. However, Schwartzman et al. (1993) argued that metastable precipitation of gypsum is likely at temperatures far above its stability field and would be consistent with a surface temperature of 73°C at 3.5 BYA.

Because the isotopic data of Knauth and Lowe (1978) give the same temperature change with geologic history as those of Schwartzman et al. (1993), because it is becoming increasingly difficult to explain the oxygen isotope data by temporal changes in isotopic composition of seawater or diagenesis, and because Knauth and Lowe (1978) interpret the changes in isotopic ratios to reflect climatic variations, we adopt the surface temperature change. A key feature of this curve, for this study, is the 100°C temperature at 3.75 BYA that declines to 73°C by 3.5 BYA and defines the onset of cooling of the near-surface environment.

**The onset of cooling of the atmosphere**

We now consider the cause of the delay in cooling of the atmosphere until 3.75 BYA and the implications of this delay for microbial evolution. Earth's atmosphere during its first few hundred million years, just after condensation of the oceans, may have contained up to 20 bars of carbon dioxide. Kasting and Ackerman (1986) modeled the dependence of surface temperature of Earth's early atmosphere on such an atmospheric carbon-dioxide partial pressure, and they concluded that the temperature was in the range of 85-110°C. We combine the inferred surface temperatures after 3.75 BYA (Schwartzman et al. 1993) with a temperature of 100°C at 4.4 BYA. The temperature history suggests that from approximately 4.4 BYA to 3.75 BYA the surface temperature was constant at approximately 100°C. This prolonged constant surface temperature has important implications for cooling of atmospheres of bombarded planets and for microbial evolution.

We propose that the cause of the delay in the onset of cooling from the probable 100°C surface temperature at 4.4 BYA until 3.75 BYA was due to three effects of the early heavy bombardment of Earth that lasted until approximately 3.8 BYA. First, some period of time was required to form the continents so that continental rocks could have been available for chemical weathering and associated cooling from removal of carbon dioxide from the atmosphere. Large-impact cratering events, occurring during the early heavy bombardment, would have required some time to produce the first topographic dichotomy of continents and ocean basins (Frey 1980) so that weathering and cooling could occur. Second, the heavy bombardment also would have caused release of carbon dioxide back to the atmosphere by impact volatilization of carbonate rocks and thus sustained the high level of initial greenhouse heating until 3.8 BYA. Finally, the heavy bombardment could also have periodically sterilized Earth and delayed continuous habitation of microbial organisms on land masses and prevented acceleration of chemical weathering of silicate rocks and associated cooling until the end of the bombardment.

Consider the formation of the oceans and the time needed to form the silicate rocks of the continents. Chyba (1987) calculated that, if comets contained 10% water, then during the early bombardment comets supplied enough water to Earth to account for the current ocean volume. Frey (1980) argued that, as early as 4.4 BYA, Earth was differentiated into a sialic igneous outer crust with a higher-density mantle and a global ocean. In his view, large asteroid impacts had, by 4.0 BYA, produced the topographic dichotomy of continents and oceans. Isotopic data also suggest the continents first existed at some time between 4.3 and 4.0 BYA (Bowring et al. 1989). The initial global ocean, a product of comet impacts, would have prohibited removal of atmospheric carbon dioxide because rocks were not exposed to weathering. However, if land masses were in place after 4.0 BYA, cooling of the atmosphere by removal of carbon dioxide through weathering of surface siliate rocks could, in the absence of other factor, have been started at this time.

One such adverse factor is impact volatilization of carbonate rocks, which would have recycled carbon dioxide to the atmosphere and helped to sustain the high initial surface temperatures on Earth through greenhouse heating until the end of the heavy bombardment. This mechanism might have kept enough carbon dioxide in the atmosphere of Mars until the end of the heavy bombardment of the inner planets at 3.8 BYA, so that surface temperatures remained above freezing, liquid water existed, and surface runoff channels formed (Carr 1989). Still another impact mechanism could have retarded cooling until 3.8 BYA.

Microbial enhancement of continental rock weathering would have facilitated surface cooling (Schwartzman and Volk 1989) if microbes were present on continents at 4.0 BYA under the suggested habitable temperature conditions. However, another adverse effect of impacts may have delayed the weathering of continental rocks in place by 4.0 BYA. Recent developments in the field of impact catastrophism indicates that microbial life may not have existed continuously or for significant periods of time on continental land masses until 3.8 BYA. Maher and Stevenson (1988) advanced the notion that impacts of large planetesimals, asteroids, and comets during the heavy period of bombardment of Earth, until 3.8 BYA, could have periodically sterilized Earth, frustrated the origin of life, or killed any existing organisms.

Sleep et al. (1989) calculated that the last impacts that completely vaporized the ocean could have occurred as early as 4.4 BYA and as late as 3.8 BYA. Oberbeck and Fogleman (1989b, 1990) found that the last ocean-vaporizing impact could have occurred at 3.8 BYA. Therefore, life may not have been continuously present, even in the deep oceans, until after 3.8 BYA. If so, the lack of continuous life on Earth's land surfaces until after 3.8 BYA, due to the heavy bombardment, may be another cause for the delay in onset of cooling of the near-surface environment.

Heterotrophic and Chemoautotrophic organisms could have existed continuously in the deep oceans after 3.8 BYA, because there would then have been no ocean-vaporizing impacts that killed deep sea life but only ones that sterilized the surface., Therefore, microbes could have continuously evolved and colonized the land surfaces after 3.8 BYA from these continuously available stocks. The sustained decline in Earth's surface temperature may have been delayed until 3.75 BYA if ocean-vaporizing impacts occurred until the end of the bombardment, if intermittent but extensive land ecosystems were created only after 3.8 BYA, and if these microbial ecosystems were effective in accelerating removal of carbon dioxide from the atmosphere.

All of the impact mechanisms for delay in cooling seem consistent with the onset of cooling at 3.75 BYA as inferred here from the isotopic data (Knauth and Lowe 1978), inferences from biology (Schwartzman et al. 1993), and the probable temperature 4.4 BYA (Kasting and Ackerman 1986). Impacts initially produced a global ocean that prohibited rock weathering and cooling; they could also have been responsible for a variety of other processes that kept the 10-20 bars of carbon dioxide present in the atmosphere until 3.8 BYA. Then, 50 million years later, cooling began when continents existed, no ocean-vaporizing impacts occurred, continental microbial ecosystems were possible, and there was minimal impact recycling of carbon dioxide.

**Implications for microbial evolution**

If the earth's temperature was constant at 100°C until 3.75 BYA and then cooled rapidly to 73°C by 3.5 BYA, and if continental microbes accelerated this cooling of the surface (Schwartzman et al. 1993), organisms must have occupied the surface of continental land mass soon after the impact bombardment ended 3.8 BYA. The implication is that the surface organisms evolved rapidly from deep-sea organisms that would only have been present continuously after the last ocean-vaporizing impact no later than 3.8 BYA.

No direct evidence exists for evolution of life before 3.8 BYA, leading to controversy over the sequence of microbial evolution. We now link the impact history of early Earth (Chyba et al. 1990, Maher and Stevenson 1989, Oberbeck and Fogleman 1989a,b, 1990, Sleep et al. 1989), 16S rRNA phylogenetic trees (Woese and Pace 1993), climatology (Kasting and Ackerman 1986), and the hypothesis that the upper temperature limit for growth of microbial groups corresponds to the actual surface temperature of Earth at the time of the groups' first appearance (Schwartzman et al. 1993) to show that each of these methodologies independently suggests the same sequence of microbial evolution on Earth from 4.4 to 3.5 BYA. We propose that this sequence of evolution is: anaerobic heterotrophs, followed by chemoautotrophs, and finally photoautotrophs.

The early bombardment of Earth before 3.8 BYA prevented life from gaining a foothold at Earth's surface and constrained it to the dark, deep ocean (Maher and Stevenson 1988, Oberbeck and Fogleman 1989a,b, 1990, Sleep et al. 1989). On the other hand, the supply of organic mass from impacting comets (Clark 1988, Oro 1961) was also greatest during this period (Chyba et al. 1990, Oberbeck and Aggarwal 1992b, Oberbeck et al. 1989). We propose that comets were the source of organic nutrients that sustained a small biomass of organisms, limited by the distribution of organics, in the hot, deep ocean just after the heavy bombardment. Thus, we suggest that the earliest microbial forms were thermophilic anaerobic heterotrophs.

After 3.8 BYA, the rate of impacts decreased and the surface could be inhabited. This development had two effects: it decreased influx of organic matter, and it allowed the surface of Earth to become continuously inhabited. The limited supply of organic matter in the presence of abundant carbon dioxide at the surface then gave a selective advantage to chemoautotrophs over heterotrophs. According to the temperature at which these types of organisms now live (up to 110°C; Kristjansson and Stetter 1992), this change in advantage may have occurred at approximately 3.75 BYA.

Removal of carbon dioxide from the atmosphere through chemoautotrophy then allowed Earth to cool more rapidly than by abiotic processes alone. We suggest that chemoautotrophy evolved earlier than photoautotrophy because chemoautotrophs can live in deeper, dark regions of the ocean, which would have been habitable before surface environments, and chemoautotrophs can tolerate higher temperatures than photoautotrophs. Thus, impacts together with chemoautotrophy may have determined the time of onset of cooling of Earth's atmosphere at 3.75 BYA. Then photoautotrophy evolved.

Surface-sterilizing impacts capable of vaporizing the photic zone of the oceans occurred after the last ocean-vaporizing impact, perhaps as late as 3.5 BYA (Oberbeck and Fogleman 1989a,b, 1990, Sleep et al. 1989). Thus, photoautotrophs could have been intermittently present after 3.8 BYA and continuously present at Earth's surface after 3.5 BYA. This presence was also implied by the microbial evolution scheme put forward by Schwartzman et al. (1993), in which the temperature tolerance of existing organisms reflects Earth's surface temperature when that type of organism first evolved. The 73°C maximal tolerance observed in existing photoautotrophs (Meeks and Castenholz 1971) is thus the temperature of Earth's surface at 3.5 BYA, just after the last surface-sterilizing impact. It would not be surprising if modern photoautotrophs would have the same upper limit of tolerance to temperature as the cyanobacterialike organisms of 3.5 BYA represented by the fossil record (Schwartzman et al. 1993).

We propose that the sequence of microbial evolution, which is consistent with impact constraints and the hypothesis that temperature constrained microbial evolution (Schwartzman et al. 1993), is as follows: anaerobic heterotrophs followed by chemoautotrophs (extant examples include *Pyrococcus*: maximal temperature, 105°C; and *Pyrodictium*: maximal temperature, 110°C; Kristjansson and Stetter 1992), with the photoautotrophs arising later (e.g., *Synechococcus lividus*: maximal temperature, 73°C; Meeks and Castenholz 1971). The branching order depicted by 16s rRNA phylogenetic trees also suggests that photoautotrophy arose after heterotrophy and chemoautotrophy heterotrophy and chemoautotrophy (Woese and Pace 1993). We suggest that heterotrophy arose no later than 3.8 BYA, chemoautotrophy arose approximately 3.75 BYA, and photoautotrophy arose approximately 3.5 BYA.

The climate history and path of evolution of the biosphere of any habitable planet may be fixed, in part, by its unique early-impact history. The decline in surface temperature may have been determined by the impact history of Earth and by microbial evolution.

Consider the possibilities if the impact history had been different. If the heavy bombardment of Earth had been more prolonged, the plateau of constant temperature may have extended in time. In that event, the surface temperature would not have reached habitable levels for many types of organisms until much later. If so, the evolution of chemoautotrophy and photoautotrophy from the heterotrophs may have proceeded later, which could have delayed the evolution of more complex life forms. The implied synergism between astrophysical environment and the biosphere suggests that global change in climate and the biosphere has its roots in astrophysics.

**References cited**


Bowring, S.A. 1989. 3.966a gneisses from the Slave province, Northwest Territories, Canada. *Geology* 17: 971-975.

Carr, M.H. 1989. Recharge of the early atmosphere of Mars by impact-induced release of $CO_2$. *Icarus* 79:311.

Chyba, C.F. 1987. The cometary contribution to the oceans of primitive Earth. *Nature* 330: 632-635.

Chyba, C.F., P.J. Thomas, L. Brookshaw, and C. Sagan. 1990. Cometary delivery of organics to the early Earth. *Science* 249: 366-373.

Clark, B.C. 1988. Primeval procreative comet pond. *Origins Life Evol. Biosphere* 18: 209-238.

Frey, H. 1980. Crustal evolution of the early earth: the role of major impacts. *Precamb. Res.* 10:195-216.

Holmden, C., and K. Muehlenbach. 1993. The $^{18}O/^{16}O$ ratio of 2-billion-year seawater inferred from ancient oceanic crust. *Science* 259: 1733-1736.

Karhu, J., and S. Epstein. 1986. The implication of the oxygen isotope records in coexisting cherts and phosphates. *Geochim. Cosmochim. Acta* 50: 1745-1756.

Kasting, J.F., and T.P. Ackerman. 1986. Climatic consequences of very high carbon dioxide levels in the Earth's early atmosphere. *Science* 234:1383-1385.

Kasting, J.F., and O.B. Toon. 1989. Climate evolution on the terrestrial planets. Pages 423-449 in S.K. Atreya, J.B. Pollack, and M.S. Matthews, eds. *Origin and Evolution of Planet and Satellite Atmospheres.* University of Arizona Press, Tucson.

Knauth, L.P. and S. Epstein. 1976. Hydrogen and oxygen isotope ratios in nodular and bedded cherts. *Geochim. Cosmochim. Acta* 40: 1095-1108.

Knauth, P.L. and D.R. Lowe. 1978. Oxygen isotope geochemistry of cherts from the onverwacht group (3.4 billion years), Transvaal, South Africa, with implications for secular variations in the isotopic composition of cherts. *Earth and Planetary Science Letters* 41:209-222.

Kristjansson, J.K., and K.O. Stetter. 1992. *Thermophilic Bacteria* CRC Press, Boca Raton, FL.

Maher, K.A., and D.J. Stevenson. 1988. Impact frustration of the origin of life. *Nature* 331:612-614.

Marshall, J.R., and V.R. Oberbeck. 1992. Textures of impact deposits and the origin of tillites. *American Geophysical Union EOS Suppl.* 73(43):324 (abstract P31B-4).

Meeks, J.C., and R.W. Castenholz. 1971. Growth and photosynthesis in an extreme thermophile, *Synechococcus liviclus* (Cyanophyta). *Arch. Mikrobiol.* 78:25-41.

Oberbeck, V.R., and H. Aggarwal. 1992a. Impact crater deposit production on Earth. Pages 1011-1012 in *Lunar and Planetary Science Conference Abstract.* Lunar and Planetary Science Institute, Houston, TX.

—. 1992b. Comet impacts and chemical evolution on the bombarded earth. *Origins Life Evol. Biosphere* 21:317-338.

Oberbeck, V.R., and G. Fogleman. 1989a. Estimates of the maximum time required to originate life. *Origins Life Evol. Biosphere* 19:549-560.

—. 1989b. Impacts and the origin of life. *Nature* 339:434.

—. 1990. Impact constraints on the environment for chemical evolution and the continuity of life. *Origins Life Evol. Biosphere* 20:181-195.

Oberbeck, V.R., and J.R. Marshall. 1992. Impacts, flood basalts, and continental breakup. Pages 1013-1014 in *Lunar and Planetary Science Conference Abstracts.* Lunar and Planetary Science Institute, Houston, TX.

Oberbeck, V.R., J.R. Marshall, and H.R. Aggarwal. 1993. Impacts, tillites, and the breakup of Gondwanaland. *Journal of Geology* 101:1-19.

Oberbeck, F.R., C.P. Mckay, T.W. Scattergood, G.C. Carle, and J.R. Valentin. 1989. The role of cometary particle coalescence in chemical evolution. *Origins Life Evol. Bioshpere* 19:549-560.

Oro, J. 1961. Comets and the formation of biochemical compounds on the primitive earth. *Nature* 190:389-390.

Perry, E.C. Jr., S.N. Ahmad, and T.M. Swulius. 1978. The oxygen isotope composition of 3,800 m.y. old metamorphosed chert and iron formation from Isukasia West Greenland. *Journal of Geology* 86:223-239.

Rampino, M. 1992. Ancient "glacial" deposits are ejecta of large impacts: the ice age paradox explained. *American Geophysical Union EOS Suppl* 73(43):99 (abstract A32C-1).

Salop, L.J. 1983. *Geologic Evolution of the Earth During the Precambrian.* Springer-Verlag, New York.

Schermerhorn, L.J.G. 1974. Later Precambrian mixtites: glacial or nonglacial? *Am. J. Sci.* 274:673-824.

Schwartzman, D., M. McMenamin, and T. Volk. 1993. Did surface temperatures constrain microbial evolution? *BioScience* 43:390-393.

Schwartzman D.W., and D. Volk. 1989. Biotic enhancement of weathering and the habitability of Earth. *Nature* 340:457-460.

Sears, J.W., and D. Alt. 1992. Impact origin of large intercratonic basins in the stationary Proterozoic crust and the transition to modern plate tectonics. Pages 385-392 in M.J. Barthlomew, D.W. Hyndman, D.W. Moqk, and R. Masson, eds. *Basement Tectonics: Characterization and Comparison of Ancient and Mesozoic Continental Margins.* Kluwer Academic, Dordrecht, The Netherlands.

Sleep, N.H., K.J. Zahnle, J.F. Kasting, and H.J. Morowitz. 1989. Annihilation of ecosystems by large asteroid impacts in the early earth. *Nature* 342:139-142.

Walker, J.C.G. 1982. Climatic factors on Archean Earth. *Palaeogeogr. Palaeoclimatol. Palaeoecol.* 4:1-11.

Woese, C.R., and N.R. Pace. 1993. *The RNA World.* Cold Spring Harbor Laboratory Press, Cold Spring Harbor, NY.

**Questions**:

1. How did the presence of continents on the early Earth affect climate?

2. What three effects of heavy bombardment influenced the onset of cooling of the Earth?

3. How did chemoautotrophs promote cooling of the Earth?

Answers on page 301

# 45

*The Earth's climate is strongly influenced by atmospheric "greenhouse" gases, such as carbon dioxide, that trap heat. Increased amounts of these gases will cause global warming. We hear a lot about global warming today, but nature has been altering the amount of greenhouse gases in the atmosphere for millions of years. Increased volcanic activity will increase carbon dioxide emissions, for example. Plants affect carbon dioxide in many ways. One way is their absorption of carbon dioxide through photosynthesis, removing it from the atmosphere. For these and other reasons, many experts have suggested that the evolution of land plants must have greatly diminished global temperature. But the complex effect of animals on plants has often been neglected in this scenario. An interesting hypothesis discussed in this article is that the appearance of highly successful, widespread plant eaters on land has sometimes reduced plant activity, including weathering, leading to warmer climates. If true, this has worrisome implications for humans who are rapidly removing wild vegetation and perhaps enhancing the greenhouse effect caused by fossil fuel burning.*

## The War Between Plants and Animals

### Its stakes are high, says paleontologist Paul Olsen. Does carbon belong in the atmosphere or the deep ocean? Should Earth be hot or cold?

by Carl Zimmer

To most people the sight of antelopes grazing placidly in a meadow would epitomize the peaceful balance of nature. But Paul Olsen, a paleontologist at the Lamont-Doherty Earth Observatory in New York, has come to see things differently. To him that meadow is just the latest battlefield in the primordial struggle between plants and the animals that eat them. And that war, Olsen claims, has had global repercussions: as its front line has surged back and forth over the past 450 million years, it has dragged Earth's climate along with it by changing the amount of carbon dioxide in the atmosphere.

Over geologic time Earth's climate depends first and foremost on carbon dioxide, which traps heat and creates a greenhouse effect. The molecules of $CO_2$ in the air at any moment are part of a vast cycle that involves the entire planet. Volcanoes steadily emit $CO_2$, which dissolves in water droplets in the atmosphere to form carbonic acid and falls to the ground as rain. The carbonic acid reacts with rocks, releasing ions of bicarbonate that flow into rivers and on to the ocean—a process known as weathering. In the ocean, plankton use the bicarbonate to build shells of calcium carbonate. When they die and settle onto the seafloor, their shells turn to limestone, and the carbon returns to the solid Earth.

Plants on land can alter this cycle by speeding up weathering. They do so in two ways. As they make organic matter through photosynthesis, they take up $CO_2$, thus providing a separate avenue by which it leaves the atmosphere. Some of the $CO_2$ gets pumped out through the plants' roots when they respire, and the rest is released when the plants die and are decomposed by bacteria. The $CO_2$ reacts with water in the soil to form carbonic acid—the weathering agent. Most of the carbonic acid involved in weathering today comes from the soil rather than from rain. And the soil itself, which is also produced by decomposing plants, speeds up weathering in a second way: it acts as a sponge that traps the carbonic acid and keeps it in close contact with the rock.

Although weathering can and does take place in the absence of plants, the best guess is that plants accelerate the process by a factor of ten. For some time now many researchers have argued that the evolution of land plants must therefore have had a profound effect on climate: by removing carbon from the atmosphere and putting it in the deep ocean, plants have diminished the greenhouse effect and cooled the planet.

But five years ago, as Olsen was preparing to teach a course on dinosaurs at Columbia University, he started to wonder whether climate researchers were actually missing half the picture. "Animals are completely left out of any theoretical studies of long-term climate change," he says. The assumption has been that herbivores don't have a large-scale effect on vegetation—yet anyone who has studied a modern ecosystem, says Olsen, knows that assumption is false. Goats can turn lush islands into bare humps of dirt. Elephants eat or knock over so many trees that they can transform dense jungles into open forests.

Nor is such destructiveness a modern invention: dinosaurs were probably the greatest plant eaters that ever lived. If the success of plants has a net cooling effect on Earth's climate, Olsen realized, the success of herbivores should have the opposite effect. The more plant eaters there are, the fewer plants there are and the more slowly rocks weather. The more slowly rocks weather, the more $CO_2$ remains in the air. The more $CO_2$ remains in the air, the warmer the climate becomes. "It became obvious that this was a theory I couldn't exclude," says Olsen. "That's how you go off on these roads. You want to try to exclude it and you can't, so you have to explore it."

Olsen's exploration has now taken him to a hypothesis encompassing not just dinosaurs but the entire history of life on land. In his view, herbivores evolve to eat plants, the plants evolve ways of eluding the herbivores, and the herbivores evolve again. As each side gets the upper hand, it changes the weathering rate, which in turn steers the planet between balmy and cold climates. The evidence for this scenario, says Olsen, is the record preserved in rocks—fossils that show the abundance of animals and plants at various epochs, isotopes that indicate how much $CO_2$ was in the air, and formations that trace the advance and retreat of glaciers.

The war began quietly about 450 million years ago, when plants first invaded land. The first primitive lichens and mosslike species reproduced slowly, had no roots, and consequently didn't cause much weathering. Around 400 million years ago, though, vascular plants evolved. They were able to extract water and nutrients from the ground with roots and carry them to the top of an upright trunk. That allowed them to colonize drier land.

Leaves and seeds came next, and they helped the plants' cause even more. By 350 million years ago trees with three-foot-wide trunks forested many parts of the world. One sign of how well plants did is the vast amount of coal—plant matter that was buried and compressed in swamps—that dates back to the Paleozoic Era. The creation of coal itself helped remove $CO_2$ from the air, but plant-induced weathering removed five times more. And shortly after forests became widespread, a chill gripped Earth for 30 million years.

But *why* did plants do so well in the Paleozoic? One reason, says Olsen, is that nothing was around yet to eat them. Vertebrates first came on land about 360 million years ago, but almost without exception they were carnivores: reptiles and amphibians feeding on fish, insects, or one another. Evolving a digestive system able to handle terrestrial plants—complete with grinding teeth, symbiotic gut bacteria, and a powerful gastrointestinal tract—is no easy thing, and it apparently took reptiles tens of millions of years to do so.

Once they did, though, the reward was huge. The first herbivores appeared 300 million years ago, and by 260 million years ago fossil remains show that ecosystems had changed over to the ecological pyramid we're accustomed to, with a few carnivores at the top and huge numbers of plant eaters at the bottom. As the herbivores feasted on plants, the formation of coal all but stopped. Meanwhile, Olsen argues, the weathering cycle must have slowed drastically, leaving more $CO_2$ in the atmosphere and heating up the planet. And in fact the rock record suggests that glaciers and polar ice caps were retreating just as the number of herbivores was increasing.

For the next 200 million years, Earth basked in a heat wave as plants struggled under attacks from herbivores. At first the treetops (where trees tend to keep their reproductive organs) were spared because the reptiles were stocky, four-legged grazers. But around 220 million years ago, tall dinosaurs appeared, some bipedal and others long-necked. When plants grew higher, dinosaurs grew longer necks.

The balance began to shift again when plants invented the flower. Flowering plants (known as angiosperms) boast enclosed reproductive systems and rugged seeds, which allow them to reproduce much faster than their ancestors. At first, 100 million years ago, they were small bushes living in marginal areas. By 80 million years ago they were spreading quickly through the understory of forests. New, ground-grazing dinosaurs such as *Triceratops* evolved to feed on them, but the angiosperms continued to explode across the landscape. At the same time, the planetary hothouse began to cool.

We'll never know if dinosaurs could have beaten back the angiosperms. A comet, many geologists now believe, crashed into the coast of Mexico 65 million years ago, creating a global pall of dust and sulfuric acid droplets that killed off so many plants that the animals that fed on them starved. The dinosaurs and many other animals disappeared forever.

The impact itself would have caused only a few months of cooling, Olsen points out—yet the climate actually kept cooling for millions of years. "If you wipe out all the herbivores, you set the clock back to zero," he explains. The fast-growing angiosperms recovered from the impact, and without any herbivorous dinosaurs left to bother them, they eventually created gigantic forests. The weathering rate sped up again, and more greenhouse gas was converted to limestone.

Mammals now had to play the same game of evolutionary catch-up as the reptiles had 300 million years earlier. It wasn't until 10 million years after the death of the dinosaurs that the ancestors of such modern plant eaters as buffalo, horses, elephants, and rabbits appeared. Once they did, the climate went through a relatively brief warming. Tropical forests spread to higher latitudes even as the total amount of vegetation on the planet was decreasing.

In the meantime, though, the plant world had discovered its most recent and powerful weapon: grass. Grasses can spread rapidly by cloning or by means of seeds that are resistant to drought and cold, and some require less water and less light than other plants. Grass is also poor food because its leaves are loaded with bits of silica. Mammals have had to evolve new equipment for handling this rough food, such as wear-resistant teeth and multichambered stomachs. In Olsen's view, they haven't caught up with plants yet. When grasses became widespread 20 million years ago, Earth slid into the current cycle of ice ages.

It's an elegant hypothesis, but is it true? Collecting the evidence to find out for sure is likely to take years. Olsen himself says, "I don't *know* that herbivores suppress weathering. I only see beautiful coincidences in timing in the fossil record." Other factors that affect the carbon cycle, such as bursts of mountain building (which exposes more rock to weathering) or volcanic eruptions, may turn out to have more impact on climate than the plant-herbivore war.

But if Olsen is right, the implications for the future could be profound. After all, we humans are the best plant destroyers since the dinosaurs. In a geologic twinkling we have replaced 40 percent of the planet's wild vegetation with crops we harvest (thus preventing much of their carbon from entering the soil), pastures we graze our livestock on, and houses we build from trees we cut down. "Over thousands or tens of thousands of years," says Olsen, "I believe that the effects of our changes to weathering are bigger and more important than the increase in atmospheric $CO_2$ from burning fossil fuel." ❑

**Questions**:

1. Why did it take Paleozoic vertebrates so long to become plant eaters?
2. When did grasses become widespread? What happened to Earth's climate then?
3. How do plants accelerate weathering?

Answers on page 301

*Carbon dioxide is the major greenhouse gas in the Earth's atmosphere. It has been one of the primary controls on global temperature since the origin of the Earth. In fact, when scientists examine the history of global climate change they see a strong correlation between carbon dioxide levels and temperature. There are however a few exceptions: the Ordovician 440 million years age and the Cretaceous 95 million years ago. During the Ordovician, an ice sheet existed which was almost the size of the present Antarctic ice sheet despite the fact that the carbon dioxide level was sixteen times their present level. During the Cretaceous, global average temperature was similar to the present despite a carbon dioxide level eight times the present level. It appears that the explanation for these discrepancies lies in the location of the continents. Factors such as latitude, whether the continents are combined or separated, and continental area covered by sea all affect patterns of ocean circulation, heat transport and cloud cover. These factors in turn allow temperature to stray from that predicted by carbon dioxide levels alone and show just how complex global climate change can be.*

# Location, Location, Location

Carbon dioxide in the atmosphere warms Earth. But just how much warming you get depends on where you put your continents.

by Carl Zimmer

If adding carbon dioxide to the atmosphere creates a greenhouse effect that warms Earth, it must have happened in the past. That's why paleoclimatology, once a small and esoteric field, is such a growth industry these days, with legions of geologists trying to glean past temperatures and $CO_2$ levels from rocks, and legions of climate modelers trying to tell us what it all means—not only for the past but also for the future of Earth's climate. On the whole, the results have been what you'd expect. "When carbon dioxide levels were low, the climate was cold, and when they were high, the climate was warm," says climatologist Thomas Crowley of Texas A&M University.

But lately two glaring exceptions to that simple rule have turned up. During the Ordovician Period, 440 million years ago, there seems to have been 16 times as much carbon dioxide in the atmosphere as there is today—and yet, judging from the gravelly deposits it left behind, there was also an ice sheet near the South Pole that was four-fifths the size of present-day Antarctica. The second exception is even more troubling. The Cretaceous Period, when dinosaurs ruled the Earth and $CO_2$ levels were about eight times what they are today, has been one of the most popular case studies for global warming forecasters. And everyone knows what the climate was like during the dinosaurs' heyday: steamy. Or was it? The latest evidence, reported just this past summer by British researchers, suggests that temperatures in the tropics 95 million years ago were no higher than they are now; and while it was a lot warmer at the poles than it is today, it was still freezing cold.

What happened to Earth's greenhouse during these two periods? Climate modelers are beginning to believe the solution to both puzzles may be the same: geography. Carbon dioxide does tend to warm the planet—no one is questioning that—but the climate you actually end up with depends to a great degree on how you arrange your continents.

In the case of the Ordovician, Crowley thinks, the solution is fairly straightforward. He and his co-worker Steven Baum have spent the past couple of years recreating the Ordovician Earth in their computer and trying to understand how it could have supported glaciers. They got help from the sun: according to astrophysicists it was 4 percent dimmer 440 million years ago than it is today, which means that although the Ordovician greenhouse had 16 times as much heat-trapping $CO_2$, it also had less heat to trap. But Crowley and Baum calculated that the net greenhouse effect was still equivalent to what you'd get by quadrupling $CO_2$ levels today. In other words, an ice sheet should have stood little chance of surviving. The survival of a permanent ice sheet depends less on whether it gets cold enough to snow in winter than on whether it gets warm enough in summer to melt all the previous winter's snow. And the key to the Ordovician ice sheet, says Crowley, is the fact that most of the continents were joined together into one roughly circular land mass called Gondwanaland, whose southern edge was just over the South Pole 440 million years ago. In the interior of the supercontinent the climate was extreme Midwestern: cold winters and hot summers and no permanent ice. But along the coast, the ocean—which warms far more slowly than the land because of its tremendous heat-storage ca pacity—put a damper on this seasonal cycle. "The moderating effect of the water mutes the amount of summer warming you get," says Crowley.

People who live in places like Maine today know what "muted" summers are like; along the southern coast of Gondwanaland 440 million yearsago, the summers would have been worse than muted. The wind blowing in from offshore was blowing over water that was right at the South Pole, and it was cold indeed. The global average temperature in Crowley and Baum's simulated Ordovician was 64 degrees—14 degrees hotter than today. But along the southern margin of Gondwanaland, three feet of snow survived the summer each year, becoming part of thickening ice sheets.

Then around 430 million years ago the ice sheets disappeared. Crowley and Baum's simulations point to an explanation that sounds paradoxical only at first: Gondwanaland was drifting southward at the time, and the farther south it went, the smaller the ice sheets became. As the center of the supercontinent came closer to the South Pole, the winters in the interior became colder—but the land still warmed up enough in summer to melt the snow. Meanwhile, what had been the southern, glaciated edge of Gondwanaland moved north into warmer waters. The glaciers soon vanished.

Thus the Ordovician mystery no longer seems so mysterious. The Cretaceous Period, though, is another story. Over the years, various analyses of the carbon locked in Cretaceous limestones—including the great chalk beds, formed from the corpses of countless plankton, that gave the period its name—have convinced researchers that the $CO_2$ level in the Cretaceous atmosphere was eight times what it is today. Analyses of oxygen isotopes, meanwhile, have suggested that the Cretaceous climate was appropriately toasty—perhaps 20 degrees warmer than today. A hothouse climate would explain why paleontologists have found fossils indicating that tropical plants and reef corals survived at higher latitudes in the Cretaceous than they do today.

Yet the evidence for the hothouse, says geologist Bruce Sellwood of the University of Reading, has never been conclusive. As a snapshot of global climate history, the rocks that went into the isotopic analyses were far from perfect. For one thing, they differed in age by as much as 30 million years. For another, most of them came from the northern mid-latitudes, and almost none from the tropics. "That just reflects the availability of chalk outcrops in Europe and North America," explains Sellwood.

He and his colleagues Greg Price and Paul Valdes set out to get a sharper picture of the Cretaceous climate. They limited their analysis to sedimentary rocks dating from a relatively small window of time around 95 million years ago. "We still have 7 million years to play with, but it's a hell of a lot better than the generalizations you got before," says Sellwood. The Reading workers also got some of the first solid information on tropical temperatures in the Cretaceous by analyzing 95-million-year-old sediments drilled recently from the ocean floor.

The results suggest that the Cretaceous was much cooler than previously thought. Temperatures at the poles were supposed to have averaged around 50 degrees, but Sellwood's group claims they actually hovered around freezing.

And the tropics, previously thought to have been 10 degrees warmer than today, were no warmer at all.

Why was the planet so cool, given all that $CO_2$ in the atmosphere, and why was there so little difference between the poles and the tropics? Like Crowley and Baum, Valdes is looking for answers in computer models. And like them, he is finding that geography matters.

Earth 95 million years ago was a world of shallow seas. The landmasses no longer formed a supercontinent, but they remained close together, and because sea levels were high during the Cretaceous, the oceans flooded into the interiors of many continents. One of these shallow seas, for example, cut North America in half from Canada to Mexico. And what all of them did, according to Valdes, was carry humid air into the heart of continents, where it created heavy cloud cover that blocked sunlight and cooled the Earth. "There are a lot of arguments and uncertainties about clouds," Valdes acknowledges—clouds can also trap heat rising from Earth's surface—"but in our model the clouds increase and create cooling."

Geography may also help explain why polar temperatures in the Cretaceous were so close to tropical ones. In today's world, warm water is carried north in the Atlantic to just south of Greenland, where it cools, sinks to the bottom of the ocean, and flows back south. Valdes thinks that in the Cretaceous this conveyor belt circulation may have been shallower—which would have made it faster at transporting heat out of the tropics. "In the Cretaceous, the gap between America and Europe was very small," he says. "There wasn't much of an Atlantic. So maybe the water didn't go down deep, it just recirculated and warmed up more and more, and so you didn't have such a big temperature gradient. I'm waving my hands around, but it's a strong hypothesis you can test"—which is what Valdes hopes to do in the near future with a new model that incorporates realistic ocean currents.

The computer models that have been used to forecast global warming don't incorporate realistic ocean currents, and their geography is pretty crude, too. Sellwood thinks that's a reason to be more skeptical of their forecasts—although no less cautious about polluting the atmosphere. "This research tells us that the link between carbon dioxide and global warming isn't as secure as we all imagine," he says. "But what I would hate to see happening is governments leaping on this and saying, 'Ah, there is no link between carbon dioxide and climate, and therefore we can go on with what we're doing now or even worse.' This work shows the whole climate system is much more complex than we imagined. That's actually a word of warning. To pump out huge amounts of carbon dioxide when we don't know how the system works is a pretty dangerous thing." ❑

**Questions**:

1. What caused the Ordovician climate to be colder than expected?
2. What caused the Cretaceous climate to be colder than expected?
3. What is the problem with climate models used to predict future global warming?

Answers on page 301

47

*With the present concern over carbon dioxide and global warming, scientists have become increasingly interested in determining changes in carbon dioxide levels throughout the Earth's history. Because no actual air samples have been preserved from more than about 250,000 years ago, carbon dioxide levels must be estimated using various means such as global climate models or stable isotopes. Recently a new method has been found: the stomatal density of fossil leaves. Tiny holes on leaves, called stomata, allow the transfer of gases in and out of the leaf. Research has shown that the density of stomata depends on the concentration of carbon dioxide in the atmosphere. At high levels of carbon dioxide, leaves have a lower density of stomata because less air needs to be drawn in to get the same amount of carbon dioxide. The carbon dioxide concentrations estimated from stomatal density of fossil leaves correlates well with expected climates based on other data such as fossil pollen. As another proxy for past carbon dioxide concentrations, stomatal density of fossil leaves may help us to understand the processes causing changes in carbon dioxide concentration in the atmosphere.*

# The Leaf Annals

## Fossil foliage puts scientists on the scent of ancient air

by Richard Monastersky

It is the breath of the planet. Volcanoes exhale it in flaming belches, while plants and eroding rocks quietly suck it out of the air, ultimately storing it underground.

Carbon dioxide flows into and out of the Earth so much that its atmospheric concentration hardly remains constant. Over millions of years, it varies whenever the planet's metabolism kicks into high gear or downshifts. Over much shorter time spans, carbon dioxide concentrations rise and fall in concert with modulations of ocean currents or vegetation patterns. On top of such natural cycles, humans have overlaid their own imprint by burning fossil fuels, pouring billions of tons of carbon dioxide into the air each year.

As a greenhouse gas, carbon dioxide helps set the planet's thermostat, so any change in its atmospheric abundance amounts to a twist of the temperature dial. Knowing how such gas variations tweaked the ancient climate can help earth scientists predict what lies in store for the next century as pollution drives carbon dioxide levels skyward. But to understand the past climate, researchers need to find a means of gauging how much carbon dioxide the air held long ago.

Though they would love to discover actual samples of the atmosphere from millions of years ago, scientists have little hope of finding anything that old because gases, unlike bone and wood, do not fossilize. The oldest air ever found comes from bubbles trapped in ancient ice, but these date back only 250,000 years—a brief tick in geologic time. The carbon dioxide exhaled by the human ancestor Lucy and her kindred australopithecines 3 million years ago has long since disappeared from the atmosphere, leaving only distant clues in its place. Air today, gone tomorrow.

But a new technique offers hope for tracing carbon dioxide levels back in time. In the June 18 SCIENCE, paleobotanists report that fossil leaves can open a window for probing the ancient atmosphere.

"This is a pioneering attempt to get a handle on the relationship between carbon dioxide fluctuations in the atmosphere and climate change," says David L. Dilcher of the University of Florida in Gainesville. Dilcher collaborated on the leaf study with Henk Visscher, Johan Van Der Burgh, and Wolfram M. Kürschner of Utrecht University in the Netherlands.

Led by Visscher, the research team developed a technique for gauging ancient carbon dioxide levels by counting the number of pores on the underside of fossilized leaves. These holes, called stomata, serve as miniature mouths, providing openings through which gases can enter and leave a leaf.

Research on vegetation growing in special greenhouses has shown that the density of stomata depends on the atmospheric concentration of carbon dioxide, which plants use during photosynthesis. Trees and shrubs grown in air with extra carbon dioxide develop fewer stomata, presumably because they can obtain the needed carbon dioxide without as many pores. Cutting down on the density of stomata helps a plant because it minimizes the amount of moisture that can escape when the pores open.

Outside the greenhouse, plants exhibit the same relationship between carbon dioxide and stomata—a fact made clear in the years following the Industrial Revolution. Over the last two centuries, carbon dioxide concentrations in the atmosphere have climbed from 280 to 355 parts per million (ppm). During that same period, trees in European temperate forests have decreased their stomatal density by 40 percent, according to findings reported in 1987 by F.I. Woodward of the University of Cambridge in England. Woodward made his discovery by examining preserved plants from the last 200 years that are kept in special repositories called herbaria.

Visscher and his colleagues leafed their way even further back in the climate chronicles by studying fossilized plants dating back 2.5 million to 10 million years. The mummified leaves, found in northwestern Germany and the Netherlands, belong to the European species *Quercus petraea*, or durmast oak, which currently grows from the Mediterranean to southern Scandinavia. To compare leaf samples, the researchers calculated what they call the stomatal index—a value that depends on the number of stomata and the total number of epidermal cells in a particular patch of leaf.

Charting the changes through time, Visscher's group found that the index averaged 10 for the oldest leaves, rose to a high of 16 a few million years later, and then dropped back to 9.5 for leaves from 2.5 million years ago. Because a high index means less carbon dioxide in the air, the leaf record suggests that concentrations of the gas were relatively high 10 million years ago, declined around 7 million years ago, and later rose toward their previous values.

The paleobotonists compared this scenario with a climate tale told by fossilized pollen, fruit, and seeds found in the same region. These indicate that northern Germany 10 million years ago had a warm-temperate to subtropical climate similar to that of Georgia and Florida, says Dilcher. The region cooled over the next few million years, causing a temperate forest to replace its predecessor. But by 2.5 million years ago, temperatures had rebounded, and warm-temperate conditions returned.

Because carbon dioxide concentrations and temperature followed each other closely through this span, the researchers proposed that the gas shifts played a role in causing the climate changes.

To flesh out the story, they sought to determine how much carbon dioxide the atmosphere held. This involved a leap-frogging technique, starting in the present and then jumping into the past. The researchers used living and preserved plants to devise an equation relating the stomatal index to carbon dioxide concentrations for modern times. That equation served to translate stomatal indexes into gas concentrations for times in the distant past. This technique suggests that gas levels 10 million years ago averaged around 370 ppm, then dropped to 280 ppm during a cool spell, and later rose back to 370 ppm during a warmer time.

If correct, these findings carry some sobering implications. Carbon dioxide concentrations are now nearing the top of the range calculated from the fossil leaves. In the past, northern Germany has had a warm-temperate to subtropical climate when carbon dioxide concentrations reached such high levels. But at the present the region maintains a temperate climate, which is characteristic of periods with less carbon dioxide. According to Visscher's group, this mismatch suggests that the climate has not yet responded to the rapid buildup of greenhouse gases since the Industrial Revolution.

"This is one more piece of evidence suggesting that, given the recent changes in carbon dioxide, we would expect to have a climate that reflects the kind of climate we reconstruct for the past, which was warmer than today," Dilcher says.

The researchers may have ventured too far out on a limb, however, by drawing such conclusions from fossil leaves, some fellow researchers contend.

Fakhri A. Bazzaz, a plant physiologist at Harvard University, says many factors aside from carbon dioxide influence stomatal density. Variations in humidity or light conditions, for instance, might spur plants to alter their leaf structure and number of pores.

Bazzaz cites another complicating factor. In his own experiments, he finds that individual species respond quite differently when growing in air enriched with carbon dioxide. Bazzaz says physiologists need to test a variety of species in different environments to see how each responds to various gas levels.

"Before this issue is laid to rest and this technique is extensively used to make inferences about the past climate, I would like to see more elaborate and detailed experimental studies done," Bazzaz contends.

Earth scientists, as well, question the reliability of stomatal studies, but the technique has attracted their attention because it could potentially fill an important gap in studies about the past atmosphere.

"If this would work, it would be very, very nice," says Thure Cerling, a geochemist at the University of Utah in Salt Lake City. Cerling has estimated atmospheric carbon dioxide concentrations for the past by examining the ratio of two carbon isotopes in ancient soils. But this technique has a large margin of error—about 400 ppm—so it works best for times in the distant past when carbon dioxide concentrations ranged in the thousands of parts per million. For more recent periods, when carbon dioxide levels resembled those of modern times, it can only offer broad hints.

John M. Hayes, a geochemist at Indiana University in Bloomington, is working on a process for gauging atmospheric carbon dioxide from isotopic information stored in sea sediments. While this approach can offer insight into why carbon dioxide concentrations vary, it is a round-about means for calculating those concentrations, says Hayes. "The stomatal density is a more direct monitor of the atmosphere," he says.

While each method has its problems, researchers hope that a combination of the available approaches will eventually yield an accurate picture. Says Cerling, "We'd like to have a bunch of different techniques that are independent of each other, because I don't think we have the same confidence with any of them that we have when we can actually measure the air itself. If all of these techniques are telling the same story and they're completely independent of each other, then we can have a lot more confidence in that tale." ❑

**Questions**:

1. Why would plants have a decreased density of stomata during high levels of carbon dioxide?

2. Since the industrial revolution, carbon dioxide concentrations in the atmosphere have risen. What has happened to the stomatal density of European trees?

3. How does Thure Cerling estimate past carbon dioxide levels?

Answers on page 301

*Throughout the Mesozoic, the age of the dinosaurs, the Earth was a much warmer place than it is today. Throughout most of the Cenozoic, starting about 40 million years ago, the Earth has been in a general cooling trend. Why did this happen? We know that the concentration of carbon dioxide in the atmosphere is one of the most important controls on the Earth's temperature, but we are still trying to work out all of the factors determining carbon dioxide concentrations. What major change occurred on the Earth about 40 million years ago that would have affected the concentration of the Earth's atmosphere? This was when India collided with Asia and began raising up the Himalayas. The presence of such a high mountain chain may have altered circulation patterns in the atmosphere and removed carbon dioxide through increased precipitation. This would further cause a complex interaction between weather patterns, erosion, and uplift, which have continued to affect carbon dioxide levels for the last 40 million years.*

# Did Tibet Cool the World?

When India and Asia collided more than 40 million years ago, the high plateau of Tibet was forced up between them. This remarkable feature could be responsible for today's climate

by David Paterson

For 250 million years, throughout the age of the dinosaurs, the Earth was warm and wet, almost tropical. Then around 40 million years ago, the planet began to cool. North America drifted away from Europe, India slammed into Asia, and new habitats were created. Antarctica iced up, temperate climates took over from some previously luxuriant tropics, and grassland and desert replaced temperate zones. Around 15 million years ago, the Earth cooled even more, leading to the glacial periods that have dominated life in the northern hemisphere ever since. What was it that reset the global "thermostat"?

The Victorian father of geology, Charles Lyell, was optimistic about finding an answer to just such a question. Writing in 1875, in the 12th edition of his *Principles of Geology*, he reflected that, in the 45 years since he had first "tried to account for the vicissitudes of climate by reference to changes in the physical geography of the globe", geologists' knowledge of the subject had greatly increased. He speculated that if the land had been subdivided more, existing as islands, the Earth's climatic history would have been more uniform. If there had been mountains higher than the Himalayas, especially in the higher latitudes, "there would be a greater excess of cold", he wrote.

With hindsight, Lyell's hypothesis seems remarkably prescient. For as geologists and climatologists come ever closer to answering the "thermostat" question, it seems that one of the world's most prominent topographical features, the vast high plateau of Tibet, may have played a large part in reducing the Earth's temperature. Some geochemists now suggest that the rising plateau, forced up from the seafloor as India crashed into Asia, was responsible for reducing the concentration of carbon dioxide in the atmosphere. The gas traps heat radiated from the Earth's surface, so as the level of carbon dioxide dropped, they argue, the Earth radiated more of the energy absorbed from the Sun and began to cool. The idea remains controversial, because it relies on a unique event to explain the "thermostat" question.

During the past decade, scientists have devised mathematical models that replicate the "vicissitudes" of the global climate. In 1983, Robert Berner of Yale University showed that there is a close relationship between the carbon dioxide in the atmosphere, the carbon stored in the oceans and on land, and the gas released by volcanic and other mantle activity.

**Dynamic Balance**

Berner argued that there had to be a dynamic balance between the processes that remove carbon dioxide from the atmosphere and those that restore it. Without such a balance, the concentration of carbon dioxide in the atmosphere might increase and lead to a hot climate similar to that of Venus, where a high level of carbon dioxide traps almost all the incoming solar energy. Or the carbon dioxide could have disappeared altogether, making the Earth cool down dramatically.

In modelling the Earth's carbon cycle over 600 million years,

Berner found that he had also produced a mathematical version of the Earth's climate history, which was generally endorsed by the geological record over this period. In the cycle, carbon dioxide in the atmosphere dissolves in rainwater to form a mild acid (carbonic acid), which reacts with limestones and granites to produce calcium, magnesium and bicarbonate ions and silica. These are eventually deposited as sediment on the seafloor and can become embodied in magma as a result of tectonic activity at the Earth's crust. The carbon dioxide locked away in these transformed sediments is eventually released into the atmosphere during processes such as seafloor spreading and when rocks under stress undergo metamorphic changes. Carbon dioxide is also released by the decomposition of buried organic remains.

The model also takes into account changes in land area and mean elevation, changes in the atmospheric concentration of carbon dioxide, the effects of plants, and the rate at which sediments on seafloors release the gas.

After constructing a dozen equations from these variables, Berner explored how levels of atmospheric carbon dioxide changed over the past 600 million years; he was even able to calculate short-term variations over just tens of millions of years. According to his model, the concentration of carbon dioxide more than 500 million years ago was 18 times higher than at present. Then over the next 200 million years, the level dropped to one that is similar to today's, with the climate becoming glacial. Next followed a period of global warming, accompanied by or perhaps driven by an increase in atmospheric carbon dioxide. Then came yet another net downswing around 120 million years ago. The model's predictions agree broadly with what the geological record of the past 600 million years suggests about the Earth's climate.

Despite its overall success, however, the model fails to predict the timing of the downturn of the global thermostat 40 million and 15 million years ago. In Berner's model, the greatest fall in carbon dioxide happens between 100 and 50 million years ago, yet the cooling in this period was small. Instead, the evidence derived from sediments suggests that the Earth cooled substantially between 50 million years ago and the present—something not predicted by Berner's model. This discrepancy is so significant, argue Maureen Raymo of the Massachusetts Institute of Technology (MIT) and William Ruddiman of the University of Virginia, that a more subtle model is now needed. They believe that the explanation behind the current ice age must lie in something other than a simple balance between volcanic outgassing and weathering.

In the mid-1980s, when Raymo was a postgraduate at Columbia University in New York, she put forward a simple yet far-reaching proposal. When India and Asia collided, she argued, the rising up of the Tibetan plateau somehow altered climatic conditions so that the Earth radiated more energy than before, setting off global cooling. The idea was originally regarded with suspicion, but computer simulations show that this theory does produce a worldwide cooling effect that corresponds well with climate changes over the past 50 million years.

The Tibetan plateau lies between the Himalayas in the south and the Kunlun Mountains in the north. It covers about 2.2 million square kilometres, or 0.4 per cent of the Earth's total surface area, with an average height of about 5 kilometres above sea level. Raymo likens the plateau to a giant boulder thrust into the atmosphere, so large that it profoundly disturbs the atmospheric circulation patterns in the whole of the northern hemisphere. In the late 1980s, John Kutzbach of the University of Wisconsin in Madison ran global climate models with and without the plateau and discovered that if there were no plateau, there would be no Indian monsoon.

This is the key to Raymo's hypothesis. She suggests that the uplift of the plateau created the patterns of air circulation that bring water-laden air off the Indian Ocean in summer and deliver monsoon rains to the Indian subcontinent, including the Tibetan plateau's southern ramparts. Dissolved in the torrential rain was carbon dioxide, forming weak carbonic acid. The higher the uplift, the greater is the volume of water-laden air drawn off the ocean by the stack of rising warm air, and the greater the rainfall. And the greater the rainfall, the more carbon dioxide is removed from the atmosphere. A combination of chemical weathering by the acid rain and fast physical erosion of the plateau's rock delivers the carbon dioxide, in the form of bicarbonate ions, to the oceans.

In essence, Raymo saw the plateau as a giant carbon dioxide extractor, pumping the gas out of the atmosphere through rainfall and then dumping the by-product in the oceans. But, as she admits, there is one flaw in her theory: it does not explain why there is any carbon dioxide left in the atmosphere—her extractor pump would have emptied the atmosphere of the gas in less than 100,000 years.

**Oversimplification**

In the course of her work, Raymo found that Berner's original model oversimplified the influence of the height of continents on the erosion caused by rain. The model assumed the erosion varied simply with land area—the greater the area, the greater the erosion. Raymo and Ruddiman drew on research done at MIT in the early 1980s, which surveyed the world's largest rivers, such as the Amazon, to gauge their weathering effects. This work showed that the mountainous relief of a continent has a greater influence on weathering than its surface area. Most of the Earth is covered by well-weathered, mainly flat regions that are drained by large rivers carrying little dissolved material. In mountainous regions, by contrast, heavy rainfall and melting snows cause extensive physical erosion: not only is solid matter carried downstream, but new surfaces become exposed to chemical erosion by acid rain. The steeper the mountains, the greater the wash-down effect of the torrents, and the greater the chemical erosion.

According to Raymo and Ruddiman, the rising up of the Tibetan plateau is an extraordinary event in the history of the world's climate. Very high, with steep sides, the plateau is also close to a warm ocean capable of providing huge quantities of rain. Eight major rivers drain it, including the Ganges, Brahmaputra, Yangtze, Indus and Mekong. Together, they carry 25 per cent of the dissolved material that reaches the world's oceans—yet the area they drain, including the plateau, is less than 5 per cent of the Earth's land area. So the effect of the plateau's weathering should be greater than that suggested by its area alone. Also, because India is still pushing up against Asia, there is a positive-feedback effect: the higher the plateau rises, the greater the monsoon rains.

Raymo cites other evidence to support her idea. Pollen analysis shows that the vegetation of Tibet, and therefore the climate, has changed dramatically since the plateau was formed, in line with cooler temperatures. More importantly, studies based on the isotopes of strontium, osmium and oxygen have indicated how levels of carbon dioxide in the atmosphere have changed over the past 600 million years. These studies confirm that temperature-enhanced weathering effects coincided with the collision of India and Asia.

Strontium behaves chemically like calcium. It undergoes the same chemical weathering cycle as calcium and magnesium carbonates and silicates, and is also deposited on the seafloor. Strontium has two isotopes, strontium-86 and strontium-87, which occur in subtly different abundances in different rocks. The ratio of strontium-87 to strontium-86 can be measured to an accuracy of 1 part in 100,000. When the strontium eventually arrives at the seafloor, deposits from different rock sources are thoroughly mixed up. But if one source predominates, then the ratio of the abundance of strontium-87 to strontium-86 in the sediment moves towards the ratio of strontium isotopes in the source rock. So, argued Raymo, the strontium isotope record should reveal where the sediment had come from. And this in turn could show whether the weathering history of the plateau tied in with Raymo's theory that there was rapid cooling 15 and 40 million years ago.

**Evidence from Isotopes**

Strontium enters marine sediment from land from two main classes of rocks—silicates and limestones. Limestones have a low ratio of strontium-87 to strontium-86, ranging from 0.706 to 0.709, while silicates have a high one, with some granites having ratios greater than 1. So enhanced weathering of granites would increase the ratio in ocean sediment.

Strontium also enters the oceans through the leaching of newly created rock emerging from places such as the mid-Atlantic ridge. This strontium has a low ratio of strontium-87 to strontium-86. But according to Frank Richter of the University of Chicago, the process accounts for only around 25 per cent of the total deposits of strontium in the ocean floor; so the rest must be delivered by rivers because the ratio of strontium isotopes in the river sediment is high, high enough to account for the net ratio of strontium isotopes in the ocean. Richter found that the strontium ratio in the oceans began to increase rapidly around 40 million years ago, having been steady for about the previous 60 million years. It then rose even higher about 20 million years ago. Richter's analysis is supported by the dating of rapid erosion in some parts of Tibet using argon isotopes.

Although the strontium ratio in the sediment could have been altered by processes outside the Tibetan plateau, such as erosion of the Canadian landmass or even glacial erosion in the Antarctic, the picture painted by this isotope pattern is supported by that of another isotope system, that of osmium. In continental rocks, the ratio of osmium-187 to osmium-186 is roughly ten times greater than the ratio of these isotopes in mantle material. Karl Turekian from Yale University suggests that if osmium-containing rocks were being leached from land into the sea, sediment records should contain evidence of changes in rates of deposition of its isotopes into the oceans. Indeed, the osmium records show similar net changes to the strontium record over the past 60 million years.

Raymo's theory is also supported by a third isotope system, that of oxygen-18 and oxygen-16. Oxygen data for the past 55 million years shows a marked change over time. As global cooling occurs, the ratio of oxygen-18 to oxygen-16 in marine calcite sediments rises. Nick Shackleton from the University of Cambridge has shown that over the past 55 million years the ratio has tended to increase, with most rapid changes occurring about 36 million and 15 million years ago.

Other evidence in favour of Raymo's theory comes from climate models worked out by Kutzbach, along with Warren Prell of Brown University, Rhode Island, who were investigating which factors most influenced the Indian monsoon. Last year, in computer experiments carried out at the National Center for Atmospheric Research in Boulder, Colorado, they looked at how the presence of mountains affects the monsoon. The results were sensational: with no mountains, the temperature over southern Asia would have been 12 °C higher. But with half mountains and full mountains, the land temperature falls and the precipitation increases as the elevation increases. This increases soil moisture and water runoff, which in turn encourage chemical weathering.

How does this affect the rest of the northern hemisphere? In summer, hot dry air rises over the plateau and draws in moisture-laden air from the Indian Ocean, causing the monsoon rains. The column of hot, dry air then cools and sinks over nearby regions: for example, Mediterranean and Central Asian summers are caused by this dry air. Meanwhile, Southeast Asia and India become wetter and stay warm. The desert landscape that lies to the north and east of the Tibetan plateau is also a result of the extraordinary uplift of the plateau.

In Europe, the models predict colder winters and colder summers and, as Kutzbach notes, the record of European vegetation over the past 20 million years does indicate a fall in temperature. We may have the Tibetan plateau to thank for the Gobi and Sahara, for the evolution of the grasses that were domesticated into wheat 9000 years ago, and for the ice ages that accompanied the evolution of mankind.

**Which Came First?**

Some researchers, however, are still not convinced by Raymo's ideas. Peter Molnar of MIT and Philip England of the University of Oxford pose the most radical question: did climate change bring about the uplift of the mountain ranges rather than the other way around? They suggest that as material is removed from an uplifted area by weathering, the feature becomes lighter and more buoyant, floating higher on the mantle. The peaks of the uplifted area may thus appear to "grow" as a result of their sides becoming eroded. So some phenomena, such as sharp incisions made by rivers into the rock that were used to infer that the uplift was recent, may themselves have been caused by climate change, rather than by the uplift itself. A gradual uplift driven by a plate collision may have been enhanced by weathering effects caused by climate change. However, Molnar and England admit they have no explanation for why the climate changed 40 or 50 million years ago, bringing about the erosion that cut deep valleys in the Tibetan region.

Raymo's response is to suggest a further positive feedback mechanism involving weathering. As the Tibetan plateau rose and cooling took place over the whole globe, she says, glacier activity on other mountainous regions increased and added to the worldwide level of erosion, causing these regions to also ride higher on the mantle. The greater the erosion, the greater the global cooling effect.

However, the most serious criticism of Raymo's theory came from Berner and Anthony Lasaga of Yale and Ken Caldeira and Michael Arthur of Pennsylvania State University. If the Tibetan plateau is such an effective remover of carbon dioxide from the atmosphere, they asked earlier this year, why is there any carbon dioxide left? Why has there been no runaway "icehouse effect"?

In Berner's model, carbon dioxide is restored to the atmosphere by volcanism, in which the gas emerges from new rock created by seafloor spreading, and by metamorphism, in which rocks under the kind of stress present when India collided with Asia release significant amounts of the gas. As far as is known, the rate of seafloor spreading has not changed appreciably in the past 40 million years. So either increased metamorphism, or some other form of carbon dioxide restoration is coming into play to balance the carbon dioxide budget—albeit at a lower carbon dioxide level.

But there may be no need to find a new source to balance the carbon dioxide budget. Caldeira suggests that as global temperatures fell due to lower atmospheric levels of carbon dioxide, the rate of chemical weathering on mountainous regions well beyond the Tibetan plateau fell too. This created a new, stable, worldwide cycle of chemical weathering and carbon dioxide replacement.

Initially there was some tension between carbon cycle modellers such as Berner and geochemists such as Raymo and Ruddiman who sought evidence from sediment core data. But some modellers now seem to be won over. Berner considers Raymo's central idea to be a useful one: in a new version of his model for atmospheric carbon dioxide (as yet unpublished) he has included the uplift ideas together with data drawn from studies of strontium isotopes.

We know that 70 million years ago Tibet lay under the sea, and we know its present state. But almost nothing is known of what happened in between. The last word on the issue of Tibet's effect on our modern climate only awaits further study of its geology. Stratigraphers apply now. ❑

**Questions**:

1. Why does reducing the level of carbon dioxide in the atmosphere cause global temperatures to drop?

2. How does increased rainfall influence carbon dioxide levels, and what happens to this carbon dioxide?

3. How could strontium isotopes be used in determining the amount of weathering that is occurring?

Answers on page 301

49

*The Ice Ages were not a time of continuous cold. The Earth alternated between cold and warm periods. Even within the warm, interglacial periods temperatures did not remain steady but rose and fell within time scales as short as a few years. Scientists are trying hard to determine what factors were controlling these rapid fluctuations in global temperatures. Some of the factors which may be important include the insulating abilities of the ice sheets themselves, the circulation patterns of the oceans, and the sudden release of large numbers of icebergs into the ocean. Future answers to the climate puzzle may even be found in the tropics. Understanding what caused these rapid fluctuations in temperature may be important to understanding how our modern climate will respond to the pollution-induced warming which we may experience in the coming decades.*

# Staggering Through the Ice Ages

## What made the planet careen between climate extremes?

by Richard Monastersky

Climate experts once viewed Earth as a well-mannered dancer moving to the stately and predictable beat of the ice ages. The world cooled and warmed with a slow, steady rhythm governed by the well-understood wiggles of its orbit. Vast ice sheets spread across the northern lands and then gradually retreated. Sea levels dipped and rose. Forests moved south and later returned. The changes kept a pace as measured as that of Earth's voyage around the sun.

But discoveries about the most recent ice age are transforming the image of a climatic waltz to something resembling a drunken lurch. During the last glacial epoch, Earth repeatedly swayed from extremely frigid conditions to warmth and back again with startling speed. As part of these shifts, the great North American ice sheet vomited huge numbers of icebergs that filled the North Atlantic. The new revelations have left scientists reeling, because the steady orbital cycles thought to control the ice ages cannot easily account for the evidence of quick climatic jitters.

"It's clear that the climate theory is not complete. The quest is now to develop the rest of the theory to explain the shorter-term events that—when we look at them in the context of our lifespan—are much more dramatic," says Gerard Bond, a marine geologist at Columbia University's Lamont-Doherty Earth Observatory in Palisades, N.Y.

Without knowing what made the ice age climate so intemperate, scientists cannot tell whether today's interglacial period is immune to the sudden swings so common when ice covered large swaths of the globe.

Much of the news about climate instability has emerged from two drilling projects in the middle of Greenland, where hardy crews bored through the 3-kilometer-thick glacial cap. Like counting tree rings, scientists can look back through the annual ice layers to trace how temperature, gas concentrations, and other factors varied during the last glacial epoch, which persisted from 115,000 until 10,000 years ago.

Although the term "ice age" conjures up a picture of unrelenting cold, temperatures during this period actually rose close to their interglacial values several times, only to plummet back toward full glacial conditions after a respite of a few hundred to a few thousand years. Willi Dansgaard of the University of Copenhagen and Hans A. Oeschger of the University of Bern in Switzerland first uncovered hints of these swings in Greenland ice cores drilled in the late 1960s and early 1980s. But the questionable quality of the oldest ice undermined confidence in their findings.

The two new Greenland cores, which match each other almost perfectly, provide dual confirmation that so-called Dansgaard-Oeschger events did indeed send temperatures bouncing up and down. In fact, the jumps from one extreme to the other occurred over decades, and in some cases over a few years—much faster than previously deemed possible.

But what does the weather over Greenland have to do with conditions in the rest of the world? As long as the evidence of climate blips remained confined to the top of Greenland, researchers could only speculate whether the fast shifts they detected there had affected other parts of the globe.

Unknown to the ice-core workers, however, a German oceanographer had already reported evidence that would draw the oceans into the picture. In 1988, Hartmut Heinrich, working in the eastern North Atlantic, described finding six unusual layers of seafloor sediments containing abnormally high percentages of rock grains.

To explain these bands, Heinrich, who works at the Federal Maritime and Hydrographic Agency in Hamburg, surmised that huge armadas of icebergs sailed across the Atlantic a half dozen times during the last ice age, dropping pulverized rock as they melted. The bergs apparently calved off of the Laurentide ice sheet, which covered Canada and parts of the United States.

Heinrich's discovery made few waves until 1992, when Lamont-Doherty's Wallace S. Broecker finally drew attention to the iceberg discharges and named them Heinrich events after their German discoverer. Bond, Heinrich, Broecker, and several others then demonstrated that the telltale sediment layers spread across much of the North Atlantic, from Newfoundland to Portugal, bearing witness to distinct iceberg flotillas 16,000, 23,000, 29,000, 39,000, 52,000, and 65,000 years ago.

The trail of debris pointed clearly to an origin along the Hudson Strait, north of Labrador, because sediments collected near there contained the greatest concentrations of rock fragments, with amounts decreasing to the east.

Last year, Bond and his colleagues married the disparate discoveries from the ocean mud and the Greenland ice. The Heinrich events, they noted, occurred during the coldest and longest spans of the Dansgaard-Oeschger cycles. Immediately after each iceberg discharge, air and sea temperatures jerked upward toward balmy conditions. Then the climate started the roller-coaster ride toward another frigid period.

Although the Heinrich events first seemed confined to the North Atlantic, their fingerprints are showing up in other parts of the world. George H. Denton of the University of Maine in Orono has tracked climate conditions in Chile by dating moraines—the rocky deposits left by receding glaciers. These moraines indicate that cold temperatures caused mountain glaciers to spread downslope at the same time as the three most recent Heinrich events, suggesting that western South America cooled in step with the North Atlantic.

Scrambling to explain Heinrich events, researchers have tended to seek answers in the regions they know best. Glaciologists first looked toward the Laurentide ice sheet, which spit out the icebergs that drifted into the cold waters of the North Atlantic. Douglas R. MacAyeal of the University of Chicago has theorized that ice sheets such as the Laurentide can grow unstable and pass through what he calls a binge-purge cycle of growth and collapse.

In this model, an ice sheet starts off stably frozen to the ground, building up layer upon layer. Eventually, the ice grows thick enough to insulate its base from the cold air temperature above.

As heat escaping from Earth warms the bottom of the sheet, the base melts, permitting the giant glacier to flow quickly across the land surface. Large fractions of the ice sheet plunge into the ocean in a Heinrich event, raising global sea level by 2 meters within the short span of a few centuries. Once thinned, the ice then refreezes to the ground and starts thickening until it next collapses.

When MacAyeal feeds rough estimates of ice-age atmospheric conditions into his binge-purge model, the Laurentide ice sheet produces Heinrich events 10,000 years apart, on average, in rough agreement with the actual cycle.

While the ice model appeals to glaciologists, some geologists who study seafloor sediments are looking for a broader explanation. In late May, at a meeting of the American Geophysical Union in Baltimore, Bond reported evidence that many different regions were discharging ice at about the same time as the northern section of the Laurentide ice sheet. Iceland and the southern lobe of the Laurentide also left their marks in the sediment layers of certain Heinrich events. According to Bond, MacAyeal's binge-purge model has a difficult time explaining why two very different parts of the Laurentide ice sheet, as well as the Iceland ice sheet, should have collapsed contemporaneously.

Denton's data from South America presents an even greater threat to the simple binge-purge model. "We have a tough time explaining one observation—that the effects of the Heinrich events appear to be felt globally, including in the Southern Hemisphere," says MacAyeal, who is now trying to alter the model. "The theoretical explanation of Heinrich events is about to go through a sea change," he says.

The sea may indeed hold the answer, according to some researchers. Because Heinrich events dumped tremendous amounts of freshwater ice into the North Atlantic, they could have altered major ocean currents, thereby spreading their influence to other parts of the globe.

The North Atlantic occupies a critical climatic position. As warm water from the tropical Atlantic heads north, the surface layer cools and loses moisture through evaporation. The cold, salty water eventually grows dense enough to sink into the deep ocean, where it flows south and rounds the southern tip of Africa.

By carrying warmth from the tropics toward the poles, this conveyor-belt system of currents helps equalize temperatures in the modern world. But melting icebergs during the Heinrich events would have stalled the conveyor belt by adding freshwater to the North Atlantic, making the water too light to sink. With this major current slowed or even halted, heat would no longer have spread around the globe as efficiently as before, leading to a sharp cooling in regions distant from the tropics.

Such scenarios assume that discharges from the Laurentide ice sheet triggered global climate shifts. But Denton and others wonder whether this theory has it backward: a case of the tail wagging the dog. What if the globe cooled first, spurring the glacial advances in Chile at the same time as they caused Heinrich events in the North Atlantic?

If so, researchers must search for some factor that would cause the planet to warm and cool every 10,000 years or so. It might seem tempting to look for answers in the vagaries of Earth's orbit, a favorite explanation for the long-term ice-age cycle, and indeed some investigators are trying that path. But the shortest known orbital oscillation—the wobbling of Earth's spin axis—has a period of 20,000 years, apparently too long to explain the Heinrich events and the even more frequent Dansgaard-Oeschger swings .

Those seeking answers to the ice-age problem may have to shift their attention away from the polar regions toward warmer climes. Jerome Chappellaz of the Laboratory of Glaciology and Geophysics of the Environment, located in St. Martin d'Heres Cedex, France, and his colleagues put the spotlight on the tropics last year when they reported on a study of methane in the Greenland ice core. From 40.000 to 8,000 years ago, methane concentrations in the atmosphere jumped in concert with local Greenland temperatures, they announced in the Dec. 2 NATURE. Because tropical wetlands emitted much of the methane in the ice-age atmosphere, Chappellaz's team suggests that the lower latitudes may have triggered both the methane changes and the Greenland temperature rises.

Lamont-Doherty's Broecker also looks toward the tropics now. In the past, he viewed the Atlantic's conveyor-belt current as the cause of most climate flipflops during the ice ages. But the newer evidence of worldwide temperature shifts has convinced Broecker that the tropical ocean might represent a critical and as-yet-overlooked force of change during the glacial epochs.

With so many new findings emerging each month, scientists are trying to make sense of the disparate evidence of iceage lurches. In the back of their minds, they recognize that the ancient events bear on the pressing issue of global warming. Might the modern climate harbor an inherently unstable element that could trigger wild swings in response to the atmospheric buildup of greenhouse gases? Or was there something unique about the ice-age Earth—the existence of the Laurentide ice sheet for example—that made the climate so susceptible to major shifts during that period only?

Bond, for one, has little hope of a quick answer to the ice-age problem. "It's getting so complicated," he says, "I sometimes think we're going backwards."

## How stable is the current climate?

Compared to the wild temperature shifts of the most recent ice age, the steady conditions of the Holocene period—the last 10,000 years-seem boring. But that stability—apparently made all the difference for human society. Only after climate quieted down did early agriculturists start domesticating crops and animals, sowing the seeds of the first civilizations.

Scientists originally thought that the relatively icefree Holocene world enjoyed some protection against the severe temperature swings seen during the last ice age. But European researchers undermined this comforting notion last year when they reported results from their ice core drilled through Greenland.

Looking within the deepest layers of ice, the scientists found evidence that the climate had gone through a series of malarialike fevers and chills during the previous interglacial period, the Eemian stage, lasting from 135,000 to 115,000 years ago (SN: 7/17/93, p.36).

Because the Eemian represents an analog of the Holocene, the ice-core evidence raised concern. To some, the findings suggested that even the modern climate might behave erratically if pushed far enough by greenhouse gas pollution.

In December, however, U.S. researchers raised a warning flag about the Eemian data. Their own ice core, drilled near the European camp in Greenland, contained folded and overturned layers near its base. Caused by ice flowing over Greenland's hilly topography, the folds essentially scrambled useful information about the Eemian stage in the U.S. core. The team wondered whether Eemian data in the European core suffered similar problems (SN: 12/11/93, p.390).

Although the European crew defended their original interpretations, they may soon have to recant some portion of their Eemian conclusions. U.S. scientists recently visited their colleagues across the Atlantic to examine the core in question for themselves.

They can't release their findings until European researchers have a chance to study the U.S. core this August, at which time the two groups will put together a joint statement. But those familiar with the controversy say the U.S. team found more evidence of folding in the European core than had previously been identified. Their work could invalidate some or all of the proposed temperature swings in the Eemian.

Stay tuned for the sizzling conclusion to the ice-core drama.

**Questions**:

1. How might the thickness of an ice sheet affect the climate?

2. What happened to the global climate immediately after large numbers of icebergs were released into the ocean?

3. How does the circulation of ocean currents in the North Atlantic affect global climate?

Answers on page 301

*During the last few years the rate at which the major greenhouse gases have been accumulating in the atmosphere has decreased. During the same time period human input of these gases has not decreased, with the sole exception of CFCs. This means that the changes may be the result of natural processes. If that is the case, what are these processes and will this trend continue.*

*CFCs are the only greenhouse gases for which we have significantly decreased our input into the atmosphere. They may also be one of our few genuinely successful attempts to stop a major source of pollution, with 128 countries signing an agreement to phase out the production of CFCs by 1996.*

*The other greenhouse gases may be slowing in their buildup due to a natural phenomena: the eruption of Mount Pinatubo. But how this may be occurring is complex and not entirely understood. Unfortunately, if Mount Pinatubo is responsible we can expect the greenhouse gases to return to their previous trends as the effects of Mount Pinatubo begin to wear off.*

# Good News and Bad News

On the one hand, greenhouse gases aren't piling up in the atmosphere as fast as they used to. On the other hand, it can't last.

by Carl Zimmer

Something surprising has happened: the accumulation of greenhouse gases in Earth's atmosphere, the gases that trap heat and warm the planet, has slowed down. Carbon dioxide, the most important greenhouse gas, has in recent years been increasing at around half its previous rate. Methane, less abundant than carbon dioxide but 20 times as effective (molecule for molecule) at absorbing heat, may be on the verge of leveling off. The same is true of chlorofluorocarbons, which are better known as ozone destroyers but are the most efficient heat absorbers of all.Why aren't climate mavens shouting joyfully from the rooftops? Because they don't believe the good news will last.The recent changes, they say, are mostly natural fluctuations—the eruption of Mount Pinatuba is one possible cause—or unintentional side effects of human actions. Thus the welcome trend is unlikely to persist.

Chlorofluorocarbons are the one exception. Spurred by the dramatic growth of the ozone hole over the Antarctic in the mid-eighties, 128 countries have ratified an agreement, known as the Montreal Protocol, to phase out the production of the two main CFCs, CFC-11 and CFC-12, by 1996. The protocol has been remarkably effective. The National Oceanic and Atmospheric Administration maintains a network of monitoring stations around the world, and James Elkins and his colleagues at NOAA have been using it to measure the CFCs in the atmosphere since 1977. In the first 11 years of study, they found that the concentration of these compounds climbed steadily higher. But as production of CFCs fell off, their growth rate in the atmosphere quickly dropped. CFC-12 is still growing, but slowly; CFC-11 has now almost completely leveled off. "We were predicting that levels would start to fall in 2000, but it looks now like we're actually a couple of years ahead of schedule," says Elkins.

As gratifying as the CFC slowing may be, it's not terribly surprising. CFCs are synthetic chemicals and thus comparatively easy to control: you just stop making them. Their concentration in the atmosphere is measured in parts per trillion. The other greenhouse gases are far more abundant. They flow into the atmosphere from many different sources, both natural and artificial, and are removed from the atmosphere in just as many ways. All these factors affect the levels of the gases in the air at any moment, and many are barely understood.

Carbon dioxide in particular is confusing atmospheric scientists these days.The leading $CO_2$ trend watcher is Charles Keeling, of the Scripps Institution of Oceanography in La Jolla, California, who started amassing a record of carbon dioxide on Mauna Loa in Hawaii in 1958. From that time to 1992 he found that the concentration of $CO_2$ in the atmosphere increased from 315 parts per million to 356 parts per million. And the rate of increase itself increased, from about .7 parts per million per year in the 1960s to 1.4 parts per million in the 1980s. On the whole this acceleration was steady—as steady as the expansion of human industry and pollution—but every few years it was punctuated by sudden bursts of even greater $CO_2$ input.

The bursts coincided with a phenomenon known as El Niño, a periodic shift in the direction of winds and ocean currents in the equatorial Pacific that has global effects on weather. Among those effects are droughts in parts of the tropics, such as Indonesia. Keeling and others believe that when vegetation in the drought-stricken areas dies and decays, it pumps about a billion metric tons of $CO_2$ into the atmosphere. Each time El Niño disappears and the rains return, plants grow back, capturing more carbon dioxide than usual and slowing the gas's accumulation in the atmosphere.

The latest El Niño began in 1990 and started to bring with it the usual $CO_2$ acceleration,. But then something baffling happened: although El Niño was still under way, the $CO_2$ growth rate abruptly dropped from 1.4 parts per million per year back down to .5 in 1992. It was the most dramatic slowdown Keeling has ever seen. NOAA's global network revealed that the slowing trend was more pronounced in the Northern Hemisphere than in the Southern.

Humans can take no credit for the slump. We've been steadily injecting about 7 billion metric tons of carbon dioxide annually into the atmosphere during the past few years, mostly through the burning of fossil fuels. The only hope researchers now have of explaining the $CO_2$ slowdown is a climatological coincidence: the change came just after Mount Pinatubo erupted in June 1991. The volcano has already been fingered as a possible cause of the acceleration of ozone destruction and the cool temperatures that have since gripped the planet. Researchers wonder if yet another atmospheric change can be pinned on it.

There are at least two ideas about how the volcano might have affected $CO_2$ levels. The rock Pinatubo hurled into the air contained a small percentage of iron, and iron is know to act as a fertilizer for plankton in some parts of the ocean. As the metal settled onto the sea surface, it may have caused these microscopic plants to grow rapidly, drawing $CO_2$ out of the atmosphere. Unfortunately there is no evidence that this actually happened.

Alternatively, Pinatubo's cooling shade of sulfuric acid droplets may have slowed the growth of soil bacteria on land. Since the bacteria eat dead plant matter and return carbon dioxide to the atmosphere, slowing them down would keep more $CO_2$ out of the atmosphere. The fact that the $CO_2$ slump was greater in the Northern Hemisphere fits this picture: there is more land and thus there are more soil bacteria in the north. That the little critters have actually been eating at a slower rate, however, has not been shown.

The reasons for the methane slump are almost as uncertain. Methane comes from a dizzying range of sources. It is produced by bacteria that live in oxygen-free environments such as wetlands, rice paddies, landfills, and the guts of cows. It is also released with vegetation burns, and it is the principal component of natural gas. Since the beginning of the industrial revolution its concentration in the atmosphere has more than doubled, from less than 800 parts per billion 200 years ago to about 1,700 parts per billion in 1992.

Four years ago two researchers made a surprising announcement: the increase in methane has been sagging for at least a decade. Like the NOAA researchers, Aslam Khalil and Rheinhold Rasmussen of the Oregon Graduate Institute have been measuring atmospheric levels of gases at stations around the globe since 1978. In 1981, they found, methane concentrations rose by 20 parts per billion. In 1990 the increase was only 11 parts per billion. These trends, Khalil says, probably reflect changes in agricultural practices that produce methane. "For a variety of reasons, the total number of acres of rice being harvested hasn't increased much in a decade," he point out. "And while the world cattle population has doubled since the turn of the century, it hasn't grown at all in the past eight years."

Since Khalil and Rasmussen reported their findings, however, the methane growth rates has really nose-dived. NOAA researchers now report that methane concentrations increased by only 4.7 parts per billion in 1992. In the Northern Hemisphere the levels increased by only 1.8 parts per billion. And while the climatologists are still putting together the numbers for 1993, they've found no evidence so far of a general reversal of this downward trend; at some stations the methane concentration didn't increase at all last year.

Ed Dlugokencky, the atmospheric chemist in charge of NOAA's methane measurements, believes that something other than cows and rice paddies must be behind this collapse. "The changes were mostly happening in the industrialized latitudes," he notes, which suggest they might be attributed to changes in fossil fuel production. Dlugokencky points specifically to a pipeline network built by the former Soviet Union in the 1970s to carry natural gas from Siberia to cities in western Russia. The shoddily constructed pipes, he says, may have leaked untold amounts of methane—until 1989, when an explosion prompted the Russians to enlist Western corporations to fix the leaks. Dlugokencky thinks those repairs alone could have substantially curtailed the methane growth rate. Since he has no hard numbers to back up this hypothesis, though (NOAA has no sampling sites in Russia), he's not yet ruling out other causes.

Pinatubo is one alternative; the cool temperatures it produced may have stunted the growth of methane-producing bacteria in wetlands as well as that of $CO_2$-producing bacteria in soil. But there is also a weirder possibility: human beings may be slowing the growth of methane by destroying the ozone layer. With less ozone in the stratosphere, more ultraviolet radiation has penetrated into the lower atmosphere. The UV rays destroy ozone there just as they do in the stratosphere, releasing excited oxygen atoms. Those atoms, in turn, react with water to form hydroxyl (OH)—a highly reactive molecule that seeks out and destroys many atmospheric compounds, including methane. As ozone is depleted, some researchers suggest, more UV rays penetrate the stratosphere, producing more OH that destroys more methane.

Everything is connected, in other words, which is one reason climate researchers are in a bit of a muddle these days when it comes to understanding the recent changes in the atmosphere. Yet most of them agree on one thing: at best the changes will only slow down global warming. Pinatubo's shade is lifting, and with it any possible effects on carbon dioxide and methane. CFCs will continue to decline, but as the ozone layer slowly heals itself, any downward push it may be exerting on methane will fade as well. Methane may level off anyway, or its concentration may start increasing again at an accelerating rate—no one knows for sure.

But the situation with carbon dioxide, by far the most significant greenhouse gas, is fundamentally different. Its concentration in the atmosphere will continue to rise. Indeed, its rate of increase seems to be rebounding upward again after the mysterious plunge of the last few years. As long as we keep putting 7 billion tons of $CO_2$ into the atmosphere each year, a habit we show no signs of abandoning, the long-term forecast for planet Earth will remain reasonably clear. The outlook is for a warmer planet in the next century." ❑

**Questions**:

1. How does El Niño affect $CO_2$- levels in the atmosphere?

2. What are the two ways in which Mount Pinatubo may have led to a decrease in $CO_2$ in the atmosphere?

3. What are the two possible causes for the recent drop in methane input?

Answers on page 301

*Geologists, in their various areas of research, are often concerned with past climates. Almost everyone seems to be concerned with the climate of the present and near future. But very few people would say that they are concerned with the Earth's climate in the distant future. What will happen to the Earth's atmosphere millions or billions of years from now? So, just in case you're wondering . . .*

# A Vision of the End

## Two and a half billion years hence, long after we are gone, Earth will lose its water, and the hardiest organisms will succumb to the sun.

by Carl Zimmer

Five billion years from now the sun will balloon into a red giant star and destroy Earth like a mote of dust in a blowtorch. By the time it does, though, our planet won't be much to cry over. It will have spent its last 2 billion years not as a blue gem of life but as a dry, overheated ball of stone, like Venus.

Such are the grim forecasts coming out of the computer sitting on Ken Caldeira's desk at Penn State. Caldeira and his colleague James Kasting have been simulating how the evolution of the sun will affect Earth's atmosphere. On its way to red-gianthood, the sun is getting brighter and hotter; according to astronomers' star models, it's about 25 percent brighter now than when Earth was born 4.6 billion years ago. According to climate modelers like Caldeira and Kasting, the Earth has managed to maintain an equable temperature so far by turning down its greenhouse effect as the sun has turned up the heat.

The scheme works like this: More solar heat means more water evaporating from the ocean, more rain causing carbon dioxide to leave the atmosphere, and more carbon getting locked up in rocks. Although volcanoes return some heat-trapping $CO_2$ to the atmosphere, there is still a steady decline in its concentration—which means Earth's temperature stays stable and life can flourish.

But plant life needs atmospheric $CO_2$ to photosynthesize, and animals like us need plants to eat. In coming eons, as the sun continues to warm, there will come a point at which so much $CO_2$ has been pulled out of the atmosphere that plants and everything that eats them will die. In 1982 the British atmospheric chemist James Lovelock ran a simple computer simulation of this process. He concluded that the biosphere will expire from $CO_2$ starvation in 100 million years. (It won't be saved even by our pollution; the man-made greenhouse effect will be a mere blip on the geologic time scale.)

A decade after Lovelock's forecast, Caldeira and Kasting decided to look into the crystal ball again. They changed a fundamental assumption of Lovelock's: that all plants die when the atmospheric concentration of $CO_2$ is less than 150 parts per million. In fact, some plants, such as grasses, use $CO_2$ so efficiently that they can get by on less than 10 parts per million. They could take over the planet when other plants die off.

As a result, Caldeira and Kasting found, the biosphere's demise can't happen until at least a billion years from now. A few tough plants and animals may cling to existence even longer. Caldeira and Kasting kept their simulation running to see how the hardy survivors would meet their end. It turns out they will be baked and dried.

With no $CO_2$ in its atmosphere, Earth will no longer be able to regulate its temperature, and it will begin to warm in tandem with the sun. A billion and a half years from now the temperature will hit 120 degrees, at which point everything but microbes will have died out. Two hundred million years later, the temperature will reach 212 degrees—boiling—and then all but the toughest microbes, of the kind that now live in seafloor hot springs, will hear time's winged chariot hurrying near.

Even they will meet their end as Earth dehydrates. The atmosphere will be full of steam, some of which will rise to the stratosphere. There high-energy radiation from the sun will split the molecules into hydrogen and oxygen, and the light-weight hydrogen will fly away into space. By 2.5 billion years from now, all of Earth's water will have been destroyed, leaving the planet sterilized.

Without water, the carbon cycle will screech to a halt. Volcanoes will continue belching up $CO_2$, and now, in the rainless atmosphere, it will once again begin to build up, creating a massive greenhouse effect on an already scorched planet. Temperatures will race to 500 degrees and above. In 3 billion years Earth will resemble dead-hot Venus, with another 2 billion years left to sit around thinking about its salad days before the sun actually swallows it.

By then we'll be long gone. "If a species lives 10 million years, it's long-lived," says Caldeira. So our own little blip of a greenhouse effect is much more worth worrying about than Caldeira's apocalypse. Still, as he points out, "every culture has a vision of creation and ending." Perhaps this is the ending for us. ❑

**Questions**:

1. How has the Earth maintained an equable temperature as the sun has gotten brighter?

2. What type of plants will survive past 100 million years from now, and why?

3. Around 2.5 billion years from now, the level of $CO_2$ will rise again. Why will this happen?

Answers on page 301

# Part Six

# The History of Life—Origin, Evolution, and Ecology

*One of the most important unanswered questions in science is "How did life begin?" Many people have thought this question unanswerable by science, and many scientists thought that actual research in this area was a waste of time. Then in 1953 Stanley Miller ran an experiment that created organic molecules, including amino acids, in an environment that was thought to represent the conditions of the early Earth. Since that experiment, the study of life's origin has become increasingly popular, bringing together evidence from chemistry, geology, molecular biology, and other fields of research. Our views of the Earth's early atmosphere have changed, from one containing methane, ammonia, and hydrogen to one dominated by carbon dioxide. Many questions still remain unanswered. Where did life start? Charles Darwin thought life may have begun in a "warm little pond," while more recent proposed settings include hydrothermal vents and bubbles in ocean foam. Where did the building blocks come from? Recent studies have shown that comets and some meteorites contain organic molecules that could be the source for life's building blocks. And possibly the most difficult and important question: How did organic materials come to life? Such Earth materials as clays and pyrite have been proposed as catalysts in the transition from lifeless organic matter to self-replicating organisms. Many scientists are now finding that we have sufficient information to bring the answers to these questions within reach.*

# How Did Life Begin?

In bubbles? On comets? Along ocean vents? Scientists find some surprising answers to the greatest mystery on earth.

by J. Madeleine Nash

The molecule was not alive, at least not in any conventional sense. Yet its behavior was astonishingly lifelike. When it appeared last April at the Scripps Research Institute in La Jolla, California, scientists thought it had spoiled their experiment. But this snippet of synthetic RNA—one of the master molecules in the nuclei of all cells—proved unusually talented. Within an hour of its formation, it had commandeered the organic material in a thimble-size test tube and started to make copies of itself. Then the copies made copies. Before long, the copies began to evolve, developing the ability to perform new and unexpected chemical tricks. Surprised and excited, the scientists who witnessed the event found themselves wondering, Is *this* how life got started?

It is a question that is being asked again and again as news of this remarkable molecule and others like it spreads through the scientific world. Never before have the creations of laboratories come so close to crossing the threshold that separates living from nonliving, the quick from the dead. It is as if the most fundamental questions about who we are and how we got here are being distilled into threadlike entities smaller than specks of dust. In the flurry of research now under way—and the philosophical debate that is certain to follow—scientists find themselves confronting anew one of the earth's most ancient mysteries. What, exactly, is life, and how did it get started?

Science's answers to these questions are changing, and changing rapidly, as fresh evidence pours in from fields as disparate as oceanography and molecular biology, geochemistry and astronomy. This summer a startling, if still sketchy, synthesis of the new ideas emerged during a weeklong meeting of origin-of-life researchers in Barcelona, Spain. Life, it now appears, did not dawdle at the starting gate, but rushed forth at full gallop. UCLA paleobiologist J. William Schopf reported finding fossilized imprints of a thriving microbial community sandwiched between layers of rock that is 3.5 billion years old. This, along with other evidence, shows that life was well established only a billion years after the earth's formation, a much faster evolution than previously thought. Life did not arise under calm, benign conditions, as once assumed, but under the hellish skies of a planet racked by volcanic eruptions and menaced by comets and asteroids. In fact, the intruders from outer space may have delivered the raw materials necessary for life. So robust were the forces that gave rise to the first living organisms that it is entirely possible, many researchers believe, that life began not once but several times before it finally "took" and colonized the planet.

The notion that life arose quickly and easily has spurred scientists to attempt a truly presumptuous feat: they want to create life—real life—in the lab. What they have in mind is not some monster like Frankenstein's, pieced together from body parts and jolted into consciousness by lightning bolts, but something more like the molecule in that thimble-size test tube at the Scripps Research Institute. They want to turn the hands of time all the way back to the beginning and create an entity that approximates the first, most primitive living thing. This ancient ancestor, believes Gerald Joyce, whose laboratory came up with the Scripps molecule, may have been a simpler, sturdier precursor of modern RNA, which, along with the nucleic acid DNA, its chemical cousin, carries the genetic code in all creatures great and small.

Some such molecule, Joyce and other scientists believe, arose in the shadowy twilight zone where the distinction between living and nonliving blurs and finally disappears. The precise chemical wizardry that caused it to pass from one side to the other remains unknown. But scientists around the world are feverishly trying to duplicate it. Eventually, possibly before the end of the century, Joyce predicts, one or more of them will succeed in creating a "living" molecule. When they do, it will throw into sharp relief one of the most unsettling questions of all: Was life an improbable miracle that happened only once? Or is it the result of a chemical process so common and inevitable that life may be continually springing up throughout the universe?

Of all the riddles that have stirred the human imagination, none has provoked more lyrical speculation, more religious awe, more contentious debate. No other moment in time, aside from the Big Bang that began the universe, could be more central to the understanding of nature than the instant that life began. "Scientific" theories on the subject are as old as civilization. The ancient Egyptians believed frogs and toads arose from silt deposited by the flooding Nile. The Greek philosopher Aristotle taught that insects and worms were born of dewdrops and slime, that mice were generated by dank soil and that eels and fish sprang forth from sand, mud and putrefying algae. In the 19th century, electricity, magnetism and radiation were believed to have the ability to quicken nonliving matter.

It took the conceptual might of Charles Darwin to imagine a biologically plausible scenario for life's emergence. In an oft quoted letter, written in 1871, Darwin suggested that life arose in a "warm little pond" where a rich brew of organic chemicals, over the eons of time, might have given rise to the first simple organisms. For the next century, Darwin's agreeable hypothesis, expanded upon by other theorists, dominated thinking on the subject. Researchers decided that the "pond" was really the ocean and began trying to figure out where the building blocks of life could have come from.

In 1953 University of Chicago graduate student Stanley Miller provided the first widely accepted experimental evidence. In a glass jar he created a comic-strip version of primitive earth. Water for the ocean. Methane, ammonia and hydrogen for the atmosphere. Sparks for lightning and other forms of electrical discharge. One week later he found in his jar a sticky goop of organic chemical, including large quantities of amino acids, Lego blocks for the proteins that make up cells. Case closed, or nearly so, many scientists believed.

Now this textbook picture of how life originated, so familiar to college students just a generation ago, is under serious attack. New insights into planetary formation have made it increasingly doubtful that clouds of methane and ammonia ever dominated the atmosphere of primitive earth. And although Miller's famous experiment produced the components of proteins, more and more researchers believe that a genetic master molecule—probably RNA—arose before proteins did.

Meanwhile, older and older fossils have all but proved that life did not evolve at the leisurely pace Darwin envisioned. Perhaps most intriguing of all, the discovery of organisms living in oceanic hot springs has provided a Stygian alternative to Darwin's peaceful picture. Life, says microbiologist Karl Stetter of the University of Regensburg in Germany, may not have formed in a nice, warm pond, but in "a hot pressure cooker."

If scientists have, by and large, tossed out the old ideas, they have not yet reached a consensus on the new. The current version of the story of life is a complex tale with many solid facts, many holes and no shortage of competing theories on how to fill in the missing pieces.

**Once Upon a Time**

Some 4.5 billion years ago, the solar system took shape inside a chrysalis of gas and dust. Small objects formed first, then slammed into one another to create the planets. Early on, the energy unleashed by these violent collisions turned the embryonic earth into a molten ball. For a billion years thereafter, the young planet's gravitational field attracted all sorts of celestial garbage. Icy comets screamed in from the outermost reaches of the solar system, while asteroids and meteorites spiraled down like megaton bombs.

Some of these asteroids could have been the size of present-day continents, says planetary scientist Christopher Chyba, a White House fellow, and the asteroids' impact would have generated sufficient heat to vaporize rock, boil the oceans and fling into the atmosphere a scalding shroud of steam. Such a cataclysm would have obliterated all living things.

Yet after a billion years, when the solar system was swept nearly clean and the primordial bombardment ended, life was already flourishing. UCLA's Schopf has identified the imprints of 11 different types of microorganisms in the 3.5 billion-year-old rocks of Western Australia. Many of the fossils closely resemble species of blue-green algae found all over the world today. Still older rocks in Greenland hint of cellular life that may have come into existence a few hundred million years earlier—perhaps 3.8 billion years ago.

At that time, scientists believe, life-threatening asteroids were still periodically pummeling the planet. Verne Oberbeck and colleagues at NASA Ames Research Center estimate that the interval between major impacts could have been as short as 3 million to 6 million years—much too brief a time to give life a leisurely incubation. This means, says Oberbeck, that the chemistry needed to green the planet must have been fast, and it must have been simple. That being the case, he asks, why wouldn't life have arisen more than once?

**The Point of Origin**

Where could life have sprouted and still been relatively safe from all but the largest asteroids? For the answer, many researchers are looking to strange, chimney-like structures found in the depths of oceans. These sit atop cracks in the ocean floor, known as hydrothermal vents, that lead to subterranean chambers of molten rock. The result is an underwater geyser: cold water plunges down through some of the cracks, and hot water gushes out through others. Fifteen years ago, when scientists began using submarines to explore these seemingly hostile environments, they were startled to discover extensive ecosystems filled with strange organisms, including giant tube worms and blind shrimp. Even more interesting, according to analysis of their RNA, the sulfur-eating microorganisms that anchor the food chain around the vents are the closest living link to the first creatures on earth. The only other life-forms that archaic are microbes living in surface steam baths like Yellowstone's Octopus Spring.

Could these overheated spots have been the places where life on earth got started? This "hot world" hypothesis has won many converts. Norman Pace, a microbiologist at Indiana University, speculates that the thin crust of primitive earth, as prone to crack as an eggshell, would have made hydrothermal vents far more common than they are today. Geochemist Everett Shock of Washington University calculates that at high temperatures organisms can get extra energy from nutrients. "The hotter it is," says Shock, "the easier life is." (Up to a point. No one has yet found a microbe living in conditions hotter than 235°F.)

Still, the question remains: Did life originate in the vents, or just migrate there? The vents may not have been a cradle but an air-raid shelter for organisms that originated near the ocean surface then drifted to the bottom. There, protected by thousands of feet of water, these lucky refugees might have survived a series of extraterrestrial impacts that killed off their relatives near the sunlit surface.

**The Ingredients**

Stanley Miller's glass-jar experiment 40 years ago suggested that the components of life were easily manufactured from gases in the atmosphere. The conditions he re-created in his laboratory faithfully reflected the prevailing wisdom of the time, which held that the earth was formed by a gradual, almost gentle convergence of rock and flecks of dust under the influence of gravity. According to this model, the earth started out cold. Its deepest layers did not catch fire until much later, after the decay of radioactive elements slowly turned up the thermostat in the core. Thus, heavy elements such as iron did not immediately melt and sink to the core, but remained close to the surface for hundreds of millions of years.

Why is this important? Because iron soaks up oxygen and prevents it from combining with carbon to form carbon dioxide. Instead, the carbon, and also the nitrogen, spewed into the atmosphere by ancient volcanoes would have been available to interact with hydrogen. The serendipitous result: formation of methane and ammonia, the gases that made the Miller experiment go.

It was, says Chyba, "a beautiful picture." Unfortunately, he adds, it is probably wrong. For the violent collisions now believed to have attended earth's birth would have melted the iron and sent it plummeting to the depths. As a result, the early atmosphere would have been composed largely of carbon dioxide—and organic compounds cannot be so easily generated in the presence of $CO_2$.

Where, then, did the building blocks of life come from? Quite possibly, many scientists think, organic compounds were transported to earth by the very comets, asteroids and meteorites that were making life so difficult. At the University of California at Davis, zoologist David Deamer has extracted from meteorites organic material that forms cell-like membranes. He has also isolated pale yellow pigments capable of absorbing energy from light—a precursor, Deamer believes, of chlorophyll, the green pigment used by modern plants.

But the amount of organic matter that can be carried by a meteorite is exceedingly small—too small, many scientists believe, to have spawned life. For this reason, Chyba argues that a far more important source may have been interplanetary dust particles floating around in the era when earth was forming. Even today, he notes, countless tiny particles—each potentially carrying a payload of organic compounds—fall to earth like cosmic snowflakes, and their collective mass outweighs the rocky softball-size meteorites by a ratio of 100,000 to 1. Comets, black with carbon, could also have flown in some raw material. Whether it would have helped to spark life no one knows, since the chemical makeup of comets remains largely a mystery.

And there's another possibility: bit objects smashing into earth could have changed the composition of the atmosphere in significant—albeit temporary—ways. "Plow a big iron asteroid into earth," argues Kevin Zahnle of NASA Ames Research Center, "and you will certainly get interesting things happening, because all that iron is going to react with all the stuff that it hits." Such conditions, Sahnle speculates, might have briefly created the methane-filled atmosphere that Miller envisioned.

**The Primordial Chemistry Lab**

Life's beginnings did not have the benefit of Miller's glass bottles, test tubes and vials. So how did nature bring the right ingredients for life together in an orderly fashion? One possibility recently suggested by Louis Lerman, a researcher at Lawrence Berkeley Laboratory, is that bubbles in the ocean served as miniature chemical reactors. Bubbles are ubiquitous, Lerman notes; at any given time, 5% of the ocean surface is covered with foam. In addition, bubbles tend to collect and concentrate many chemicals essential to life, including such trace metals as copper and zinc and salts like phosphate. Best of all, when bubbles burst, they forcibly eject their accumulated molecules into the atmosphere, where other scientists feel the most important chemistry takes place.

Biologist Harold Morowitz of George Mason University in Fairfax, Virginia, suspects that life arose in a less ephemeral chemistry lab than a bursting bubble. He focuses on Janus-faced molecules found in nature called amphiphiles. These molecules have one side with an affinity for water and another side that is repelled by water. Bobbing in the primitive oceans, the molecules would have hidden their water-hating sides away by curling into tiny spheres. These spheres, known as vesicles, would have provided an ideal setting for chemical reactions and could have been precursors to the first cells. "Once you have these little vesicles," says Morowitz, "you're on the way to life."

Which came first, though, the membrane or the metabolism? Günter Wächtershäuser, a patent attorney from Munich who also happens to be a theoretical chemist, believes that what we call life began as a series of chemical reactions between certain key organic molecules. Instead of being enclosed in a membrane, he says, they might have been stuck like pins in a cushion on the surface of some accommodating material. Wächtershäuser's surprising candidate for this all-important material: pyrite, or fool's gold. Since the shiny crystal carries a positive electrical charge, it could have attracted negatively charged organic molecules, bringing them close enough to interact. Wächtershäuser thinks these reactions could have led to the development of something similar to photosynthesis.

Still unanswered is the riddle of how these molecules came to reproduce. Chemist A.G. Cairns-Smith of the University of Glasgow thinks the answer may lie not in glittery fool's gold but in ordinary clay. The structure of certain clays repeats the same crystalline pattern over and over again. More important, when a defect occurs, it is repeated from then on, rather like a mutation in a strand of DNA. While few scientists believe such inorganic materials are actually alive, a number take very seriously the idea that clay or mineral crystals could have served as molecular molds that incorporated life's building blocks and organized them in precise arrays.

**Molecular Ancestors**

Even if one accepts the fact that organic molecules can spontaneously organize themselves and, further, that these molecules might spontaneously reproduce, there remains a fundamental chicken-and-egg problem. Modern cells are made of proteins, and the blueprints for the proteins are contained in long strands of DNA and RNA. But DNA and RNA cannot be manufactured without an adequate supply of proteins, which act as catalysts in the construction process. How, then, could nucleic acids get started without proteins, or vice versa?

One solution was put forward a decade ago, when researchers discovered that certain RNA molecules can act both as blueprints and catalysts, stimulating reactions between themselves and other molecules. Up to that point, scientists had thought of RNAs as merely molecular messengers carrying genetic instruction from DNA to the cell's protein factories. Suddenly RNA was seen in a totally different light. If RNA could catalyze reactions, perhaps at some point in the past, it spurred its own replication. Then it could have been much more than DNA's intermediary: it could have been DNA's ancestor. According to this line of reasoning, the first organisms lived in an "RNA world," and DNA did not develop until life was speeding down the evolutionary turnpike.

While searching for that ancient precursor of life last April, Scripps Research Institute's Joyce stumbled on the molecule that so tantalized him. A bit of synthetic RNA sloshing around in a test tube suddenly attached itself to a piece of protein and embarked on a course of nonstop replication. For a moment, this molecular upstart seemed close to the breakthrough Joyce had been seeking.

The molecule, he acknowledges, is not alive. Magical as it seems, it cannot replicate without a steady supply of prefabricated proteins. To qualify as living, a molecule would need to have the ability to reproduce without outside help. An important step in this direction was recently taken by Harvard molecular biologist Jack Szostak and his graduate student David Bartel, who mimicked the prolific chemistry of primitive earth by randomly generating trillions of different strands of RNA. Eventually the scientists came up with a good five dozen that were able to join themselves to other strands suspended in the same test tube. The process of linkage, explains Szostak, is critical to the formation of complex molecules from simple building blocks. What's exciting, he says, is this part of the origin-of-life puzzle does not look quite so daunting as before.

One of these days, both Joyce and Szostak believe, when someone fills a test tube with just the right stuff, a self-replicating molecule will pop up. If that happens, the achievement could be as upsetting as it is amazing. For it would challenge the most fundamental conceptions of what life is all about. Life, to most people, means animals or plants or bacteria. Less clear cut are viruses, because they are nothing more than strands of nucleic acid encased in protein, and they cannot reproduce outside a living cell.

As scientists close in on life's origins, the working definition of life will be pondered, debated and perhaps even expanded. If a sliver of fully functional RNA arises in a test tube and starts building its own proteins, who is to say it is any less alive than the strand of RNA doing the same thing inside a cell?

Some people will always hold to the belief that it is a divine spark, not clever chemistry, that brings matter to life, and for all their fancy equipment, scientists have yet to produce anything in a test tube that would shake a Fundamentalist's faith. The molecule in Joyce's lab, after all, is not as sophisticated as a virus and is still many orders of magnitude less complex than a bacterium. Indeed, the more scientists learn about it, the more extraordinary life seems. Just as the Big Bang theory has not demystified the universe, so progress in understanding the origin of life should ultimately enhance, not diminish, the wonder of it. ❑

**Questions**:

1. What evidence suggests that life may have originated around hydrothermal vents?
2. What other extraterrestrial source, besides comets and meteorites, could have supplied organic molecules to the Earth, and what are its advantages?
3. How might clays have helped in the origin of life?

Answers on page 301

53

*Cyanobacteria are some of the oldest organisms on Earth. The oldest identifiable fossils belong to this group of organisms. But even more interesting is the fact that the fossil cyanobacteria in many cases look just like the cyanobacteria that exist today. It appears that they have evolved little if any over billions of years. This has led some researchers to propose that they may be able to undo mutations as they occur. Other researchers disagree with this idea, believing that cyanobacteria participate in the evolutionary processes just as other organisms do. Mutations may have occurred but they would have been selected against. In either case, the similarities between modern cyanobacteria and their ancient counterparts are truly fascinating.*

# Static Evolution

## Is pond scum the same now as billions of years ago?

by Elizabeth Pennisi

To understand evolution, researchers from all walks of science typically search for signs of change. Paleontologists seek fossils that link extant organisms with ancient ancestors. Biologists tracking populations of plants and animals get excited when they detect one species splitting into two. Chemists and molecular biologists cheer when alterations they induce in some laboratory brew mimic the processes by which life originated (SN: 8/7/93, p.90).

But what intrigues J. William Schopf most is lack of change. Schopf, a paleobiologist at the University of California, Los Angeles, was struck 30 years ago by the apparent similarities between some l-billion-year-old fossils of blue-green bacteria and their modern microbial counterparts, which often form a living film on stagnant water.

"They surprisingly looked exactly like modern species," Schopf recalls.

At the time, researchers had unearthed very few fossils dating that far back, so for all they knew, the similarity was nothing more than a fluke. "The question then became: How widespread, or general, was this observation?"

Now, after comparing data from throughout the world, Schopf and others have concluded that modern pond scum differs little from the ancient blue-greens. "This similarity in morphology is widespread among fossils of [varying] times;" says Schopf.

As evidence, he cites the 3,000 such fossils found; these represent about 300 species, some 90 of which have modern look-alikes. Exquisitely preserved specimens have the same sizes, shapes, organization, even colonial structures as modern bacteria, he reported in late February at the annual meeting of the American Association for the Advancement of Science, held in San Francisco. Billions of years ago, these bacterial cells even divided as they do today—asexually

"Species after species, I find remarkable identity," Schopf emphasizes.

This lack of change prompted him to rethink how evolution occurs.

Almost 50 years ago, evolution researchers divided organisms on the basis of how quickly they evolved, suggesting that different mechanisms might underlie differences in rate of change. The most slowly evolving species included horseshoe crabs, crocodiles, and coelacanth fish, all of which still resemble their earliest fossils, from 100 million years ago.

Those scientists didn't know about the lineages of blue-green bacteria, also called cyanobacteria. "The fossil record is now seven times longer," says Schopf (SN: 5/1/93, p.276). He thinks these bacteria belong to an even slower evolutionary category—one, perhaps, of arrested evolution. "As far as I can tell, I think they've stopped," he told SCIENCE NEWS.

Genetic changes underlie evolution. Because sexual reproduction involves the mixing of genes between two organisms, new combinations of DNA arise more readily, speeding up the evolutionary process. As asexual organisms that reproduce without mating, blue-green bacteria depend solely on mutations and therefore would evolve more slowly.

But Schopf contends that these microbes went one step further to guard against alterations. As evidenced by their ability to withstand X-ray, ultraviolet, and even gamma-ray exposure, these microbes possess incredibly efficient DNA repair mechanisms, Schopf maintains. Thus they fix mutations that do occur.

"Evolution is economical, and evolution is conservative," he explains. "The object of an organism is not to change; the object is to be well adapted to many different environments. What DNA does is establish a mechanism so that when [genetic] changes do occur, they are corrected ."

True enough, cyanobacteria have learned to live almost anywhere. The harshness of Earth's early atmosphere demanded that they become largely self-sufficient. They use light to convert carbon dioxide into usable energy and possess the biochemical expertise to carry out this process using either water, like modern plants, or hydrogen sulfide.

Moreover, they can fix nitrogen from the air. During their early history, the ability of blue-green bacteria to use water—and to produce oxygen as a by-product—gave them a competitive edge by interfering with the metabolism of the other microbes existing then. "It was microbial gas warfare," says Schopf. "They [were] the first polluters of the world's environment."

In those days, cyanobacteria dominated all environments, having become the perfect ecological generalists, says Schopf. Even today, these microbes exist in hot springs and on snow fields; in incredibly acidic, basic, salty, or pure water; inside rocks; in deserts; or with little, no, even excess oxygen present. Some can survive long periods with no light to drive their photosynthetic machinery.

"I don't think they have to evolve, because as a group, they don't go extinct," Schopf adds.

Not everyone agrees that these microbes simply undo mutations in their genetic code. "The fact is that whatever changes did occur were selected against," counters Stjepko Golubic, an evolutionary biologist at Boston University. "That doesn't mean that they didn't participate in the evolutionary process—that [their] genotypes are frozen [in time]." In support of his position, he points to evolving properties of blue-green bacteria grown for a long time in a laboratory.

Golubic and his colleagues have compared fossil and modern stromatolites, chalky disks and cylinders produced by certain cyanobacteria. These comparisons show that at both times, these organisms thrived along shorelines in intertidal zones and produced pigments on their upper surfaces to protect them from light. "They didn't undergo ecological changes, and they didn't undergo morphological changes," concludes Golubic. "But how much the genotype has changed is still a question."

Schopf admits that appearances can deceive. These bacteria could have kept their general structure and yet evolved into very different organisms biochemically. But he doesn't think so.

Blue-greens thrive now in environments very much like those that hosted pre-Cambrian blue-greens. The carbon isotopic record indicates that the type of photosynthesis practiced by blue-greens today went on in that period. Moreover, modern species retain the ancient metabolic pathways, showing little sign of innovation in their biochemistry.

Schopf's work on cyanobacteria has led him to view the period from 2.5 billion to 500 million years ago as a distinct evolutionary era with its own tempo and mode of change.

Until now, researchers had based most of their ideas about evolution on investigations of the most recent 500 million years. However, "this is a time of relatively rapid evolution," says Schopf. Most species last no more than 8 million years. During that span, they and other extant organisms partition the environment, specializing to fit a particular niche and changing quite a bit. Unfortunately, these organisms become very good at using resources under a particular set of conditions and can't adjust quickly enough to survive when conditions change. They become extinct, making room for other species.

In contrast, before 500 million years ago, species came on the scene and stayed. Asexual organisms prevailed, particularly the oxygen-generating cyanobacteria. These slow-growing generalists could handle Earth's unforgiving conditions. Eventually, however, changes wrought by these organisms primed the ancient environment for more complex life forms (SO: 12/19/89, p.376).

These new life forms could outstrip the plodding blue-green bacteria and took many habitats away from them, says Schopf. Consequently, cyanobacteria today exist primarily in less hospitable places, such as hot springs.

Golubic interprets the data about blue-green bacteria quite differently. Just because these organisms have persisted 2 to 10 times longer than other "living fossils"—crocodiles and the like-doesn't mean they follow a different set of evolutionary rules, he argues.

Nor does he think that blue-greens have been relegated solely to extreme environments. True, they often thrive in those places. But in 1979, other researchers discovered that huge numbers of oval and round blue-green bacteria populated the open ocean, sometimes reaching densities of 10 million cells per liter. They live everywhere except the cold polar regions and have adapted to both nutrient-rich and nutrient-poor waters, Golubic notes.

By themselves, these microbes account for 20 percent of the primary production in the seas, providing fodder for the rest of the marine food chain. Their presence means that cyanobacteria continue to play an important role in sustaining life on this planet, Golubic stresses.

He argues, too, that these microbes have not forsaken other amenable environments. Rather, they have allowed themselves to become integrated into more complex creatures. Genetic studies indicate that all chloroplasts—whether from trees, grass, spinach, or algae—carry the genetic signature of blue-green bacteria (SN: 2/4/89, p.71). Thus he views blue-greens as still existing everywhere, but in such close partnership with higher organisms that they have become one with them. They may exist solo in extreme habitats simply because higher life forms could not follow them there.

Anything which is green in this world is cyanobacteria in origin," he points out. "So who says they did not evolve?"

**Questions**:

1. Out of 300 fossil species of cyanobacteria, how many have modern counterparts?
2. What is one piece of evidence that cyanobacteria may be able to repair damage to their DNA?
3. What is the relationship between cyanobacteria and plants?

Answers on page 301

*The origin and initial diversification of nearly all known animal phyla occurred very rapidly over 500 million years ago. This initial radiation of animals has become known as the Cambrian explosion and was thought to have taken somewhat less than 40 million years to occur. Recent radiometric dating of volcanic rocks at the beginning and at the end of this period have compressed it to less than 10 million years. This is an incredibly short time span for such a variety of life to have evolved and focuses more attention on the possible causes of this rapid period of evolution. Theories explaining the Cambrian explosion include increases in atmospheric oxygen levels, the evolution of new physiological features, and tectonic processes. The Cambrian explosion is one of the most important events in the history of life, and these new findings should encourage more research on this critical period of time.*

# Evolution's Big Bang Gets Even More Explosive

by Richard A. Kerr

Life on Earth was a slow starter. Single-celled organisms made their debut in the geologic record nearly 3.5 billion years ago, but until less than 600 million years ago, early in the Cambrian Period, evolution dragged. Then something clicked, and animals burst into a multimillion-year frenzy of evolutionary innovation like nothing ever seen before or since. "The lid was off; evolution was going full tilt. Never again in the history of marine life do we see so many phyla, classes, or orders appearing so rapidly," says paleontologist Jack Sepkoski of the University of Chicago. "Virtually every phylum of [marine invertebrate] animals comes in, plus things that can't be related to modern forms. They're the fastest rates of evolution we ever see." And now this "Cambrian explosion" of life has gotten even more spectacular.

On page 1293 of this issue of SCIENCE, geochronologist Samuel Bowring of the Massachusetts Institute of Technology (MIT) and his colleagues report that two lumps of volcanic rock from Siberia have yielded a new, more recent date for the beginning of the explosion, shrinking its duration to a mere 5 million to 10 million years—less than a third as long as paleontologists had traditionally assumed. "This Big Bang in animal evolution happened faster than we imagined," says Sepkoski. And that means paleontologists will have to reconsider past explanations of what caused the Cambrian explosion and possibly come up with new ones. They had invoked such mechanisms as changes in atmospheric composition, sudden shifts of continents, and new features of the animals' own physiology. But whether those explanations will hold up in light of the new date is anyone's guess.

Until now, almost every wall-chart of the geologic time scale put the beginning of the Cambrian explosion at 570 million years ago and its end 20 million to 40 million years or more later. But those dates were never really firm. The order of events was clear enough from the sequence of fossils laid down stratum by stratum: Simple, mostly unskeletonized animals appear in the Precambrian, before 600 million years ago, then give way in the Cambrian explosion to arthropods such as trilobites, shelled mollusks, starfish-like echinoderms, and bizarre creatures such as the aptly named *Hallucigenia*. The absolute timing and therefore the duration of events could not be measured accurately, however.

The problem was that Cambrian fossils and most of the sedimentary rock in which they are found can't be given an absolute age, since they lack sufficient amounts of the radioactive elements needed for dating. Volcanic rock, which often does have the right elements, is scarce in Cambrian strata. What dates were available were suspect because the existing techniques relied on elements such as rubidium that could be leached out of minerals over hundreds of millions of years.

Beginning in the 1980s however, refined isotopic dating methods based on the decay of uranium to lead were beginning to narrow the uncertainties. Researchers had more confidence in these techniques because the key uranium and lead isotopes can be tightly locked in crystals of zircon, a mineral commonly found in volcanic rocks. And because each sample contains two different clocks—two radioactive isotopes of uranium and their lead decay products—investigators can double-check how well the sample has retained the key elements and even correct for losses.

Last year geochronologist William Compston of the Australian National University in Canberra and his colleagues measured uranium and lead isotopes in zircons from Cambrian strata in Morocco. The uranium-lead dating revealed that the explosion had ended about 525 million years ago, tens of millions of years later than many time scales had assumed. If the 570-million-year age for the explosion's beginning were correct, that would have implied a longer evolutionary frenzy than was thought. But Bowring and colleagues have now used a different version of uranium-lead dating to determine a dramatically younger age for the beginning of the Cambrian explosion—533 million years or less. That result squeezes the explosion down to a few million years.

The discovery is one of the fringe benefits of the end of the cold war. Geologist John Grotzinger of MIT found the crucial samples—the cobble-sized remains of a volcanic eruption laid down when the evolutionary explosion began—in Siberia while doing field work last summer with Peter Kolosov of the Yakutian Geoscience Center in Yakutsk. The site on the Arctic coastline where they found the rock had been closed to foreigners because of its early-warning defense radars, and on geologic maps the exposed Cambrian layer had been labeled as simply a "conglomerate," giving no hint that it contained datable volcanic rocks. After separating zircons from two of their samples, Bowring and Clark Isachson of MIT measured the uranium and lead isotope content to get the crucial date.

With the Cambrian explosion now confined to a mere 5- to 10-million-year interval around 530 million years ago, paleontologists will be revisiting earlier notions about what could have set off such an intense burst of evolutionary creativity. An often-proposed idea is that multicellular animals couldn't proliferate until atmospheric oxygen, which had been rising gradually, crossed a threshold. Until then, the theory goes, organisms had to be small and simple because there was not enough oxygen around to support the metabolic machinery of multicellular animals. The problem with that idea is timing: The first sizable multicellular animals appear in the fossil record roughly 575 million years ago, well before the beginning of the Cambrian explosion, says evolutionary biologist Andrew Knoll of Harvard University, a co-author of the *Science* paper.

Rather than a dearth of oxygen, some researchers have suggested that maybe it was animals' sheer simplicity that held them back. Until animals reached a certain level of physiological and anatomical complexity, this reasoning goes, they could not expand into many ecological niches left empty until then. Once animals achieved the complexity of modern worms, says Sepkoski, "they exploded into these ecological arenas, exploiting new resources, including each other." Indeed, some paleontologists view the sudden profusion of protective shells and armor plating during the Cambrian explosion as a response to the new-found ability of animals to prey on one another.

Not content to leave the explanation to biologists, paleomagnetician Joseph Kirschvink of the California Institute of Technology has proposed a tectonic driving force for the Cambrian explosion. He sees evidence in the paleomagnetic and geologic records for a thorough reorganization of Earth's crustal motions about the time of the explosion. That redistribution of mass in turn seems to have caused Earth's outer shell of tectonic plates to slip as a unit around the planet's insides, Kirschvink says, like the crust of a hot roasted marshmallow. The geologically sudden shifts of latitude and climate that resulted could have driven an evolutionary leap. Whether it was an arms race among emerging new predators and prey or a slip-sliding Earth, the cause of the Cambrian explosion—when it's finally pinned down—is likely to be as startling as the event itself. ❑

**Questions**:

1. What two radioactive elements were used for dating, and what mineral were they found in?

2. How much time passed between the first evidence of single celled organisms and the Cambrian explosion?

3. Why would a lack of oxygen hinder the development of multicellular animals?

Answers on page 301

# 55

*The beginning of the Cambrian saw a rapid burst of new life forms, giving it the name the Cambrian explosion. During this time most of the modern phyla came into existence. Along with the many recognizable organisms were a number of organisms that were difficult to classify. Among these was the world's first large predator, Anomalocaris. What exactly this predator was and how it related to other organisms was difficult to determine. Throughout the last hundred years Anomalocaris has had its bits and pieces classified as different organisms or parts of organisms. In the last 20 years, studies have finally sorted out this unusual organism and the most recent findings from China have even found some relatives. The largest of these anomalocaridids may have reached 2 meters in length with jaws that could swallowed a person's head—a monster even by today's standards.*

## The First Monsters

### Long before sharks, Anomalocaris ruled the seas

by Richard Monastersky

The doomed trilobite scuttled across the floor of some prehistoric ocean, oblivious to a pair of bulbous eyes projecting up out of the muck like twin periscopes. As the unsuspecting animal passed, the camouflaged predator bolted from its lair, raising a cloud of obscuring silt.

The trilobite rolled into an armored ball, its hardened shell forming a nearly impenetrable shield. But against this opponent, the defensive gesture provided little protection. Two grasping limbs plucked the trilobite off the seafloor and popped it into a circular mouth unlike any known on the planet today. A ghastly ring of teeth crushed the shell as if it were a nut.

Just as it terrorized the seas of Earth's Cambrian period a half billion years ago, this nightmarish creature called *Anomalocaris* has plagued paleontologists since the 1880s, when parts of the strange beast first surfaced. A century passed before researchers could finally piece together the different parts of this anatomical oddity. Today, *Anomalocaris* and its kin are yielding new secrets, thanks to specimens recently uncovered in southwest China and other regions.

"We have found the earliest monsters," declares Jun-yuan Chen of the Nanjing Institute of Geology and Paleontology, referring to the extreme antiquity of the Chinese animals. These anomalocaridids are the oldest-known large predators in the fossil record, and their presence in the Cambrian seas reveals that an elaborate ecosystem had developed far earlier than paleontologists previously thought.

The anomalocaridids appeared at a turning point in the history of life. Prior to the start of the Cambrian period 544 million years ago, animals had extremely simple bodies capable of limited motion. A zoo at the close of Precambrian time would have displayed a relatively mundane array of creatures related to jellyfish and coral; the star attractions would have been worm-like animals, which distinguished themselves with their abilities to slither across the seafloor.

At the beginning of the Cambrian, however, life took a sudden turn toward the complex. In a few million years—the equivalent of a geological instant—an ark's worth of sophisticated body types filled the seas. This biological burst, dubbed the Cambrian explosion, produced the first skeletons and hard shells, antennae and legs, joints and jaws. It set the evolutionary stage for all that followed by giving rise to most of the major phyla known on Earth today. Even our own chordate ancestors got their start during this long-past era.

Yet alongside the familiar arthropods, echinoderms, mollusks, and other phyla, this frenzy of innovation produced many creatures that defy imagination. The history of *Anomalocaris* research, littered with errors, demonstrates the difficulty paleontologists have experienced in trying to crack some of these conundrums.

"The unfolding story of the anomalocaridids (the group including *Anomalocaris* and its near relatives) is almost as unlikely as the animal itself," declares paleontologist Derek E.G. Briggs of the University of Bristol in England in the May 27 SCIENCE.

Through the years, researchers have misidentified almost all major parts of *Anomalocaris*, sticking them on a variety of other animals almost like a game of pin the tail on the donkey. Even the creature's name, which means "odd shrimp," has its origins in a mistake. When Canadian paleontologist J.F. Whiteaves found the front appendages of *Anomalocaris* in 1886, he identified them as the body of a shrimplike crustacean with an unknown type of head, recounts Stephen J. Gould in his book *Wonderful Life: The Burgess Shale and the Nature of History* (1989, Norton).

Charles Doolittle Walcott made the next set of errors after discovering the rich deposit of Cambrian fossils in Canada's Burgess Shale in 1909. He classified the broad *Anomalocaris* body as the squashed remains of a sea cucumber. At the same time, he put an *Anomalocaris* appendage on the head of a different creature called *Sidneyia*. As for the peculiar ring-shaped mouth, Walcott identified this as a flattened jellyfish with a hole in the middle.

For over 70 years, the giant predator lay disassembled and unknown, until Harry B. Whittington of the University of Cambridge in England and his then-student Briggs finally put the pieces of the puzzle together. During careful study of fossils from the Burgess Shale, they discovered the feeding appendages, mouth, and body of *Anomalocaris* in their bizarre but true-to-life positions.

In their reconstruction, Whittington and Briggs envisioned a 60-centimeter-long animal that raised and lowered its side flaps as a form of underwater wings. The resulting undulating motion would have resembled the way modern manta rays swim. Compared to most other Cambrian animals, which measured no longer than a finger, this flounder-sized predator qualified as a giant.

Whittington and Briggs noted that the jointed front appendages had just the right reach and flexibility to capture prey and pass it back to the strange circular mouth. Unlike most maws, this ring of 32 teeth could not close completely. Rather, it worked by crunching food and pushing the pieces to additional rings of teeth inside. Some specimens from the Burgess Shale showed three circles of teeth stacked one atop the other.

While the animal made sense anatomically, the researchers could not fit it into any existing taxonomical category. With its jointed feeding appendages and segmented body, *Anomalocaris* looked somewhat like an arthropod—the great phylum that includes insects and crustaceans. But the resemblance ended there. Neither the mouth nor body of the animal showed a connection with arthropods or any other phylum. The only group in which *Anomalocaris* seemed to belong was a grab bag of misfits with the self-explanatory title of "problematica."

A decade later, paleontologists know that *Anomalocaris* was not alone. Excavations from 1990 through 1992 outside the city of Chengjiang in southwest China have turned up three similar, but distinct, creatures belonging to the anomalocaridids, report Chen and Nanjing colleagues Gui-qing Zhou, who collaborated with Lars Ramsköld of the University of Uppsala in Sweden. They describe the Chinese finds in the May 27 SCIENCE.

Chen and his colleagues have added more parts to the body of *Anomalocaris*. The Chinese specimens reveal that the animal had a broad tail with a pair of long, trailing spines—elements missing from the creatures described by Whittington and Briggs. The tail and spines are also preserved on a complete specimen collected in 1991 by Desmond Collins of Toronto's Royal Ontario Museum, who is working near Walcott's original quarry in the Burgess Shale.

The second type of giant predator from Chengjiang resembles *Anomalocaris*, but it had a wider body and shorter front grasping limbs, each armed with daggerlike pincers.

A third type is known only by its imposing jaws: Chen and coworkers found a large ring of teeth measuring 25 cm across, which apparently belonged to another anomalocaridid. Judging from the dimensions of complete *Anomalocaris* specimens, the researchers estimate that this beast would have reached 2 meters in length, making it bigger than most people.

Unlike Whittington and Briggs, Chen's group believes that *Anomalocaris* propelled itself with fishlike movements of its body and tail instead of its side flaps. "Side-flap swimming is very energy-consuming. I don't think they would get any speed that way," Chen says.

Because the predators had flat bodies, Chen and his colleagues suggest that "anomalocaridids may have spent much time partly buried or camouflaged in the bottom sediment, with the stalked eyes protruding over the bottom and scanning the surroundings for swimming prey."

Taken together, the range of new fossils points to a wide variety of eating habits. The two new Chinese forms as well as the original *Anomalocaris* have front limbs well designed for grasping large animals, says Briggs. In the case of the gaping jaws from Chengjiang, this anomalocaridid could easily have consumed something the size of a human head.

A new type of *Anomalocaris* from Australia, however, went for more delicate fare. It had front limbs suited for straining small animals from the water. Another relative, called *Peytoia*, bore long, rakelike appendages apparently adapted for combing through seafloor sediments, Briggs says. He was the first to recognize the use of this limb back in 1979, although he did not know to what creature it belonged until 2 years later.

The anomalocaridid discoveries have forced paleontologists to redraw their earlier, simplistic image of life in the ancient oceans. Not only did the Cambrian explosion produce a diversity of different body types, it also gave birth to a remarkably complex ecology.

"Originally," says Briggs, "people regarded the Cambrian as a rather early stage in the development of ecosystems. The assumption was that predation wouldn't have been a very well developed strategy."

According to this theory, the earliest predators would have started off as relatively simple creatures that then evolved more specialized features over many millions of years. As the predators added to their offensive weaponry, prey would evolve sophisticated defense systems.

But the fossils show that the arms race accelerated almost overnight during the Cambrian explosion. Creatures with hard shells and long spines abound in the Chengjiang fauna, displaying a broad sweep of protective armor. Likewise, *Anomalocaris* appeared on the scene with an array of formidable feeding tools.

"Things like *Anomalocaris* indicate that there were very highly evolved, large, well-adapted predators even in the lower Cambrian. It indicates that the Cambrian ecosystems were not that dissimilar to the types of things that you see today. There were different organisms occupying the niches, but it was the same sort of setup," says Briggs.

Chen agrees that the early Cambrian ecology blossomed quickly. "There was a highly developed ecosystem. The food chain was as complicated as it is today," he says.

When they described *Anomalocaris* a decade ago, Whittington and Briggs regarded the creature as an evolutionary dead end, without a home in any existing phylum. But new interpretations, driven by fossil discoveries, are starting to make sense of these peculiar predators.

Chen and his colleagues see a distinct resemblance between anomalocaridids and two other Cambrian oddballs: a creature from Greenland called *Kerygmachela* (SN:7/11/92, p.22) and another called *Opabinia* from the Burgess Shale. By Chen's analysis, the Greenland creature was the closest relative of the anomalocaridids because both possessed a pair of movable front appendages, 11 body segments, side flaps and two long rear spines.

*Opabina*—which looked somewhat like a swimming vacuum cleaner because of its long, hoselike front appendage—also resembled the anomalocaridids in having large eyes on stalks, side flaps, two rear spines and a similar tail. Chen's team therefore folds anomalocaridids, *Kerygmachela*, and *Opabina* together in a taxonomic category closely related to arthropods, perhaps even with the phylum Arthropoda.

A once unclassifiable patchwork of body parts, *Anomalocaris* is finally revealing its kinship with the more familiar phyla of animal life. "We no longer need to think of anomalocaridids as something really strange or bizarre. It's beginning to look like quite an important group, one pretty close to the arthropods," says Briggs.

With that, paleontologists are finding the earlier Cambrian seas an easier place in which to navigate. ❑

**Questions**:

1. Prior to the beginning of the Cambrian, what were some of the most complex creatures in the ocean?
2. What features did Anomalocaris share with the arthropods?
3. What was the mouth like on Anomalocaris?

Answers on page 301

*Graptolites are Paleozoic fossils of colonial animals. They are especially useful in biostratigraphy (correlating rocks) because they often occur in deep water shales where other fossils are rare. As floating organisms, they often sank into poorly oxygenated sediments where few other organisms could live. Graptolites were thought to have died 300 million years ago, but a recent sample of the seafloor off New Caledonia revealed that they are apparently still alive. This find is similar to that of the coelacanth fish in 1938, which was known only from fossils over 100 million years old. And like that find, it provides a rare opportunity to study the soft tissues of creatures heretofore known only by the incomplete remains of the fossil record.*

# It's Alive, and It's a Graptolite

## From the seafloor off New Caledonia comes a strange colonial creature that was supposed to have been extinct for 300 million years.

by Kathy A. Svitil

When naturalists in the nineteenth century began to turn their attention to the deep sea, they expected to find it crawling with "living fossils." Darwin had taught them that organisms evolve in response to a changing environment, and since the environment in the deep never changed—or so it was then assumed—there was no reason for organisms there to evolve at all. The assumption was wrong, and for the most part the search for living deep-sea fossils has been a disappointment. But there have been some notable exceptions. In 1938 a South African chemist and amateur ichthyologist named J.L.B. Smith, staring at a sketch of a fish netted off Madagascar, realized he was looking at a coelacanth, a fish that had been thought to be extinct for 100 million years. And in 1992 Noel Dilly, sorting through a pile of unlovely organisms retrieved by French oceanographers from the seafloor off New Caledonia, found himself looking at a graptolite: a strange sea creature last seen alive around 300 million years ago.

"I don't want to sound arrogant," says Dilly, an amateur marine biologist and professional ophthalmologist at St. George's Hospital Medical School in London, "but I feel I must almost be in the same bracket as Dr. Smith. Here I am, a complete outsider, who by virtue of an incredible stroke of good fortune has stumbled across something that is very exciting and very interesting."

Actually Dilly is hardly an outsider in marine biology. He is considered one of the world's leading authorities on pterobranchs: colonial animals, somewhat like coral, that attach themselves to the seafloor at depths ranging from 2 to 2,000 feet. A pterobranch colony is typically an inch or so wide. It consists of a hardened mass of collagen inhabited by tens of zooids—the individual members of the colony. Each zooid is less than a tenth of an inch long and lives in its own tube or tiny depression, depending on the species. The zooids crawl out of these holes to filter microscopic plankton out of the water with their wavy tentacles. As they're eating, they're also secreting collagen from a suckerlike organ near their mouth. Thus the colonial skeleton is built up layer by layer, much like a plaster cast.

For more than 50 years some researchers have suggested that pterobranchs must be related to the graptolites, a group of marine organisms that flourished throughout the world ocean between 570 million and 300 million years ago. Like nearly all fossil organisms, graptolites are known only from their hard parts. But those hard parts look very much like pterobranch hard parts: they are colonial skeletons with individual tubes or holes, called thecae, that presumably housed zooids. And they appear to have been built in the same way as pterobranch skeletons, of layers of material secreted by the zooids. Some graptolites attached to the seafloor, and those looked either like a bushy seaweed or like a flattened bagpipe, with thecae protruding from a skeletal blob. But many graptolite species floated in the water, and their skeletons resembled hacksaw blades, with each tooth a theca.

Those little blades have been beloved of geologists and gold miners for more than a century. "Because graptolites were floating organisms, they were widely distributed by ocean currents," explains William Berry, a geologist at the University of California at Berkeley. "More important, the shape of the little hacksaw blades changed quickly over geologic time. Sometimes there are cups on one side of the blade, sometimes on both sides; sometimes the blades form a V or a Y. Once you understand the pattern of change, you can date the layers in a geologic structure relative to one another based on the relative positions of graptolite deposits. And then, if you know that a gold deposit always occurs, say, in the youngest layer in a pile, all you have to do is look for the graptolite structure that corresponds to that layer. Back in the last century, it turns out, prospectors in Australia used graptolites extensively to find gold deposits."

In spite of the physical resemblance between graptolites and pterobranchs, though, some paleontologists have balked at putting them in the same class. One part of the graptolite skeleton, a long, thin spine called the nema, could not, these researchers have argued, have been formed the way pterobranch zooids form their skeleton, through successive layering. A zooid, the argument went, could not have left its theca, climbed a spine many times longer than its body length, secreted material at the tip, and still have remained connected to the colony. In fact, in some species of graptolite it seemed the zooids couldn't leave their thecae at all, because the openings looked to be smaller than their bodies. But the linchpin of the argument against saying pterobranchs were living graptolites was simply that no one has ever found a pterobranch that built a nema.

Enter the French research submersible *Cyana* and a French marine biologist named Michel Roux, who collected a batch of pterobranchs off the coast of New Caledonia and sent them to Noel Dilly for sorting. "I remember the morning vividly," Dilly says of the day the package arrived. "I thought, 'Oh my God, not another boring collection to hack through.' And then when I took the first one out of the pot I thought, 'This is like nothing I've seen.' I saw those long needle-like processes and thought, 'I don't believe this.' There was about a 40-minute period when I was looking at these things through the microscope and waiting to wake up from a dream."

The things Dilly spied under his microscope—he called them *Cephalodiscus graptolitoides*—were orange-colored colonies of pterobranchs consisting of a flattened crust of collagen peppered with saclike cups. Each cup contained a single white zooid, between a twenty-fifth and a tenth of an inch long and, in many cases, fatter than the door to its home. Yet these zooids were clearly capable of squeezing out: near the cups were long, tapered, needle-like spines that looked for all the world like a graptolite's nema. Some of the spines were more than an inch long—nearly 30 times the body length of a typical zooid and the same length as the impossible-for-a-pterobranch-to-build nema.

When Dilly examined the spines under an electron microscope, he found that their internal structure was the same as that of a nema, with layers of collagen piled one on top of the other, from the bottom of the spine up. "I've examined electron micrographs of the ultrastructure of floating and bushy kinds of graptolites," says Berry, "and the patterns in those 400-million-year-old forms are *identical* to those of Dilly's pterobranchs. If I were shown pictures of these spines I'd say it was a graptolite."

The zooids, Dilly suspects, use spines as feeding poles. "It's no good filter-feeding down a deep, dark hole," Dilly explains, "or on the bottom in the mud. The place for a filter feeder is out in the current. I think the nema is just a gadget that allows the colony to get off the bottom." Each time a zooid crawls or squeezes out of its hole, Dilly says, it leaves a trail of collagen, much like a snail leaving a trail of mucus. When it stops to feed, the collagen lumps up underneath it and hardens. The next time the zooid crawls out, its sits on that lump and secretes more collagen on top. Gradually the lump gets taller. "The next thing you know," Dilly says, "you have a tower"—and a rather silly-looking headless organism perched precariously on top of it, its tentacles flapping in the current, looking far more like a creation of Dr. Seuss's than a hardened survivor of natural selection.

Yet pterobranchs *are* survivors: the one obstacle to linking them to graptolites has not been overcome, and zoologists can study the living organisms knowing they are also learning about the long-dead. Since his discovery, Dilly has videotaped a different species of pterobranch off Bermuda; his footage shows zooids actually crawling out of their holes and making spines, albeit smaller ones than those of *C. graptolitoides*. In the near future he plans to dive on Australia's Great Barrier Reef and in the Florida Keys, and he's hoping to find more living fossils. But he's not getting greedy. "Maybe I'll never do it again," he says. "But history will always have to put me down as the guy who said that *Cephalodiscus graptolitoides* was a graptolite. Even if it decides to debunk me." ❑

**Questions**:

1. What are pterobranchs? What is a zooid?
2. When did graptolites flourish?
3. Why did paleontologists balk at putting graptolites and pterobranchs in the same class?

Answers on page 301

*The earliest fossils certain to be vertebrates belong to the ostracoderms or jawless fish. Understanding the evolutionary relationships between these fossils is important to the interpretation of significant developments in vertebrate evolution and the relationships between modern vertebrates. The authors use cladistic methods to try to determine these relationships. Two of the most primitive of the living vertebrates, the hagfishes and the lampreys, may not be as closely related as previously thought. It appears that the lampreys are more closely related to the ostracoderms that gave rise to modern fish than they are to the hagfishes. These recent studies also address questions of the habitat in which early vertebrates evolved and the position of conodonts within the vertebrate evolutionary tree.*

# Evolution of the Early Vertebrates

## Recent discoveries provide clues to the relationships between early vertebrates and their modern relatives

by Peter Forey and Philippe Janvier

The first vertebrate probably arose about 550 million years ago in the Cambrian period, immediately after the great evolutionary explosion that produced most of the major groups of multicellular plants and animals. With one possible exception, however, these animals do not appear to have left their mark in the fossil record. The oldest undeniably vertebrate remains do not appear until the Ordovician period, about 460 million years ago. These animals, collectively and informally known as "ostracoderms," were relatively small, fishlike creatures that became extinct about 100 million years later. Some among them, however, had a promising future; in fact the ostracoderms gave rise to most of the vertebrates on the earth today, including human beings.

The 100-million-year reign of the ostracoderms was a crucial period in vertebrate history. It was during this time that many of the features we regard as important evolutionary advances arose, including true bone, pair pectoral appendages and jaws. The last feature is especially significant because jaws undoubtedly served as the impetus for many of the sophisticated vertebrate qualities that followed. This is because jaws permit the bearer to increase the intake of energy by biting off large pieces of other organisms.

The significance of the jaw is recognized by the classification of the modern vertebrates into two major groups, the gnathostomes (meaning "jawmouths") and the agnathans (meaning "without-jaws"). Modern agnathans are represented by two groups, the lampreys and the hagfishes. These jawless fishes are recognized not only for their unsavory feeding habits (lampreys rasp the flesh and suck the blood of other animals, whereas hagfish eat their way through dead or dying fishes), but also for their primitive appearance. The gnathostomes constitute all the other living vertebrates, including the bony and cartilaginous fishes and the tetrapods. The absence of jaws in all of the known ostracoderms assigns these fishes to the agnathan group, yet some share many similarities with the gnathostomes. Herein lies a central puzzle of vertebrate evolution: Who were the ostracoderms, and how are they related to modern vertebrates?

Unraveling these relationships has long plagued paleontologists because of the peculiar anatomy of the ostracoderms, and because their remains have not always been complete. Although ostracoderms have been known since the middle of the 19th century, their anatomy was not carefully appraised until the early part of this century, when the Swedish paleontologist Erik Stensiö analyzed structural details in serial sections of the fossils. Recently, many new discoveries of different kinds of ostracoderms, combined with acid-etching techniques that remove either the rock or the bone, have dramatically increased our knowledge of these creatures and their relationships to modern vertebrates. Our results are consistent with the notion that hagfishes are the most primitive vertebrates known (living or extinct) and that the lamprey is more closely related to gnathostomes than either is to the hagfishes.

### Jawless but Armored

Since they were first discovered, more than 600 species of ostracoderms have been recovered in the fossil record. Like the piscine equivalents of medieval knights,

*Reprinted with permission from AMERICAN SCIENTIST, November/December 1994, Journal of Sigma Xi, the Scientific Research Association.*

most species were clad in an external bony armor (ostracoderm means "shell-skin"). The diversity of armor types suggests that some as yet unknown selective forces were at work. Some paleontologists have suggested that giant invertebrates, the scorpion-like eurypterids, may have posed a threat. Others suggest that the bony armor may have helped to prevent the loss of water from the body tissues or that it served as storage for calcium and phosphate. In any case, the ostracoderms were also experimenting with other aspects of their new found vertebrate body, being modified with various types of bone, shields, tail shapes, gills and nasal openings.

In spite of their diversity, most ostracoderms can be classified into one of nine or ten groups. Distinctions between them are usually based on certain characteristics, such as the presence of a head shield, the location of the eyes, whether one or two nasal openings is present, and such general features as the shape of the head, the body and the tail. Although we may never discover the reason for the ostracoderms' developments, an appreciation of their appearance should provide clues to their evolution.

The osteostracans were a very successful group of ostracoderms with nearly 200 species. These fishes lived during the Silurian and Devonian periods in North America, Europe, Siberia and central Asia. Most osteostracans appear to have had pectoral fins and an upturned, flexible tail, suggesting that they would have been among the most maneuverable ostracoderms. The dorsal surface of the osteostracan head was encased in a large semicircular shield, while the undersurface was covered with tiny scales or plates surrounding the openings of the mouth and gills. Beneath the head shield, the brain and the gills were encased in a "skull" made of perichondral bone (a type of bone present in many gnathostomes, but not in lampreys and hagfishes). The top of the head had openings for the eyes and the pineal organ and a single combined nasohypophyseal opening (so named because the hypophysis [or pituitary] develops as an outpouching from this duct in modern agnathans). The dorsal nasohypophyseal opening of osteostracans is strikingly similar to that of lampreys, and has been the primary argument that these two groups are immediately related. However, it is now believed that this similarity came about as a result of parallel evolution. The most distinctive features of the osteostracans are the "sensory fields," on the lateral margins of the head shield and just behind the eyes. The sensory fields were connected to the inner ear by radiating systems of tubes. It has been suggested that they were some form of electric organ, a special sense organ or, more likely, a specialized part of the lateral line system (a sensory structure over the head and along the sides of some fishes that enables them to detect vibrations in the water).

During the Lower Silurian and Upper Devonian periods, the geographic regions of China and North Vietnam were covered by seas inhabited by ostracoderms known as the galeaspids. The galeaspids have only been known for 30 years, during which time about 60 species have been discovered. Recent work by Liu Yuhai, Pan Jiang and Wang Nienzong in Beijing and Ta Hoa Phuong of Hanoi University has resulted in a number of exciting discoveries of these fishes. The galeaspids superficially resemble the osteostracans in having a single broad shield on the dorsal surface of the head, whereas the mouth and the gill openings are on the underside of the head. The shield is usually semicircular, but may be extended into the long horn or rostrum in some species. Unlike the osteostracans, the galeaspids have a distinctive median opening in front of the eyes, just behind the anterior margin of the head shield. A similar opening is found in modern hagfishes, and it is believed that it was used for the intake of water into the gill chamber.

While the galeaspids were swimming in the ancient seas of China, the anaspids were living in the brackish and fresh waters of Europe and North America. Anaspids are poorly known but appear to be characterized by small spindle-shaped bodies and downward-turning tails. Alex Ritchie of the Australian Museum in Sydney has provided evidence for ribbon-like paired fins that extended behind the gill openings. One of us (Janvier) has suggested that the fins may have undulated, moving the fish slowly forward or backward.

There appear to be two types of anaspids, both of which are reminiscent of lampreys in having large eyes with the nasohypophyseal and pineal openings on top of the head. One group has enlarged scales along the length of the animal's back and immediately behind the gills. The members of the other group completely lack scales, or have poorly developed scales. The best known genus of these "naked" anaspids is *Jamoytius*, which resembles modern lampreys in having a circular cartilage surrounding the mouth, large eyes and a basket-like skeleton that supported the gills. These similarities have led some workers to suggest that the naked anaspids are immediately relatives of lampreys.

The heterostracans were among the first of the armored agnathans to be discovered. Although their head shields were initially identified as fossil squid shells or crustacean carapaces, in 1858 the British naturalist Thomas Huxley realized that the shields belonged to vertebrate animals. At this writing the group is known to consist of about 300 species that lived during lower Silurian to upper Devonian times in North America, Europe and Siberia. Their remains are so common in coastal, brackish and possibly freshwater deposits from the Devonian that they are used locally for stratigraphic correlations.

All heterostracans had a well-developed dermal skeleton covering the head, body and tail, and a single external gill opening. In most species the head armor was made of dorsal and ventral shields. Each shield was made of a particular type of bone, called aspidin, that did not contain bone cells. Aspidin tissue was also present in the scales or shields of some other groups of ostracoderms (including the anaspids and galeaspids). Unfortunately, little is known of the internal anatomy of heterostracans, since they appear to have lacked a bony endoskeleton. Some reconstructions of the brain and the inner ear are possible because these structures left impressions on the underside of the head shield.

The tissue structure of heterostracan bone is considered to be the primitive vertebrate type. Not only is its structure simpler than true cellular bone, but it is also found in much older groups of ostracoderms dating from the Ordovician. These Ordovician species are the earliest records of animals that are clearly recognizable as vertebrates. They differ from heterostracans in having several separate gill openings. Until the mid-1970s one of these Ordovician groups, the astraspids, was thought to have the primitive form of the vertebrate armor (made of small polygonal plates that could grow). However, some recent discoveries of heterostracan-like fishes from older deposits have changed this view.

In particular, Alex Ritchie and Joyce Gilbert-Tomlinson of the Australian Museum discovered several heterostracan-like fishes in 1977, including the genus *Arandaspis*, in the Middle Ordovician rocks of central Australia. In 1993 Pierre-Yves Gagnier, then of the National Museum of Natural History in Paris, described a very similar fish, *Sacabambaspis*, from Upper Ordovician deposits in Bolivia. More recently, the Australian paleontologist Gavin Young of the Bureau of Mineral Resources in Canberra found similar fragments of these fishes, collectively called the arandaspids, in the oldest rocks of the Australian Ordovician.

Like the heterostracans, arandaspids have large, solid head shields on the bottom and top of the head. The shields are separated by many small branchial plates (which protected the gill openings), and the body is covered with narrow, elongated scales. However, the large dermal plates of the arandaspids were made of minute polygonal units, which themselves could not grow, but were added to the margin of the shields. Being the oldest known vertebrates, the arandaspids call into the question the assumption that the original vertebrate armor consisted of relatively large, polygonal plates, or tesserae, that were capable of growth.

About 60 species of thelodonts are known to have lived in the waters of the Silurian and Devonian throughout the world. Although complete bodies are rarely found, the pointed scales that covered their bodies are so abundant they are commonly used for stratigraphic correlations. The remains of complete thelodonts that have been found represent two very different body shapes. Tiiu Märss of the Estonian Academy of Science and Susan Turner of the Queensland Museum in Brisbane, Australia, have described one group with a broad head, flattened from top to bottom, and a downturned tail. Another group of thelodonts found in Silurian deposits in the Canadian arctic have been described by Mark Wilson and Michael Caldwell of the University of Alberta as having very deep bodies. The deep-bodied thelodonts have small, pipette-like mouths, an oblique row of gill openings and very unusual, forked tails. The deep-bodied thelodonts have different shapes of scales on different parts of their bodies suggesting that some isolated scales, originally attributed to different species of thelodonts, may belong to the same animal. Despite their similar scales (with crowns of dentine and bases of aspidin), it may turn out that the two types of thelodonts are only distantly related to each other.

Perhaps the most unusual of the recently discovered agnathans were described by Gavin Young in 1991. In Lower Devonian deposits in Queensland, Australia, Young discovered a fossil ostracoderm that resembled an osteostracan in the overall shape of the head and the shield and what appear to be pectoral fins. These fishes, the pituriaspids, differed from the osteostracans in lacking a dorsal nasohypophyseal opening, and the site of the nostrils remains a mystery. The pituriaspids have an elongated shield, which is pierced by openings for the eyes and by associated openings of unknown function, as well as a very long rostrum.

At this point the pituriaspids are known from only a few poorly preserved specimens. Unfortunately, they were unearthed in a remote desert of Queensland, and a new expedition does not appear to be imminent. This is especially frustrating for vertebrate paleontologists because the pituriaspids are probably the most significant discovery of fossil agnathans since the discovery of the galeaspids in the 1960s.

**Relationships and Characters**

Given the broad range of ostracoderm features, how do we determine which characteristics came first and which were subsequently lost during vertebrate evolution? Surprisingly, there is a reasonable way of assessing the evolution of anatomical traits, or characters, based on the phylogenetic relationships between the groups involved. Here we shall provide a brief review of the phylogenetic schemes that have been proposed in the past and what our own studies suggest.

Up to the first half of this century, lampreys and hagfishes were thought to be each other's closest relatives. Both groups of fishes have smooth, scaleless skins, pouch-like gills, complex tongues and median nostrils. They were also thought to be degenerate in lacking paired fins and a dermal skeleton; it was assumed that their ancestors had paired fins and armor.

During the 1970s, a considerable amount of evidence cast doubt on this view. It became apparent that the lampreys and gnathostomes share a number of significant features, including the presence of neural arches along the rod-like notochord that supports the body during development, large eyes with associated eye muscles, osmotic regulation of their body fluids, nervous control of the heart, and a brain bearing such structures as a cerebellum, optic tectum and choroid plexus. Interestingly, many of these features appear to be associated with the active life-styles of lampreys and gnathostomes, compared to the relatively lethargic hagfishes. On the basis of these similarities, Søren Løvtrup of Umea University in Sweden proposed in 1977 that the lampreys are more closely related to the gnathostomes than either is to the hagfishes.

In some respects this view has been challenged by the recent work of David Stock of Stanford University and Gregory Whitt of the University of Illinois. Stock and Whitt examined the nucleotide sequences of the 18S ribosomal RNA molecule from various members of the phylum Chordata, which includes the vertebrates, urochordates (tunicates), and cephalochordates (represented by amphioxus). The marine-dwelling and relatively sedentary tunicates and amphioxus are filter-feeders that are considered by many zoologists to be the animals most closely related to vertebrates. In Stock and Whitt's study the vertebrates were represented by lampreys, hagfishes and gnathostomes.

When Stock and Whitt compared the sequence of 1,631 nucleotide bases from the ribosomal RNA of each organism, they found that the resulting phylogenetic relationships depended on *how* the groups were compared. For example, when amphioxus (or amphioxus grouped with the tunicates) was used as the basis for comparing the vertebrate groups, the lampreys and hagfishes were resolved as each being other's closest relatives. In contrast, when the tunicates alone were used as the basis for the comparison, the lampreys and gnathostomes appeared to be most closely related to each other. In other words, the results varied according to whether tunicates or amphioxus are considered to be more closely related to the vertebrates. It seems that more molecular analyses are necessary before we can assess the historical message within the genetic code.

Before we examine what the fossil record has to say about these relationships, we should briefly explain our technique. In recent years there has been a pronounced swing toward cladistic methods of phylogenetic analysis. The method was first introduced in 1950 by the German entomologist Willi Hennig, and later improved by several comparative biologists who employed various computer algorithms. Cladistic classifications express relationships in relative terms; for example, taxon A is more closely related to taxon B than either is to taxon C. A and B are said to be sister groups, and C is the sister group of A+B.

The goal of cladistic analysis is to recognize sister-group relationships by identifying derived characters, unique to the members of the sister-group pair, that are assumed to have arisen in the immediate common ancestor. Primative characters that are present in taxa other than the putative sister groups have no relevance to establishing the relationship. For example, the absence of paired fins and jaws in both hagfishes and lampreys is irrelevant to determining their cladistic relationship because these traits are absent in invertebrates as well. Finally, in any cladistic analysis, the taxa are grouped so that the number of character changes (across the taxa) is kept to a minimum. The "best" solution is the most parsimonious for a given set of characters.

Using such a cladistic method, we analyzed 56 characters of the living and fossil agnathans to construct a phylogenetic tree. Our results suggest that all of the ostracoderms (except for *Jamoytius*) are more closely related to the gnathostomes than either group is to the lamprey or hagfish. Among the ostracoderms, the osteostracans are most closely related to the gnathostomes, whereas the anaspids are the most distantly related to the gnathostomes. Finally, among the living taxa, the lampreys and the gnathostomes are more closely related to each other than either is to the hagfish. On the basis of these relationships we can begin to reconstruct the evolution of some vertebrate characteristics, including their earliest habitat, their nasal structures, the brain and the lateral line system.

One of the earliest vertebrate developments is the sense of smell, recognized in both living and fossil forms with the presence of nasal openings. In lampreys and hagfishes the nostrils open into a duct, the prenasal sinus, which itself opens to the outside through the nasohypophyseal opening. In gnathostomes, there is no longer a developmental connection between the nasal sacs and the hypophysis. At one time it was believed that the nasohypophyseal connection signified that lampreys and hagfishes are very close relatives. However, hagfishes differ from lampreys in that the prenasal sinus continues past the nostrils as a nasopharyngeal duct that opens into the roof of the pharynx (allowing for the continuous passage of water). We now know that the fossil agnathans had various forms of nasal apparatus, including those seen in the hagfish and the lamprey. Consequently the nasohypophyseal connection cannot be used to link the modern agnathans.

Among the fossil ostracoderms, paired nasal sacs appear to be present across the groups, including the galeaspids, thelodonts and heterostracans. The presence of paired nasal sacs is probably the primitive condition for most vertebrates, except perhaps the hagfish. The confluence of the nasal sacs, seen in lampreys, osteostracans and hagfishes, appears to be the derived condition.

The prenasal sinus also seems to be a primitive vertebrate structure, perhaps ultimately leading to the development of the median nasal sac found in lampreys and osteostracans. This is consistent with the view expressed by Hans Bjerring of the Swedish Museum of Natural History, which holds that the prenasal sinus is the remnant of the primary inhalant device in the vertebrate ancestors. Gnathostomes retained the paired nasal sacs, but lost the prenasal sinus, resulting in nostrils that open directly to the outside. In some primitive gnathostome forms, part of the hypophyseal duct is present in the form of a tiny canal that connects the hypophysis to the roof of the mouth.

Some structures appear to be embryologically and phylogenetically related—in particular, the lateral line (which is made of simple sensory buds, or neuromast organs), electroreceptors and the labyrinth of canals in the inner ear (used in the sense of balance). The form and distribution of these systems across the vertebrates suggests these structures arose through a sequence of increasing complexity. Hagfishes appear to have the most primitive complement of these structures, lacking both electroreceptor organs and lateral-line structures, yet they lack typical neuromasts. Lampreys have electroreceptors on the head and sparsely scattered on the body. Lampreys also have short lines of neuromasts on various parts of the head (but not on the body), which represent the lateral-line structure.

The next stage in the evolution of the lateral-line system may be shown by the arandaspids, astraspids and probably the anaspids. The heads of these fishes have a series of discontinuous grooves which mark the armor and part of the trunk, and may have contained neuromasts. The heterostracans, galeaspids, osteostracans and thelodonts have a well-developed lateral-line system expressed as canals throughout the head and the body that lies deep within the armor and the scales and that opens to the surface through pores. This is very similar to the condition in gnathostomes. Thus the lateral line seems to have evolved from a series of isolated neuromasts, to distinct lines and grooves on the head, to neuromasts embedded within tubes in the armor and scales of the head and the body.

It seems clear that an increase in the complexity of the vertebrate brain is associated with increases in the complexity of the visual, acoustic and lateral-line systems. For example, the hagfish has the simplest kind of vertebrate brain, with well-developed olfactory lobes that match the refined sense of smell in these fishes. However, the midbrain and the hindbrain of the hagfish are very poorly developed. The lack of an optic tectum, a cerebellum and oculomotor nerves is consistent with the presence of a simple eye that lacks muscle attachments, a simple inner ear with only one semicircular canal, and the absence of a lateral line.

The brains of lampreys, osteostracans, galeaspids and possibly the heterostracans, on the other hand, have clearly developed beyond that of the hagfish, although they are not as well developed as in gnathostomes. These fishes have a well-defined optic tectum, which is undoubtedly related to well-developed eyes that can be controlled with extrinsic eye muscles. The cerebellum of the lamprey is very small, whereas it appears to be prominent in osteostracans, galeaspids and probably the heterostracans. Lampreys, galeaspids and osteostracans have two semicircular canals that are equivalent to the two vertical canals in the gnathostome ear (which also has a horizontal canal). The heterostracans also have at least two semicircular canals. Unfortunately we know nothing about the brains of anaspids, arandaspids, thelodonts and pituriaspids.

Our phylogenetic scheme should also provide clues to the habitats of the early vertebrates. Historically it was thought that vertebrates originated in fresh water and subsequently invaded the seas. The idea was based on the belief that the earliest vertebrates were preserved in oxidized (hence freshwater) sandstone in the Harding deposits of Colorado. Since their remains were highly fragmented, it was believed that the bodies had been broken up after death and washed onto a nearshore deposit. This view was supported by the observation that nearly all modern vertebrates have lower levels of salts in their blood compared to seawater, thus requiring specializations in the kidney and the gills to regulate the osmotic composition of body fluids. The interesting exception to this is the hagfish, whose salt concentrations are nearly identical to those of seawater (not unlike marine invertebrates).

A series of recent investigations now suggest that the early vertebrates probably arose in a marine environment. A study of the Harding sandstone by Nils Spjeldnaes of Oslo University in 1967 revealed that it was deposited in a highly saline (thus possibly marine) environment. In 1982, David Darby of Minnesota University at Duluth provided good evidence that the astraspids lived in a marine environment. Fossils of complete, articulated arandaspids in marine rocks of Australia and Bolivia preclude the possibility that they lived in fresh water and that their broken remains were washed to sea. The earliest known lamprey (*Hardistiella*) was found in marine rocks of Montana by Richard Lund of Adelphi University. The oldest known hagfish (*Myxinikela*) and the lamprey *Mayomyzon* were recently discovered by David Bardack of the University of Illinois at Chicago in the Upper Carboniferous marine deposits of Mazon Creek in Illinois. It is worth noting that Mazon Creek also contains freshwater deposits; however, no agnathans have been found in them.

The marine origins of the ancestral vertebrates suggest that the hagfishes have primitively retained the marine state, whereas the freshwater habitat of some lampreys is a secondarily acquired niche. This of course does not explain the observation that most vertebrates have lower concentrations of salt compared to seawater. We propose the following possibility. The fossil record of the Silurian and the Carboniferous periods shows that the earliest members of the living fish groups (sharks, lobe-finned fishes and ray-finned fishes) often lived in marginal, brackish environments, comparable to modern mangroves or tidal flats. Although these environments have great variations in the amount of salinity, they tend to have a large food supply for these fishes. It may be that the modern descendants of these early fishes retained the physiological ability to survive in such an environment (having low levels of salt in the blood) even though most now live in the open sea.

**Who Were the Conodonts?**

The arandaspids from the Lower Ordovician are the oldest known vertebrates. According to our phylogeny, however, arandaspids are relatively derived fishes that must have been preceded by forms resembling thelodonts, anaspids, hagfishes and lampreys. These groups probably had as yet undiscovered representatives in the (earlier) Cambrian period. Curiously, with the possible exception of one group of animals, there appears to be no evidence of vertebrates in the Cambrian period.

The possible exception is the animals that bore the minute toothlike structure known as conodonts. These phosphate-bearing structures are found in marine rocks from the Upper Cambrian to the Upper Triassic and are widely used for stratigraphic zonation and correlation. For many years vertebrate paleontologists disowned conodonts because the teeth of modern agnathans are horny (not phosphatized), and because no animal had been found bearing conodonts in "life position." In 1983, however, Derek Briggs of Bristol University and his colleagues announced the discovery of a soft-bodied worm-like animal (about 6 millimeters long) bearing a conodont feeding apparatus. The animal was preserved in Lower Carboniferous sediments (from about 345 million years ago), which had been laid down in brackish water.

By 1993 sufficient remains had been found to allow Richard Aldridge of Leicester University and his colleagues to reconstruct the conodont animal's appearance. It appears to have been bilaterally symmetrical, with an asymmetric tail-fin fold, large eyes and <-shaped myomeres (muscle blocks). Aldridge and his collaborators also identified traces of a notochord. Although these qualities are hallmarks of a vertebrate, there appears to be nothing that resembles gill openings. Ivan Sansom of the University of Durham in Britain and his colleagues have also examined the hard tissues of the conodont. They note that the surface layer appears to be an enamel-like tissue that lies atop cellular "bone." As we pointed out in 1993, however, these tissues are very different from the bone and enamel of vertebrates, so it is not at all clear that conodonts are structurally comparable to vertebrate teeth.

Because of these features, it is difficult to assess where the conodont animal fits within the vertebrate phylogeny we propose in this paper. Curiously, the hard tissues of the conodont animal suggest that they have more derived characteristics than either lampreys or hagfishes. There is clearly a need for more studies on the conodonts and the various types of bone and enamel present in fossil agnathans. As has been previously suggested, it may be that the conodont animal is a larval stage of an extinct agnathan. We cannot say.

**Vertebrate Mysteries**

Although fossil agnathans have helped to resolve some of the questions concerning the evolutionary pathways that led to the origin of gnathostomes, they remain mute on some important issues. In particular, it is still not clear how jaws arose or how true bone developed.

When jaws first appear in the fossil record (in acanthodian fishes) about 430 million years ago, both the mandibular arch and the associated hyoid arch (behind the jaw) are fully formed. The hyoid already supports the mandibular, and the intervening spiracle is classically regarded as a former gill slit. There are no structures in the gill skeleton of agnathans that are the obvious precursors of gill arches of the gnathostomes. It may be that the gill structures of the agnathans have nothing to do with the gill arches of gnathostomes.

It has been proposed that the velum, a pumping device in agnathans, may have formed the mandibular arch. Indeed, the nerve supply to the agnathan velum suggests that this may be so. Moreover, in lampreys the velum develops from the same embryonic tissue as the jaws do in gnathostomes. Jon Mallat of Washington State University at Pullman has suggested that the earliest vertebrates used the velum to pump in water and food particles, much as do the larvae of lampreys. Modifications of the velum allowed them to take in larger food particles. Jawlike structures may have permitted these vertebrates to feed on whole prey, perhaps by sucking them into the mouth.

The origin of true cellular bone, present in osteostracans and gnathostomes, poses another puzzle to paleontologists. At various times cellular bone has been regarded as representing either a primitive or an advanced condition. The view that cellular bone is primitive is supported by the embryological development of some modern vertebrates. These species have acellular bone that is derived from a bony matrix in which the cells die and disappear. On the other hand, none of the ostracoderm fossils with acellular bone (all but the osteostracans) has a trace of "dying" cells. Consequently, our phylogeny supports the view that cellular bone is an advanced feature. Even so, the presence of cellular bone in the Ordovican suggests that it is an ancient trait of vertebrates.

Other evolutionary leaps in the history of early vertebrates are also problematic. The appearance of true paired fins (with muscles and skeleton) has long been regarded as arising from a fold of skin devoid of muscles, since this appears to be the way fins develop in embryonic gnathostomes. It is possible to imagine such a gradual evolution, although no supporting evidence has been found in the fossil record. In contrast, the sudden appearance of the horizontal semicircular canal in the inner ear of gnathostomes is more difficult to imagine taking place in a gradual manner. What would be the use of an incipient semicircular canal? There appears to be no trace of a precursor to such a canal in any living or fossil agnathan.

**Conclusion**

Our phylogenetic analysis of the living and fossil vertebrates allows us to make some general statements about the evolution of vertebrates. Among the living forms, it seems clear that the lampreys are more closely related to gnathostomes than either group is the hagfishes. The hagfishes are the most primitive vertebrates known, living or extinct. Thus most features of the hagfish that were considered to be degenerate are probably primitively simple, including the lensless eye, the simple hypophysis, the simple ear, the lack of nervous regulation of the heart, the absence of a lateral-line system, electroreceptors and a cerebellum, and an inability to regulate internal ionic concentrations.

It also seems to be the case that the majority of ostracoderms are more derived than either lampreys or hagfishes. This would suggest that the lampreys and hagfishes represent a primitive naked state, and that the dermal skeleton arose after these vertebrates had diverged. At first the dermal skeleton consisted of dentine and acellular bone; and only later was it made of cellular bone. A calcified endoskeleton also appeared in the ancestors of gnathostomes after the lampreys and hagfishes diverged.

Perfected paired fins appear to have arisen first in the common ancestor of the osteostracans, the pituriaspids and the gnathostomes. Osteostracans and gnathostomes have pectoral fins and cellular bone. The pituriaspids are the only other group of agnathans that might have had pectoral fins. Lampreys probably never had pectoral fins, although they might have had paired fin folds along part of their body length. Hagfishes and their ancestors never developed paired fins.

Some features of modern hagfishes and lampreys—such as the median nasohypophyseal opening, the prenasal sinus, the pouch-like gills and the complex tongue—belong to the history of gnathostomes. Indeed, these features may have been present in the common ancestor of all known vertebrates but have become modified or lost in gnathostomes. There are other viewpoints on the relationships between modern and fossil agnathans based on other methods. Erik Jarvik of the Swedish Museum of Natural History considers that hagfishes, lampreys and ostracoderms are more closely related to each other than any are to gnathostomes. He also suggests that the agnathans are more derived than gnathostomes. However, this theory assumes a number of parallel evolutionary paths for which we see no evidence.

Although many questions in the evolution of vertebrate structure remain unanswered, the ostracoderms provide information that no other source could have given. It seems that much could be gained by further excavations in Early Ordovician and Cambrian deposits where ostracoderms should be present. As the recent history of the field shows, new techniques of preparing and observing the fossils may reveal previously unnoticed details.

**Acknowledgments**

In writing this review we have freely culled and begged information from many of our colleagues who study agnathan fishes. We thank all of them who have discussed their work with us and acknowledge that they may not necessarily agree with our phylogenetic conclusions. Peter Forey would also like to acknowledge financial support from the Robert O. Bass Visiting Scholarship Fund, Field Museum, Chicago.

**Bibliography**

Aldridge, R.J., D.E.G. Briggs, M.P. Smith, E.N.K. Clarkson and N.D.L. Clark. 1993. The anatomy of conodonts. *Philosophical Transaction of the Royal Society of London*, B 340:405-421.

Forey, P., and P. Janvier. 1993. Agnathans and the origin of jawed vertebrates. *Nature* 361:129-134.

Gagnier, P.Y. 1993. *Sacabambaspis janvieri*, Vertébrés ordovicien de Bolivie. *Annales de Paléontologie*, 79:119-166.

Halstead, L.B. 1973. The heterostracan fishes. *Biological Reviews* 48:279-332.

Hardisty, M.W. 1982. Lampreys and hagfishes: analysis of cyclostome relationships. In *The Biology of Lampreys*, ed. M.W. Hardisty and I.C. Potter, pp.165-260. London: Academic Press.

Janvier, P., and A. Blieck. 1993. L.B. Halstead and the heterostracan controversy. *Modern Geology* 18:98-105.

Løvtrup, S. 1977. *The Phylogeny of the Vertebrata*. London & New York: Wiley & Sons.

Stock, D.W. and Whitt, G.S. 1992. Evidence from 18S ribosomal RNA sequences that lampreys and hagfishes form a natural group. *Science* 257:787-789.

Turner, S. 1991. Monophyly and interrelationships of the Thelodonti. In *Early Vertebrates and Related Problems of Evolutionary Biology*, ed. M.M. Chang, Y.H. Liu and G.R. Zhang. Beijing: Science Press, pp.87-120.

Wang, N.Z. 1991. Two new silurian galeaspids (jawless craniates) from Zhejiang province, China. With a discussion of galeaspid-gnathostome relationships. In *Early Vertebrates and Related Problems of Evolutionary Biology*, M.M. Chang, Y.H. Liu and G.R. Zhang. Beijing: Science Press, pp.41-66.

Wilson, M.V.H. and M.W. Caldwell. 1993. New Silurian and Devonian fork-tailed 'thelodonts' are jawless vertebrates with stomachs and deep bodies. *Nature* 361:442-444.

Young, G. 1991. The first armoured agnathan vertebrates from the Devonian of Australia. In *Early Vertebrates and Related Problems of Evolutionary Biology*, ed. M.M. Chang, Y.H. Liu and G.R. Zhang. Beijing: Science Press, pp.67-86.

## Questions:

1. Lampreys and gnathostomes share many important features. What about their life-styles might account for this?

2. What was one of the earliest vertebrate developments and what feature in the fossils show evidence of this development?

3. What is the only group of vertebrates for which we have evidence in the Cambrian?

Answers on page 301

*Traditionally it has been assumed that the rapid diversification of insects coincided with the origin and diversification of flowering plants, the angiosperms. Presumably the diversification of flowering plants created numerous new niches for insects. There are many examples of cooperation between flowering plants and insects, particularly in the fertilization of the plants. Recent work has shown that this view is incorrect. Insects appear to have diversified millions of years before the diversification of angiosperms. Additional studies have shown that social behavior in insects arose much earlier then previously thought and that insects are relatively immune to extinction, the last point being a fact that may not surprise anyone who has had to deal with roaches, mosquitoes, or any other unwanted insects. These studies are having a profound effect on our view of the early evolution of insects.*

# Insects Ascendant

Bugs don't need flowers as much as flowers need bugs: insects were fruitful and diversified long before the first flower bloomed.

by Carl Zimmer

Over the past two summers of digging in Arizona's Petrified Forest, Stephen Hasiotis and Russell Dubiel of the United States Geological Survey have unearthed numerous hunks of sandstone filled with miniature tunnels, ramps, and chambers—all trademarks of termite nests. What makes the nests remarkable is that they are 220 million years old. Until now the oldest evidence of termites, and for that matter of any insect with a complex social organization, had dated from 100 million years ago—just about the time when flowering plants began to evolve. That coincidence made sense: the abundance of food offered by flowering plants, many researchers figured, must have given insects the opportunity to create complex societies and thus to improve their chances of survival. "We screwed up that picture," says Hasiotis.

Hasiotis and Dubiel's discovery may turn out to be just one part of a massive revision in our view of insects and their relationship with flowering plants. Today insects are by far the most diverse land animals, comprising an estimated 5 million species (compared with 4,000 mammals). Most are exquisitely adapted to feeding on some particular group of flowering plant, or angiosperm. Thus many researchers have assumed that the rise of angiosperms sometime after 115 million years ago opened the way for the diversification of insects.

But recently Conrad Labandeira of the Smithsonian Institution's Museum of Natural History and John Sepkoski of the University of Chicago have followed several lines of research to just the opposite conclusion: the success of insects, they say, cannot be attributed to flowering plants. Insects diversified and flourished long before flowers did.

"I've never bought the idea that there was nothing interesting about insect life before angiosperms," Labandeira says. His gut feeling was reinforced by his dissertation project at Chicago, a study of the mouthparts of fossil insects. Insects have evolved a kitchenful of feeding utensils, including hooks for ripping into plant sacs, and siphons for sucking up nectar. If angiosperms had had such an important effect on insect evolution, the appearance of many diverse mouthparts presumably should have coincided with the appearance of angiosperms. Yet Labandeira found that 85 percent of the manifold types of insect mouthparts existed before flowering plants did.

Meanwhile Sepkoski was building a data base detailing the origins and extinctions of families of fossil insects. Labandeira helped him flesh it out. A lot of insect research, it turned out, was hiding in relative obscurity in old German and Russian journals. "It took five or six months to shake every source in all the languages," says Labandeira. Eventually the researchers raised the number of families in the data base to 1,263. That's when they began to see patterns in insect evolution that had been invisible before.

From their origin 390 million years ago, it seems, insects have steadily increased in diversity, as measured by the number of different families alive at any one time. Compared with other life-forms, insects are actually slow to evolve new families—but they are even slower to go extinct. Some 84 percent of the insect families alive today were alive 100 million years ago; for land vertebrates the figure is 20 percent. By the time angiosperms became successful, the rate of diversification in insects had in fact slowed down, possibly because they had began to saturate their habitats.

According to Labandeira, these patterns imply that insects evolved to take advantage not of angiosperms but of ferns, cycads, and other primitive plants that were already around more than 200 million years ago. It was then that they developed their eating equipment and started to diversify and form societies. Rather than being crucial to the rise of insects, angiosperms may have actually depended on the previous success of bugs—which are now the angiosperms' chief pollinators—for their own good fortune.

How could researchers have believed the opposite for so long? They lacked a complete data base on fossil insects, Labandeira says, and the early record isn't great anyway. But they may also have been biased by their surroundings. "We now live in a world dominated by one kind of plant," says Labandeira. "It's just hard to take those angiosperm glasses off and imagine a world before it." ❑

**Questions**:

1. What features in Arizona's Petrified Forest indicate social behavior in insects?

2. What features of insects were used to study their diversification? What makes these features useful for studying the diversity of insects?

3. If insects are slow to evolve, how have they reached such high levels of diversity?

Answers on page 301

59

*While the dinosaurs have been getting all of the attention from the public, another group of large animals from the Mesozoic have been getting increasing attention from scientists. These are the pterosaurs. New discoveries over the last ten years have addressed questions of how pterosaur wings attached and how they flew, what they ate, and how they socialized. How pterosaur wings attached is important not only for understanding how they flew but also for understanding how they moved around when they were not flying. Studies of their possible eating habits have shown that the pterosaurs may have filled many of the ecological niches that are occupied by birds today. Other studies have addressed the question of social interactions such as how they raised their young. The picture that emerges is one of a highly diverse group of organisms that could thrill time-traveling "bird" watchers just as much as their modern counterparts.*

# Masters of an Ancient Sky

For 160 million years reptiles flew above the Earth. For just over 200 years we humans have known something of their existence. But only recently have we really begun to understand their lives—or their looks.

by Carl Zimmer

The fossil skull sitting on Alexander Kellner's worktable makes no sense. Usually fossils are spark plugs in the engine of the imagination. One look at the long, gently curved bones of a spider monkey's arm and you see a graceful swing from branch to branch; one glance at the weighty leg bones of a mastodon and you hear the rumble of a heavy stride. But this skull greets you with cognitive dissonance: all you see is essentially a flat triangle of bone. "A paleontologist once came in here and said, 'What is *that*?'" says Kellner, himself a paleontologist at the American Museum of Natural History in New York.

Slowly, though, with Kellner's guidance, the details that will give this triangle meaning emerge. There is a pair of large eye sockets midway along the skull, staring out of a braincase the size of a child's head; the straight, smooth daggers in front now become a set of jaws. Most prominent on the skull is a crest of bone that starts at the tip of the creature's beak; it rises above the upper jaw, passes over and behind the eyes, and shoots out behind the skull like an oar. The crest doubles the length of the skull, from two feet to four. How could any creature support such a burden? "It was hollow," Kellner says simply, picking up a piece of the crest to show how the bone is only a few hundredths of an inch thick in some places. While the crest's interior is now filled with rock, in life there was only air. It's all a little less confusing now but no less bizarre.

This skull belonged to a pterosaur—an extinct flying reptile—named *Tupuxuara*, and it is one of the latest additions to an inventory of more than 100 pterosaur species known to paleontologists. *Tupuxuara* lived 110 million years ago in what is now Brazil. It has wings measuring 18 feet from tip to tip, over 6 feet longer than the wings of the largest living bird, the albatross. Yet since many of its bones were as thin as its crest, it weighed no more than 45 pounds.

For the most part, Kellner says, *Tupuxuara* soared, but from time to time it would flap its wings to stay aloft. With its sharp eyes it would scan the broad lagoons below for food. The fern-and-pine-lined shores of these lagoons were home to many other species of pterosaurs, as well as dinosaurs, turtles, crocodiles, birds, and amphibians. Most important to *Tupuxuara*, the lagoons were teeming with fish. With a few precise adjustments to its wings, it would dip down to the surface of the water, pluck a fish, and gulp it as it flew away.

If, in Kellner's telling, this pterosaur's behavior comes off as remarkably birdlike, it is no accident. According to much of the new research by the small number of pterosaur enthusiasts worldwide, these extinct reptiles in many ways mirrored the habits of their far more familiar successors in the sky. Pterosaur flight, the researchers say, was like that of birds down to subtle details; pterosaurs played many of the same rules birds do in today's ecology, and pterosaurs may even have had birdlike social structures and child-rearing techniques.

Pterosaurs haven't always been seen this way; they weren't

even originally seen as denizens of the air. In 1784 the Italian zoologist Cosimo Collini, trying to make sense of the first pterosaur fossil ever found, guessed that he was looking at a sea creature—an amphibious mammal. Early in the nineteenth century the French anatomist Baron Georges Cuvier saw drawings of this fossil and quickly interpreted the animal as a flying reptile, but his colleagues found that concept hard to swallow. Throughout the 1800s researchers suggested pterosaurs were everything from extinct bats to flying marsupials. Eventually they seem to have reached consensus on the image of a lizard that, while it didn't actually fly, did glide through the air on batlike wings.

Part of their confusion stemmed from the pterosaurs' bones. Hollow and thin, they were perfect for flight but not for preservation. Only a few sites scattered around the world have yielded many pterosaur fossils that aren't badly damaged. Generally they are found in slabs of limestone, greatly flattened and most often fragmented.

Now, however, beautiful pterosaur fossils are emerging from a scrubby plateau in northeastern Brazil called Chapada do Araripe. The site itself isn't newfound. Two wandering naturalists discovered it in 1817, and paleontologists have been hauling out fossil fish, amphibians, insects, and plants ever since. But not until 1971 was the first pterosaur discovered at Araripe, and most of the other pterosaur fossils there have been uncovered within just the past ten years. Fourteen new species, including *Tupuxuara*, have now been found at Araripe, each dramatically different from all previously known pterosaurs.

The Araripe pterosaurs are miraculously preserved. For reasons geochemists still don't fully understand, animals that died in the lagoons of Araripe sometimes began to fossilize unusually quickly. In other ancient lagoons, bones that sank to the bottom were covered in mud and gradually crushed as the mud accumulated and turned to stone. But when an animal died in the Araripe water, it was quickly coated with a layer of sediment. Layer after layer of limestone built up on the bones, encasing them in large, round nodules. How quickly the nodules formed is hard to say, but judging from the presence of fossilized muscle and sometimes even bacteria on the skin of Araripe animals, it could have happened within hours. It's as if nature were packing away fine china for 110 million years. When paleontologists unearth these pale lumps, crack them open, and soak away the limestone, they often find fossils with unprecedented three-dimensional detail.

Kellner is one of the world's experts on the Araripe pterosaurs, and he doesn't conceal his pride in them. "Paleontologists always like to say 'My fossils are the best,'" he says, "but these *really* are the best." Stafford Howse, a paleontologist at the University of London, recently finished a study of a pterosaur skull from Araripe that is preserved down to its ear bones. "If the thing had died six months ago," he says, "and been buried in the soil for the worms and maggots to get to and had then been exhumed, this is the condition it would be in. It's perfect in every respect."

As far as paleontologists can now determine, pterosaurs descended from a small, lightweight bipedal reptile that lived perhaps 250 million years ago, an animal that also gave rise to dinosaurs (and thus, most paleontologists believe, ultimately to birds) at around the same time. How pterosaurs evolved toward flight is unknown; they may have made their first faltering attempts by leaping from the ground or tumbling out of trees. All researchers can say with certainty is that 225 million years ago there were pterosaurs, which were fully qualified fliers.

The first pterosaurs were small, ranging from robin to sea gull size. They generally had long, narrow heads filled with teeth. Most notably they possessed a finger on each hand—their "pinkie"—that was longer than their entire body; this outsize appendage supported a wing. Their other three fingers were perfectly normal and tipped with claws. Trailing behind them was a long trail that, like the tail on a kite, stabilized their flight. They probably ate fish as a rule, although some may also have eaten insects. At some point on the evolutionary path from their cold-blooded ancestors they must have become warm-blooded, because otherwise they wouldn't have had enough sustained energy to fly. Bolstering this notion, and giving strong hints of what they looked like when alive, is a pterosaur fossil found in Russia in 1970 that shows signs of a thick coat of fur.

This standard pterosaur model persisted for 45 million years. But around 180 million years ago a new version made its appearance. These newer pterosaurs are called pterodactyloids (from the name given the first member of the group found, *Pterodactylus*, or "wing finger"; the older pterosaurs are referred to as rhamphorhynchoids). Pterodactyloids manifested some significant changes: Their long head grew still longer, yet, because it had lost some bones in the skull, it became even more lightweight. Their neck became flexible and birdlike. They lost some of all of their teeth. Most important, their tail shrank to a stump, making it useless for stabilizing flight. The only way that tail loss can be explained, say paleontologists, is by pterodactyloids' having developed more sophisticated brains capable of stabilizing flight with quick, small changes to the wings.

For more than 30 million years pterodactyloids remained relatively rare. But 144 million years ago, for reasons unknown, their primitive relatives vanished, and at that point the pterodactyloids exploded into a strange diversity of species. Some became enormous—one late species named *Quetzalcoatlus* was 39 feet across, making it by far the largest animal ever to fly. Other pterodactyloids developed extremely peculiar heads: one species had hundreds of bristles for teeth; another had a duck bill, another had a spoon bill. Many pterodactyloids had bizarre crests like *Tupuxuara*'s; some had crests shaped like swords, others like keels.

For all the hallucinatory appeal of pterosaur heads, however, it is the creatures' method of flight that remains the central question for paleontologists. Pterosaurs were the first vertebrates to fly, and only two other animals have since joined them in the air: birds and bats. By comparing the bones of all three creatures, paleontologists have tried to draw analogies that might suggest exactly how pterosaurs flew. Judging from the bones in the shoulder region, researchers have concluded that the extinct reptiles could have flapped their wings as powerfully as a bird or a bat.

Yet the analogies can go only so far. Bird and bat wings themselves are very different from each other. In birds, the wings are made of feathers with stiff shafts that are rooted in the arms and the fused fingers of the hands. The wings of bats are made of elastic membranes that run between four elongated fingers and reach all the way down to the feet. By moving their feet and fingers, bats control the shape and tautness of the wings.

Pterosaur wings, supported by that single outlandish finger, could not have been really close to either of these models. *Something* had to give the wings support, to keep them from flapping uselessly in the wind. Either the wings had to be stiffened internally or they had to be tied down to legs or feet. But we have no perfectly preserved soft-tissue fossil to tell us. The best clues come from 80 fossil impressions of wings left behind in the mud. Unfortunately these tracings do not provide researchers with any definitive answers. Usually the wings became distorted and were pulled into unnatural positions in death, and without exception their impressions are maddeningly ambiguous. In many of the fossils it's not even clear if the impressions were made by a wing or by some other soft tissue folded into a provocative position by accidental events. But since the nineteenth century, paleontologists, observing that pterosaur wings looked as if they were made of a membrane like that of a bat, have assumed that the creatures did in fact sport leathery wings that ran, like a bat's, from finger to feet.

For more than a decade paleontologist Kevin Padian, of the University of California at Berkeley, has been vigorously fighting this view. "If you're choosing between the bat and bird analogy," he says, "there are many more reasons to draw comparisons with birds." Batlike pterosaur wings would show a clear attachment to the foot, he says, perhaps with a notch in one of the bones, but he has seen none. A bat's wing also has a tendon running along its trailing edge, but there is no evidence of one in pterosaur fossils.

There is evidence, however, of a different sort of structural component. Paleontologists have long noted that some pterosaur fossil wings bear thin parallel ridges. Peter Wellnhofer, the curator of the Bavarian State Museum in Munich, argued in 1987 that these ridges represent tough fibers, possibly made of the protein collagen, sandwiched inside the wing to provide stiffness. Recently, Padian and Jeremy Rayner, a zoologist at the University of Bristol in England, found that the tips of pterosaur wings support this idea. Since bats have an elastic membrane that they pull tight, their wing comes to a sharp point. But while studying a particularly well preserved pterosaur in Germany, Padian and Rayner discovered that its wing tip forms a blunted curve. Such a shape requires some internal stiffening, such as could have been provided by fibers.

This pterosaur had been preserved in a belly-up position, exposing the underside of its wings. Padian and Rayner found that at the trailing edge some of the fibers had apparently separated from the wing and were lying on top of it at an angle, leaving behind corresponding grooves on the wing itself. The way the fibers had frayed indicates that rather than being sandwiched inside the wing, as many paleontologists had imagined, they lined the underside like the ribbing of an umbrella.

If fibers were indeed a structural component of the wing, then they, rather than the ambiguous impressions of the wing membrane, may be the best indicator of a pterosaur wing's shape. And according to the fibers, pterosaurs had narrow, birdlike wings that attached at the hip or the mid-thigh. Rayner, whose speciality is animal aerodynamics, says the fibers would have given the wings so much stiffness that a pterosaur wouldn't have needed to keeps its wings taut with its feet, as a bat does. Instead the fibers would have acted more like a bird's stiff feather shafts, which, when the bird brings its wing down, transfer the lift generated by the wing to the rest of the body.

The shape of pterosaur wings must have affected not only how they flew but also how they moved on the ground. Birds, with their wings free of their hind legs, can walk easily. But bats have to deal with two problems: their wings connect their arms to their legs, and their legs are designed to swing out to the side. In flight this keeps the wing membrane tight, but on the ground it makes a bat's legs too loose to support the weight of its body. As a result, bats crawl on all fours, with their long arms and flexible legs splayed out to the sides.

Just as pterosaurs' wings were designed for birdlike flight, Padian maintains, their legs were designed for bipedal, birdlike walking. He points out that the head of their femur was oriented to swing the leg forward and backward, not side to side, and their ankle bones had fused into a hinge that couldn't bend out in a sprawl. Far from crawling on all fours, he says, pterosaurs walked on two legs as comfortably as a dinosaur or a bird—which, Padian argues, shouldn't be surprising. "They really are coming from fairly similar ancestral stocks," he says, "and so a lot of what they do is going to be predisposed to be similar." Pterosaurs with wings folded up on either side also leave their hands relatively free for handling food or climbing trees.

But the traditional batlike stance still charms many paleontologists. "I find myself with the fuddy-duddies dotted around the world," says Donald Unwin of the University of Bristol. "In our humble opinion, the legs sort of stick out sideways and the animals tend to sprawl somewhat on the ground and walk around in a quadrupedal fashion." Unwin, along with Wellnhofer, has argued that the hips of pterosaurs had sockets unsuited for walking on two legs. If a pterosaur tried to walk upright, they maintain, the pelvis wouldn't bear directly down on the femur and the femur heads would tend to pop out of their sockets. "I think you can get pterosaur hind limbs into a bipedal position in the same way I can do a split," says Unwin. "It hurts, but I may be able to do it with a bit of training."

Padian counters that what might seem unlikely judging from only one part of a skeleton becomes possible when you add supporting cartilage and muscle. "What pterosaurs *can't* do," Padian adds, "is jerk the hind limb out to the side in exactly the way a bat does it."

Recently, Christopher Bennett of the University of Kansas gave some of Padian's ideas a new twist. By tilting the hip 60 degrees from the horizontal, he found that a protruding part of the socket's rim was positioned so that it could bear down on a femur head. A pterosaur with such a hip wouldn't be hunched over like a bird; it would be upright, something between a gorilla and a human. Instead of walking on its toes, it would roll back on its heels and keep its feet flat on the ground. Pterosaurs standing as tall as 20 feet might have strolled along the banks of lagoons with a humanlike gait.

Once you know how an extinct animal moved—whether it crawled or ran, flapped its wings or only glided—you can get an idea of how it fit into the ecological web of its time. If, as Padian argues, pterosaurs flew more like birds than bats, then one would expect them to play birdlike roles in their ecology as well.

Rayner and graduate student Grant Hazelhurst have found evidence for this idea in the geometry of pterosaur bodies. Previous research has shown that the ratios determined by a flying animal's mass, wingspan, and wing surface area predict what kind of flight style it will have. Forest birds share similar ratios, while soarers have their own set, as do the insect eaters, divers, and so on. Rayner and Hazelhurst found that the combination of long, thin wings and slender, lightweight bodies made large pterosaurs very aerodynamically efficient, able to soar on the weakest of rising air currents. Many of the largest pterosaurs may have been like today's frigate bird: perhaps they soared hundreds of miles over the ocean, grabbing fish or harassing other pterosaurs to surrender their catches. Perhaps, also like the frigate bird, they were helpless if they accidentally landed on water. Smaller pterosaurs probably flew like sea gulls and petrels, needing to flap more often to keep themselves aloft. Other pterosaurs were aerial hunters like falcons. And the smallest ones were the swallows of their day, chasing insects and maneuvering to match their prey's unpredictable flight.

Rayner and Hazelhurst's work is just as interesting for what it says pterosaurs *didn't* do as for what it says they did. Their long wings disqualified them from diving for fish like a duck or loon because they would have caused too much drag in the water. The researchers also say those long wings would have been unwieldy in forests. "If you've got a bird living in trees or nesting in twigs, it has a real problem if it's got very long wings because they can knock against the environment," Rayner explains.

Rayner is the first to point out that the fossils may be skewing his view. The fossils of small land animals are most often preserved in watery environments, such as oceans or lagoons, so creatures that kept to forests, deserts, or plains left behind much sketchier traces. Rayner didn't consider many pterosaurs that have left only fragmentary remains. For example, paleontologists know the giant *Quetzalcoatlus* only from the arm bones of a single animal. As a result, it's still not clear exactly what its physical proportions were, and paleontologists suspect that its wings may actually have been small for its body. *Quetzalcoatlus* is also unusual because it wasn't found by an ocean but instead in what were once seasonal wetlands far inland in Texas. The biggest birds that live in these kinds of habitats today are egrets and herons, which (possibly like *Quetzalcoatlus*) have relatively small, squarish wings that let them keep clear of the trees and other plants around them. Imagine *Quetzalcoatlus* as a spectacularly oversize egret, wading silently and delicately through the marsh, its long neck pulled back, and then suddenly plunging its head into the water to snatch fish with its chopsticks-style jaws.

Other pterosaurs apparently foreshadowed birds' feeding habits in surprising ways. One, named *Pterodaustro*, had hundreds of long bristlelike teeth in its lower jaw. Flamingos have similar ridges in their bill that they use to filter out algae and insects from water. *Pterodaustro* was exactly suited for that kind of feeding and supremely unqualified for anything else. In London, Stafford Howse has been studying the jaw of a pterosaur he refers to as the Purbeck spoonbill, after the geologic formation in England where it was found; like the living spoonbills of Africa and Asia, it had a long, narrow beak ending in two horizontal disks. Spoonbills today move their beak through mud at the bottom of streams and lakes to stir up the animals they feed on. The Purbeck spoonbill seems to have been even better adapted to this kind of feeding because it had several dozen sturdy curved teeth on both jaws. Although useless for chewing, since they jutted away from the mouth, they would have been perfect as a mud rake. Moreover, part of the roof of the Purbeck spoonbill's mouth was lined with horn; when it dug up snails and other shelled animals, it could have broken them open like a nutcracker.

In 1991 it occurred to University of Miami ecologist Thomas Fleming that some pterosaurs ought to have eaten fruit. Fruit-bearing plants now depend on bats, birds, and primates to eat their fruit and spread the seeds in their droppings. Yet paleobotanists believe that for 40 million years after fruit-bearing plants first appeared, these dispersers either hadn't yet evolved or were rare. There were plant-eating dinosaurs, of course, but they probably destroyed the seeds in their enormous, slow-moving digestive systems. However, Fleming hypothesized, the seeds would probably have survived a passage through a pterosaur's small gut. Furthermore, it made sense that pterosaurs would have taken advantage of a food that could provide them with the energy they needed for flying. The only problem with Fleming's idea was that at the time there weren't any pterosaurs known that might have been fruit eaters.

Fleming decided to publish his suggestion anyway, and only after his paper had been typeset did Wellnhofer tell him about a pterosaur from Araripe that he and Kellner were beginning to describe; they had named it *Tapejara* (meaning "the old being" in the language of the Tupi Indians). This pterosaur had an eight-inch-long head dominated by a high, thin prow in front of its nose that tapered back to a narrow prong above the eye. Its toothless jaws were like a precise pair of tweezers, sharp and slightly bowed.

The combination of crest and beak rules out many of the unusual pterosaur feeding techniques. The fish eaters had either gripping teeth or long jaws like a pelican's. *Tapejara*'s beak might have been good for picking at carrion, but its crest would have gotten in the way when it tried to poke into a carcass. Yet both its crest and its beak made *Tapejara* well suited for picking fruit. It could have used the crest to push aside thick foliage and could then have plucked fruit from the stem with its beak, which is precisely what hornbills and toucans do today. If so, then perhaps pterosaurs like *Tapejara* were crucial to the evolution of fruits like avocados and mangoes.

Flying and eating undoubtedly took up a healthy part of a pterosaur's day but not all of it. What did pterosaurs do the rest of the time? What was their social life like? Christopher Bennett has recently shed some light on the question for a pterosaur known as *Pteranodon*, which lived from 115 million to 70 million years ago along the coast of a sea that ran down the middle of North America. It was among the best of the soaring pterosaurs, with a wingspan ranging from 10 to 25 feet. *Pteranodon* fossils have been found at sites that were 100 miles or more from the coastline of the ancient sea, which suggests that these pterosaurs regularly flew great distances in search of food.

Bennett compared some 1,100 *Pteranodon* fossils, plotting on a graph the length of their individual leg bones, finger bones, and so on. In each case the measurements bunched into two distinct groups. The average individual in the small group, he calculated, had a wingspan of 12.5 feet, while the larger ones averaged 19 feet. The smaller pteranodons seemed to be more common, outnumbering the larger creatures two to one.

Beyond their size—and presumably their weight—there

were only two other differences Bennett could find between the groups: the pelvises of the large animals were proportionately narrow, while those belonging to the smaller ones were wide; the smaller animals also had small crests, while the large pteranodons had extravagant ones.

Paleontologists have long speculated on the purpose of pterosaur crests. Did they improve the aerodynamics of the body? Did they act as a rudder? Kellner points out that *Tupuxuara*'s fossilized crest was covered with the impressions of blood vessels. A dense mesh of capillaries on the crest, he proposes, might have acted like an air conditioner during flight, bringing hot blood close to the skin.

Bennett doesn't think either aerodynamics or cooling makes sense. If a small crest is sufficient to do the job, then why do a third of the pteranodons have much bigger ones? The only explanation that made sense to him was that he was looking at male and female pteranodons. The crests were primarily male mating displays, like antlers on an elk or a long tail on a bird, and the wide hips of the smaller, female pteranodons were designed to pass eggs.

To understand this combination of sexual differences in size, displays, and population ratios, Bennett turned to living animals as analogues. He found several creatures with these traits, ranging from elephant seals to boat-trailed grackles. And, he learned, they all court, mate, and raise their young in a similar fashion. Bennett thinks *Pteranodon* probably did the same. He envisions male pteranodons on crowded rookeries lining the ocean, competing for the attention of females. Usually crest and body-size display would be enough for males to ward off challengers, but occasionally they might have to fight. The winners would maintain big harems of the smaller females, but they'd offer no help in raising their young. The unlucky males could only look on enviously.

The effort of attracting females would have taken a heavy toll on the males. A wing broken in combat would have meant being grounded and eventually starving. Big crests might have made flying more difficult. While the males had a 50 percent wider wingspan than the females, they might well have been twice as heavy, and flying might thus have been far more taxing for them. This kind of built-in hazard could explain the skewed sex ratio Bennett found. "Apparently it increases the reproductive success of the ones who survive," he says, "so it can be selected for, even though in other respects it gets you into trouble."

In studying these bones, Bennett was able to get an idea of what it was like to grow up as a pteranodon. Along with measuring the size of the various bones, Bennett looked for signs of immaturity, such as unfused bones and spongy tissue at the joints where the animals still had youthful cartilage. After separating the creatures by age, Bennett found that typical bone from young pteranodons was similar to the bone of human children and other warm-blooded young in that it was full of blood vessels. Older pteranodons had bones of less uniform composition: the inner and outer surfaces of the hollow bones were covered with a layer of hard, smooth, less vascularized material. These findings suggest that pteranodons went through a massive growth spurt that ended at adulthood. Once they reached full size their bones hardened over. "In other words," says Bennett, "they weren't like living reptiles, like alligators, that continue growing for many, many years. If you could let alligators live a hundred years, they'd still be growing."

Bennett also found that all the immature pteranodons represented in the sample—about 14 percent—were already the same size as adults. "If they were going out fishing and feeding themselves and slowly continuing to grow," says Bennett, "you would expect to find young animals in the middle of the seaway, but they are just not there. There you find only animals that are virtually adult size."

This would tally with their fast growth. Pteranodons must have started life with a wingspan of a few inches at most, since their wings had to fold inside an egg that could in turn fit inside a mother. "But there's no way that a pterosaur could grow rapidly to large size and feed itself at the same time," Bennett says, because flying is so metabolically costly. After pteranodons hatched, they stayed in their nests, waiting for their mother to bring them food from the ocean—just as modern albatrosses do. "While the birds are being fed by their mother, they just sit around and grow till they are virtually adult size," Bennett explains. "And then they say, 'Okay, I'll go out and take care of myself now.'" Albatrosses need only a year to grow to full size, and judging from their bone growth, Bennett suspects that pteranodons grew as quickly (although reaching a wingspan not of 11.5 feet but up to 25).

Ironically, it might have been their similarity to birds that spelled the pterosaurs' ultimate doom. Beginning 88 million years ago the number of pterosaur species steadily dwindled, until 65 million years ago they vanished completely, along with their cousins the dinosaurs. While geologists have pretty much convinced themselves that a comet or asteroid impact did in the dinosaurs, paleontologists don't think an impact alone could have caused the extinction of pterosaurs, since they had clearly been declining for 23 million years.

One suggestion, put forward by Unwin, is that pterosaurs were slowly replaced by birds. The fossil record suggests that pterosaurs were the sole residents of the air until 145 million years ago. The earliest birds may then have taken advantage of ecological niches that pterosaurs had left empty. One of the first birds, *Hesperornis*, was an ocean diver, an occupation for which Rayner and Hazelhurst showed that pterosaurs were badly suited. Later, as a species of pterosaur or bird went extinct, both animals would compete to fill its role, and birds would win the niche more often—perhaps because feathered wings are more adaptable than fibered membranes. Indeed, beginning 100 million years ago birds made up a larger and larger proportion of species of all aerial animals.

Many of the last pterosaurs were giant soarers. Unwin suggests that pterosaurs were better suited for this kind of flight than birds could ever be and thus held on to this last corner of the sky until the end (soaring birds didn't appear until 5 million years after the extinction of pterosaurs).

Most researchers, though, aren't captivated by any hypothesis that's been put forward. It may simply be too difficult to craft a story—whether of ecological competition or some other process—from the fossil evidence. But pterosaur paleontologists are not overly concerned; right now they're too busy discovering how pterosaurs lived to be obsessed with how they died.

"In the seventies, dinosaurs had a renaissance," says Kellner, referring to research that helped transform our image of these creatures from lumbering sluggards to the intelligent, warm-blooded, socially complex ancestors of birds. "I think we're on the verge of something like that with pterosaurs. The study is really only getting started now." ❑

**Questions**:

1. How long ago did the earliest pterosaurs live?
2. What evidence supports the idea that pterosaur wings did not need to be attached to the legs or feet?
3. What ecological role were pterosaurs better at than birds, allowing some species to survive almost to the Cenozoic?

Answers on page 301

*One of the major controversies surrounding dinosaurs is whether they were warm-blooded or cold-blooded. This is a debate that began about 25 years ago and has become a particularly contentious issue in the last 10 years. One piece of evidence that has been used to determine the physiology of dinosaurs is the growth patterns of bones. Unfortunately, studies of dinosaur bones have found that growth patterns aren't exactly like cold-blooded reptiles or warm-blooded mammals. Dinosaurs may have had a physiology distinct from either reptiles or mammals. Studies of birds, which are thought to have evolved from dinosaurs, also display a confusing pattern of bone growth which suggests that the earliest birds may have had a physiology different from modern birds. Instead of falling neatly into our modern categories, dinosaurs may force us to reevaluate what it means to be cold-blooded or warm-blooded.*

# The Pulse of T. Rex

## Were dinosaurs warm-blooded? Does it even matter?

by Richard Monastersky

There's a moment in the movie *Jurassic Park*, a brief, unremarkable image, that bothers physiologist John Ruben. He doesn't mind the look of Spielberg's dinosaurs or even the way they act. Let them run, jump, stalk Jeff Goldblum, or even dance a jig if they want. What John Ruben doesn't like is their breath.

More precisely, he objects to the sight of vapor coming out of a dinosaur's snout—a small point that reveals a fundamental assumption the movie makes about the metabolism of these great beasts.

"The problem with *Jurassic Park* is that they portrayed dinosaurs as endothermic to underline the notion that they were interesting, active, and dangerous. How do I know that? Because they showed one of these dinosaurs lying there with steam coming out of its nose. The only way an animal can do that is if it's very tightly regulating its body temperature and is a lot warmer than its surroundings. That's why you never see reptile breath," says Ruben, a researcher at Oregon State University in Corvallis.

Welcome to the blood feud over dinosaur physiology—a debate about endothermy versus ectothermy, about warm-bloodedness versus cold-bloodedness. Paleontologists launched this war 25 years ago, and it has since spread from museums into magazines, best-selling books, and even public television shows.

The zealousness of some combatants and the derogatory comments that have flown back and forth led paleontologist James O. Farlow to upbraid a few colleagues when he addressed the question of physiology in *The Dinosauria* (1990, University of California Press).

A dinosaur researcher from Indiana University-Purdue University at Fort Wayne, Farlow wrote: "Unfortunately, the strongest impression gained from reading the literature of the dinosaur physiology controversy is that some of the participants have behaved more like politicians or attorneys than scientists, passionately coming to dogmatic conclusions via arguments based on questionable assumptions and/or data subject to other interpretations."

The debate still simmers, in large part because researchers have failed to find any means of resolving the issue. Lacking conclusive evidence one way or the other, paleontologists have had the freedom to argue ad nauseam, driven by nothing more substantial than faith in their own theories.

New discoveries, however, are providing solid information that offers the hope of ending this long-standing controversy. "We are on the edge of finding out what the metabolic physiology of dinosaurs was really like," says Ruben.

When early paleontologists looked at the oversized femur of a *Triceratops* or a *Tyrannosaurus rex*, most of them envisioned a large reptile of a physiology to match. Reptiles typically take their temperature cues from the outside environment.

In the cold, their metabolism slows and they grow lethargic. When the air warms or the sun comes out, reptiles arise from their torpor and resume an active life. In contrast, birds and mammals keep their body temperatures continuously elevated, which requires them to consume more food than a reptile of similar size.

The idea of cold-blooded, slow-moving dinosaurs came under fire in the 1960s, after John H. Ostrom of Yale University discovered a sickle-clawed terror called *Deinonychus*. With its slashing weaponry and flexible skeleton, this dinosaur had the look of an active, agile predator—an image inconsistent with the concept most paleontologists had of typical reptiles. Ostrom suggested that dinosaurs may have had physiologies more like mammals and birds.

Since then, other researchers have argued that the sheer bulk of the larger dinosaurs would have kept them from cooling down during the night. By dint of their dimensions, these giants could have maintained a constant body temperature.

Paleontologist Robert T. Bakker pushed the physiology argument further than other scientists in his popular book THE DINOSAUR HERESIES (1986, William Morrow). Bakker claimed that all dinosaurs were "automatic" endotherms—animals with fast, stable metabolisms that supply enough internal heat to keep their body temperatures constant. Ectotherms are just the opposite, having body temperatures that fluctuate with outside conditions.

Among his different points, Bakker contended that the structure of dinosaur bone closely resembles that of mammals and birds and does not match that of modern reptiles. He concluded that the internal structure of dinosaur bone proved such animals grew quickly, as modern endothermic mammals do.

New work on dinosaur bones, however, paints a more complex picture, suggesting that these once ruling reptiles do not fit neatly into any physiological category.

Although researchers have examined slices of dinosaur bone for decades, Anusuya Chinsamy of the University of Pennsylvania in Philadelphia advanced this work recently by comparing the bones of young and old animals from a single species. Charting the variations among the samples enabled her to reconstruct how dinosaurs grew.

Chinsamy's work has yielded a confusing array of results. When she examined cross sections of femurs from one type of dinosaur called *Syntarsus*, she found growth rings—dark bands similar to those that appear in trees. Growth rings develop when an animal temporarily stops building new bone, most often during a difficult season such as wintertime. Because the bones of mammals and other endotherms do not show growth rings, their presence in *Syntarsus* links that animal with ectotherms.

Buy Chinsamy also found evidence that this small predatory dinosaur stopped growing when it reached adulthood—a pattern typical of endotherms and not of ectotherms. What's more, the structure of the *Syntarsus* bone indicated that the animal grew rapidly, another characteristic of endotherms.

Seen from one side, this dinosaur looked like a typical reptile; seen from another angle, it seemed more like a mammal or a bird.

Another type of dinosaur, *Dryosaurus*, revealed the opposite pattern. This ornithopod did not show any growth rings, meaning it grew at a fairly high, sustained rate, much as mammals do. But unlike pandas and people, a mature *Dryosaurus* did not stop growing.

David J. Varricchio of Montana State University in Bozeman found a similarly complex picture when he analyzed a small, big-eyed dinosaur called *Troodon*. This animal also grew quickly, possibly reaching adult size within 3 to 5 years, Varricchio calculates. However, rings in the bone suggest that *Troodon* stopped growing occasionally—more like an ectotherm than an endotherm.

Chinsamy and Varricchio both conclude that dinosaurs may not fit neatly into the physiological categories occupied by modern animals. Instead, they may have existed somewhere in the middle. "They could have been souped-up cold-blooded animals," Varricchio offers.

Chinsamy recently took a different look at the physiology question by examining early birds from the Cretaceous period, the culmination of the dinosaur's reign on Earth. Contrary to what many paleontologists might have expected, she found that ancient birds had growth rings, suggesting that they had not yet achieved the endothermy of their modern offspring. Chinsamy and her colleagues reported their findings in the March 17 NATURE.

Many paleontologists think that birds must have been endothermic from the start because active flight consumes tremendous amounts of energy, something only endothermic animals can provide for long periods of time. Furthermore, when it became fashionable to call dinosaurs endotherms, ancient birds automatically gained similar footing because they evolved from dinosaurs.

Chinsamy's findings now punch a hole in that argument by suggesting that ancient birds were not endotherms. "If she is right about that, it seems unlikely that their [dinosaurian] ancestors would have had the full suite of endothermic features either," Farlow says.

Although he once firmly supported the idea of endothermic dinosaurs, Farlow now describes himself as a militant agnostic. "A lot of people, including myself, have done work that suggests maybe it's not quite as simple as all that. Some people, who were at one point very big enthusiasts of endothermy, have back-pedaled quite furiously," says Farlow. Most academic paleontologists have taken a similarly conservative view, he says, leaving only a tiny, but vocal, minority to champion the theory of warm-blooded dinosaurs, principally in the popular media.

Ruben describes the situation in even harsher terms. "The amount of misinformation that is out there on dinosaur endothermy is so unbelievable. Most of the people who have worked on dinosaur physiology don't know what they're talking about," contends the Oregon State physiologist.

According to Ruben, studies done to date have focused on factors only indirectly related to metabolism and therefore could not reveal whether an animal was ectothermic or endothermic. In fact, he thinks that paleontologists have been looking in the wrong place. Instead of examining slices of a femur, researchers might find more information by looking up a dinosaur's nose.

Ruben's former student, Willem J. Hillenius, recently demonstrated that endothermic animals have a special set of nasal bones directly related to their metabolism. Called the maxilloturbinals, the bones form thin, folded sheets inside the nasal passages of birds and mammals.

Animals with a fast metabolism need such protection because they inhale and exhale 5 to 10 times more frequently than reptiles. Without these bones, mammals would lose far too much water simply through breathing, says Hellenius, now a researcher at the University of California, Los Angeles.

The maxilloturbinals work as a humidifier-dehumidifier system. When a mammal inhales, air moving through the nasal passage evaporates water from the tissue covering the maxilloturbinals. That process not only provides necessary moisture for lungs, it also cools the maxilloturbinals. Then, as the animal exhales, the cool tissue reabsorbs moisture from the breath, drying the air before it leaves the body.

Looking back in the fossil record, Hillenius traced the evolution of endothermy in mammals by searching for maxilloturbinals or the internal ridges to which they attached. In the April EVOLUTION, he reports that the initial development of a mammal-like metabolism began as far back as 260 million years ago, some 45 million years before the first mammals appeared.

If mammalian ancestors did indeed take such a long time to evolve a faster-revving engine, Hillenius' discovery offers some hints to explain the physiological transition. Many researchers have speculated that endothermy developed because it helped animals maintain a high body temperature. But the reptilian ancestors of mammals wouldn't have received such thermal benefits because their metabolism, while slightly elevated, was still too low to provide enough heat.

Instead, Hillenius' findings support a different idea: Endothermy evolved because it enhanced an animal's ability to maintain strenuous activity. Even a modest increase in aerobic fitness would have given these creatures an edge.

Having tackled the mammal question, Hillenius now plans to address the dinosaur problem. By looking for maxilloturbinal bones in dinosaur skulls, paleontologists may eventually determine whether these creatures had a fast metabolism, he thinks. "We've still got to do our homework, but I think we've got a potential test," he says.

If dinosaurs don't have such bones, that doesn't necessarily relegate them to some physiological backwater where lethargic animals doze the day away. Paleontologists are now learning what physiologists have know for years—that some modern ectotherms can grow rapidly and have extremely active lifestyles. In particular, many lizards can sprint just as fast as, if not faster than, mammals of equivalent size, says Ruben. What ectothermic animals lack is the endurance of mammals and birds.

"It's hard for a lot of people to imagine that you can have an animal that's fast growing, fast, and interesting unless it's like we are. It's sort of a chauvinistic perspective," Ruben says. "I think in the end we're going to find out that dinosaurs were probably fairly typical ectotherms, metabolically. But that doesn't mean that they were sluggish or uninteresting.

**Questions**:

1. Who first proposed the idea that dinosaurs may have been warm-blooded?

2. What feature of a bone indicates how an animal grew during its life?

3. What feature of a mammal's skull increases a mammal's endurance relative to reptiles?

Answers on page 301

# 61

*Nearly everyone has at one time been fascinated with dinosaurs. Over the years Hollywood has repeatedly tried to recreate dinosaurs, with varying levels of success. In the summer of 1993, the movie* Jurassic Park *so convincingly recreated dinosaurs that it even impressed paleontologists, who normally complain that movie makers play fast and loose with science. The big difference between* Jurassic Park *and previous movies is that it is based on current theories in science, particularly the idea of extracting actual samples of dinosaur DNA. Extracting DNA fragments from fossils has become a hot area of research. Although it is currently impossible to clone dinosaurs from fossil DNA, fossil DNA fragments have two important uses to scientists: to sort out evolutionary relationships and to study migration patterns of humans. Because variations and similarities in the DNA of two species tell us how closely related they are, the use of fossil DNA fragments could help solve some important controversies in evolution where other fossil evidence is ambiguous, such as the origin of birds.*

## Dino DNA: The Hunt and the Hype

### Several groups are racing to get the first DNA out of dinosaur bones, but other researchers say their efforts are taking attention away from the real scientific value of ancient DNA

by Virginia Morell

Eighteen months ago, Mary Schweitzer, a biology graduate student at Montana State University's Museum of the Rockies, was examining a thin section of *Tyrannosaurus rex* bone under her light microscope, when she noticed a series of peculiar structures. Round and tiny and nucleated, they were threaded through the bone like red blood cells in blood vessels. But blood cells in a dinosaur bone should have disappeared eons ago. "I got goose bumps," recalls Schweitzer. "It was exactly like looking at a slice of modern bone. But, of course, I couldn't believe it. I said to the lab technician: 'The bones, after all, are 65 million years old. How could blood cells survive that long?'"

Yet even as she was doubting the evidence before her eyes, Schweitzer was moving ahead to another daring thought. If red blood cells had survived fossilization, it should be possible to get at the dinosaur's DNA. "The minute I saw those structures, getting the DNA became my goal," she says. "It was obvious that was the thing to try for."

Does that mean the real Jurassic Park is right around the corner? Maybe, and then again, maybe not. From the interior cavities of the *T. rex*'s bone (which her colleague, paleontologist Jack Horner, provided), Schweitzer has already extracted a molecule that might be dinosaur DNA. And their discovery has set off a furious race with other labs to be the first to publish on dinosaur genes. But before you start thinking that next summer's big amusement will be live dinosaurs at your local zoo rather than a Steven Spielberg movie, be warned that all these efforts are plagued with technical problems. No one knows enough about developmental genetics even to consider cloning a dinosaur should any genetic material from the age of reptiles be recovered. More important, most researchers think it's unlikely that any such material has survived in the first place. "Given what we know about the decay of the DNA molecule, the onus remains on those who are searching for dinosaur DNA to prove that they've found it," says molecular biologist Noreen Tuross of the Smithsonian Institution.

Beyond the technical problems lies a deeper concern about the attempt to retrieve dinosaur DNA, an effort some researchers disparage as "Disco Science." That is the possibility that this summer's intersection of science and Hollywood (which has already landed Horner on the cover of *U.S. News and World Report*) will overshadow some less dramatic—but potentially more fruitful—attempts by other scientists to get DNA from extinct animals and people that are hundreds or thousands of years old, rather than millions. These less glamorous specimens could provide answers to questions about population movements and biodiversity that dino DNA can't even begin to address.

None of this criticism means researchers won't be able to extract DNA from dinosaurs. In fact, they may well succeed. In a typical dinosaur fossil all the bone's original organic structures have been replaced by minerals and silica. But in a few cases, such as Horner's and Schweitzer's *T. rex* skeleton, the interior cavities of the bone were not completely mineralized and appear to contain organic matter. Horner suspects the bones are thick enough to protect their internal cellular structure—and perhaps their genetic material—from water and oxygen, two elements lethal to the survival of DNA.

What is more, the route into the bones isn't the only chance for success. Other scientists hope to sidestep the skeleton and go for the extinct animals' blood in a move taken straight from the pages of Michael Crichton's novel. At least two research groups are attempting to remove blood from the gut of amber-preserved biting insects, whose last meal might have been a dino. The net result of these efforts is that "someone's going to get dinosaur DNA, and soon," says the leader of one competing group, Raul Cano, a molecular geneticist at California Polytechnic State University.

Cano's team, not to miss a trick, is pursuing both bone and blood possibilities. Cano and his graduate student Hendrik Poinar, who recently smashed the age record for ancient DNA by recovering a gene fragment from a 120-million-year-old amber-preserved weevil, say they have—like Schweitzer and Horner—already managed to extract miniscule amounts of DNA from a fossilized dinosaur bone. But, again like Schweitzer and Horner, they can't say whether it is dinosaur DNA or DNA from something else. Because the polymerase chain reaction (PCR) technique that enables scientists to study the bits of DNA they harvest is exquisitely sensitive, it can pick up any stray DNA—a human technician's, for instance, or DNA from a fungus on the original bone—and amplify that instead. Consequently, neither group is rushing to claim victory. Says Horner: "It's easy to extract DNA from a dinosaur bone. It's trying to prove that it's from the dinosaur—and not from some contaminant—that's hard."

In spite of these roadblocks, big game DNA hunters keep pushing forward, and not just for the glory of being first, they say. Horner and Cano both say they want to use the genes to reconstruct dinosaur evolutionary history. That, says Horner, "is the big question: Where do dinosaurs fit?" They could be most closely related to either birds or crocodiles—or both or neither. Michael Parrish, a paleontologist and biochemist at Northern Illinois University, argues that with the molecule in hand they will be able to "answer a lot of questions about phylogeny." In addition, he points out that if they could get DNA from more than one species, researchers could then go on to study questions about the classification of specific dinosaurs.

But many other researchers think this is a highly unlikely scenario, arguing that DNA, a notoriously unstable molecule, would be hard put to survive in any boney material, no matter how well preserved, for 65 million years or more. It will be tough enough to get DNA from one dinosaur specimen, they say, let alone from the numbers needed to do any sort of species comparison. "There is no evidence—none—that even a very common, stable macromolecule like collagen has survived for that long," says the Smithsonian's Tuross. "It takes a leap of faith to believe that something like collagen has been destroyed and yet DNA persists. "Backing Tuross up, biochemical studies of the rate of decomposition of DNA in solution show that under normal conditions, with typical exposure to air and water, the molecule is seriously degraded if not entirely broken down after 40,000 to 50,000 years.

Yet there are hints that organic molecules can survive the millennia. Schweitzer has seen fibers on her *T. rex* bone that look like collagen, although she has not tested them. And biochemist Gerard Muyzer, from the University of Leiden in the Netherlands, reported last fall that he and his colleagues had extracted a bone protein, osteocalcin, from the bones of several dinosaurs including ceratopsians and hadrosaurs. Muyser's group used two types of antibodies that reacted specifically to osteocalcin from modern bones, and found that the antibodies also labeled the ancient protein. The researchers noted that burial in an impermeable matrix and the absence of high heat made it possible for the material to survive; a similar burial might preserve delicate DNA.

But preservation is only one worry. A more persistent concern among researchers in the young field of molecular paleontology is that the current dinosaur-sized hype may squash their own less glamorous but scientifically more rewarding investigations. "It's the new Disco Science question: Who's going to be the first to get dinosaur DNA?," says Bob Wayne, an evolutionary biologist at the University of California, Los Angeles, and an editor of the *Ancient DNA Newsletter*. "But the trouble is that these very topical questions tend to obscure other research on more recent materials—such as mammal pelts in museum collections—which are much more likely to contain real DNA from the original source."

Kelley Thomas, a molecular evolutionist at the University of California, Berkeley, agrees that "the real value of this work is not so much in the age or type of DNA you can get, but in what it has to offer the study of conservation biology and genetic continuity."

For example, graduate students in Thomas' lab are attempting to recreate the genetic history of the Alaskan and Northwest coast Steller sea lions and various salmon species—organisms whose populations have crashed in the past and are under pressure again. "The museum specimens are snapshots of the genetic diversity that was once found in those populations," explains Thomas. Using DNA taken from the museum specimens, researchers can see if there is a relationship between genetic diversity and a healthy population, Thomas says. "We've assumed that because the species has passed through a bottleneck we've hurt it. Now we can determine if those assumptions are true." The answer could be important for many endangered species.

Wayne, for his part, is extracting DNA from museum pelts of a now extinct wolf-like animal that inhabited the Falkland Islands until the latter part of the last century. "Charles Darwin wrote the first description of this species," says Wayne, "but about 30 years after he reported it, hunters wiped it out." Since then, various scientists have argued that the animal was actually a wolf, a large fox, or some type of domestic dog that had gone feral. "By extracting the DNA from hairs on the pelts," Wayne says, "this can be resolved definitively.

The questions being addressed aren't limited to animal species; molecular geneticists have had good luck retrieving viable DNA from human remains and using these to address questions of migration patterns. In the May 1993 *Proceedings of the (UK) Royal Society*, Erika Hagelberg and John Clegg of Cambridge University published initial results of a 3-year study of the prehistoric colonization of the South Pacific islands of Oceania. Previous efforts to trace the settlement patterns of the early island-hoppers relied on linguistic and archeological evidence—and indicated that the original settlers were voyagers from Southeast Asia. But Hagelberg sampled mitochondrial DNA sequences taken from human skeletons at a variety of sites in Oceania (some dating as far back as 2700 years before present). By comparing these skeletal genes to mitochondrial DNA from modern inhabitants, Hagelberg has shown that some of the earliest settlers probably came from Melanesia, and not just Southeast Asia.

Other investigators are looking for 13,000-year-old DNA from ground sloths to untangle their family tree. Still others are taking genetic material from 800-year-old skeletons of early Americans—the Hohokum people—to learn if they are related to current Native Americans. Though none of these projects are as romantic as the hunt for dinosaur DNA, together they constitute what Wayne terms "the real nuts and bolts research" of ancient DNA studies. "What we really want to do is reconstruct the historical variability of species, to get a feel for the variation in gene frequency over time. It's opened up a whole new avenue of research for museum collections, a new way of studying evolutionary history," he says.

And while these cautious researchers may never inspire a Michael Crichton novel or Steven Spielberg film, they are already in a position to do what the dinosaur DNA hunters are still struggling to accomplish: get hard data and replicate it. ❑

**Questions**:

1. What are the two possible sources of dinosaur DNA?

2. DNA has been extracted from dinosaur bones. Why hasn't anyone claimed that it is dinosaur DNA?

3. What is the oldest DNA extracted so far, and what did it come from?

Answers on page 301

*One of the most famous fossil mammals, other than hominids, is the saber-toothed cat,* Smilodon. *The best fossils have been obtained from the Rancho La Brea tar pits in southern California. So far, at least 2,500 individuals have been found. Because the fossil bones are so well preserved, studies can identify many of the injuries and diseases these animals suffered from. This is the science of paleopathology. Of the 166,000* Smilodon *bones at the George C. Page Museum, 5,000 belong to the paleopathology collection. The most important information that these injuries give us is in how* Smilodon *may have lived. Evidence of healing of some injuries indicates that the animal lived for quite a while after it received the injury. This suggests that* Smilodon *was a social animal, behavior that is rare among modern cats. Other injuries tell how* Smilodon *may have hunted. These malformed and broken bones may give us as much information as the perfect specimens chosen for museum displays.*

# Saber-Toothed Tales

Perfect fossils from a smelly tar pit tell us what the saber-toothed cat looked like. But it's the pit's broken and mangled bones that tell us how the cat actually lived.

by Rosie Mestel

In days of old, the cats were gold, and their teeth were used for stabbing: that's the picture paleontologists once painted of *Smilodon fatalis*, the saber-toothed cat that roamed North America until its extinction 11,000 years ago. Hunting alone, the tan-colored cat raced its hapless prey across grassy plains and then plunged its vicious, daggerlike fangs into the neck of its quarry.

These days, though, paleontologists paint a very different picture of *Smilodon* (whose name means "saber tooth" and has nothing to do with smiling). The new view is based largely on fossils that collections manager Christopher Shaw tends in a back room of the George C. Page Museum in Los Angeles. These fossils were pulled out of the museum's accompanying Rancho La Brea tar pits, shallow ponds of sticky asphalt that trapped late Ice Age mammals like flypaper. Over a 25,000-year period, from the time of the first deposits in the tar pits 36,000 years ago, at least 2,500 saber-toothed cats were trapped. From this collection of 166,000 *Smilodon* bones, the largest such collection in the world, paleontologists are slowly piecing together a picture of the life and times of the ferocious feline.

They know, for instance, that *Smilodon* could roar like a lion, because its throat bones—the hyoids—are shaped like those of a roarer. And they're now pretty sure the cat did not have the dry-grass coloration of a lion, as earlier paintings of *Smilodon* depict, but was spotted like a leopard, as the Page Museum mural now presents it. The evidence for the spots is indirect, because no one has yet unearthed a fossilized flap of saber-tooth skin. But pieces of leaves found in the tar or stuck to the teeth of fossilized herbivores suggest that the Pleistocene Los Angeles basin was a plain dotted with sagebrush, buckwheat, and clumps of oak and pine. In short, it was a place where spotty camouflage would be just the ticket for a lurking predator.

And lurk it did, if *Smilodon*'s bones are being interpreted correctly. Overall, its skeleton is closer to a lion's than to any other cat's, but it differed from the lion in size and bulk. At three feet high, *Smilodon* was a foot shorter than the lion. Nor was it as long, measuring about five and a half feet from snout to rump, compared with the lion's seven feet. But it was almost twice as heavy: recent estimates put it at 750 pounds. The saber-toothed cat also sported a bobtail, in marked contrast to the long tails that help provide balance for such distance runners as cheetahs, leopards, and lions. Short legs, bulky body, puny tail—all these add up to an animal that didn't run far while hunting. That same bulk means the cat could topple large, lumbering prey that didn't run far either, like a juvenile mastodon or mammoth. So it's most likely that *Smilodon* waited for its prey to draw near and then launched a surprise assault.

This much about *Smilodon*'s life was pieced together from the best, near-perfect bones of the museum's cat collection. But perfection isn't always a plus in paleontology. Shaw actually prefers the bad bones—the mangled, misshapen specimens that are housed in a special set of shelves draped with protective plastic. This is the paleopathology collection: more than 5,000 bones, every one a mess in one way or another. There are dislocated hips, their sockets burnished smooth by the incessant scraping of bone on bone. There are bite wounds, some suspiciously saber-toothed in shape, others hideously bubbly from infections that persisted for months. There are femurs, tibias, and fibulas that are blemished by lumpy, bony projections—evidence of the body's attempts to reinforce overexerted muscles that just kept on ripping. And each bone has something to tell Shaw and company about the way the long-dead cat lived its life.

Many of the injuries, for instance, were so dramatic that the cat would surely have dragged its limbs or limped badly, yet the amount of infection or scraping means that the animal was around for a while after it ran into trouble. How could it have survived its injury if it couldn't hunt? A solitary hunter, as most cats are, would have been a goner. This suggests that *Smilodon* was a social creature, like the lion. "This is just an awesome injury," mutters Shaw, showing off a mashed-up pelvis with a thighbone to match. "Now, this animal was so crippled that it could not have gotten enough food if it were a lone hunter. So this and other examples suggest that these animals had a social structure that actually encouraged the nurturing of injured individuals. Perhaps their families brought them food, or at least protected them from other carnivores." At the very least, injured animals may have been allowed to hang around the site of a kill and eat after everyone else was done.

Of course, you could argue—and paleontologists used to—that the tar pits, with their writhing masses of trapped ground sloths, dire wolves, and mastodons, were a year-round smorgasbord of captive treats. Injured animals, no longer able to fend for themselves, could limp over and dig in. But that popular picture is wrong. Sure, the museum houses more than a million bones, but that's a million bones piled up over 25,000 years. "It wasn't a giant orgy of death here," says Shaw. "There wasn't a lot of carrion for crippled animals to come in and make a livelihood out of. We're talking one large herbivore per *decade:* that's not very good on the pickings."

Another specimen, a femur with a nasty, knobby growth on it—the tell-tale mark of a repeated muscle injury—provides additional indirect information about *Smilodon*'s habits. "Obviously the cats won't be stressing their muscles when they're sleeping," Shaw says. "They're going to be stressing their muscles when they're using the most violent and jerking motions—like attacking." He and fellow paleontologist Fred Heald used that starting point—simply tallying which muscles were most commonly injured—to figure out which muscles the cat used most when grappling with prey. The collection is littered with such injuries. The particular femur Shaw is holding is knobbed at the place where an adductor muscle joined the thighbone to the pelvis, pulling the thigh forward and sideways when flexed. A few trays down, there's a foreleg bone with a hummock where the deltoid muscle attached from the shoulder blade, elevating the leg on demand.

"Overall, we found that the most common position for these animals to be stressed in was with their forelegs out, forward, and slightly bent and their front paws in a grasping position," says Shaw. "Their hind legs would be crouched in a pushing or pulling stance. So it seems that these animals were most often injured trying to acquire prey either by pushing animals, bowling them over, or pulling them back toward themselves." *Smilodon*'s traumatic injuries—fractured lower limbs, dislocated hips, crushed chests—fit the pushing-pulling scenario, too, because these are injuries you'd expect from an animal wrestling bulky prey and occasionally ending up underneath it. (The larger, less-studied dire wolf paleopathology collection has many more hoof-dented skulls, as befits a hunter who goes in headfirst for the hamstring or the throat.)

This fits with another *Smilodon* theory, put forward by William Akersten, former curator at Rancho La Brea and now curator at the Idaho Museum of Natural History. Instead of going for the neck as a lion does, Akersten's *Smilodon* would have bitten a chunk out of its upended victim's belly. (That's more like the method of the modern-day Komodo dragon. These reptiles are also saber-toothed, suggesting that a saber would be just the tool for belly-biting.) "The saber teeth are very strong in the forward-backward direction but not nearly as strong side-to-side," explains Shaw. "If these things were going into the neck of a living animal, where there are lots of big bones and everything's dynamic, everything's moving, it's more likely we would have found more broken teeth in these animals. But out of 700 skulls, we have only two specimens with broken canines that have wear on them, meaning ones that were broken during the animal's lifetime."

What's more, Akersten argues that the ten-inch-long saber teeth were actually blunter than butter knives and that the animal's lower jaw would have gotten in the way of any stabbing stroke. Akersten thinks *Smilodon* bit its victims, driving its sabers deep into the belly with powerful head-depressing muscles and gripping a chunk of flesh with its other, interlocking teeth. It ripped out the chunk, then hung around until the animal bled to death.

Akersten's idea is not universally accepted, though. "All of us crazy paleontologist, saber-toothed-cat types get together once in a while to argue about the new stuff and the old stuff," says Shaw. "And it's kind of neat, because as it's an extinct animal we'll never really answer all the questions. So you can be forever right—or forever wrong. Some of the arguments might be settled if we ever find a saber-toothed cat frozen in a glacier someplace. I'd sure love to find a frozen saber-toothed kitty so I could dissect the little guy." ❑

**Questions**:

1. How do scientists know that *Smilodon* could roar?

2. What features of *Smilodon*'s skeleton suggest that it waited for its prey to come near before attacking? What was its most likely prey?

3. What injuries suggest that *Smilodon* obtained its prey by either pushing or pulling it over?

Answers on page 301

63

*Of all the questions regarding the evolution of life, the origin of modern humans remains the most fascinating and the most controversial. Because of the rarity of relatively complete hominid fossils, each new find stands a good chance of challenging the dominant theories of human evolution and generating much controversy. Recent investigations have done just that. These studies have challenged our views on when Homo erectus first left Africa suggesting that the initial migration may have occurred much earlier than previously thought. This early migration has also rekindled the debate over whether modern humans evolved in Africa and then migrated out, or if modern humans evolved simultaneously in different parts of the world. At present this debate has not been resolved and may not be for some time.*

# How Man Began

New evidence show that early humans left Africa must sooner than once thought. Did *Homo sapiens* evolve in many places at once?

by Michael D. Lemonick

No single, essential difference separates human beings from other animals—but that hasn't stopped the phrasemakers from trying to find one. They have described humans as the animals who make tools, or reason, or use fire, or laugh, or any one of a dozen other appealing oversimplifications. Here's one more description for the list, as good as any other: Humans are the animals who wonder, intensely and endlessly, about their origin. Starting with a Neanderthal skeleton unearthed in Germany in 1856, archaeologists and anthropologists have sweated mightily over excavations in Africa, Europe and Asia, trying to find fossil evidence that will answer the most fundamental questions over our existence: When, where and how did the human race arise? Nonscientists are as eager for the answers as the experts, if the constant outpouring of books and documentaries on the subject is any indication. The latest, a three-part *Nova* show titled *In Search of Human Origins*, premiered last week.

Yet despite more than a century of digging, the fossil record remains maddeningly sparse. With so few clues, even a single bone that doesn't fit into the picture can upset everything. Virtually every major discovery has put deep cracks in the conventional wisdom and forced scientists to concoct new theories, amid furious debate.

Now it appears to be happening once again. Findings announced in the past two weeks are rattling the foundations of anthropology and raising some startling possibilities. Humanity's ancestors may have departed Africa—the cradle of mankind—eons earlier than scientists have assumed. Humans may have evolved not just in a single place but in many places around the world. And our own species, *Homo sapiens*, may be much older than anyone had suspected. If even portions of these claims prove to be true, they will force a major rewrite of the book of human evolution. They will herald fundamental changes in the story of how we came to be who we are.

The latest shocker comes in the current issue of *Nature*, where Chinese scientists have contended that the skull of a modern-looking human, found in their country a decade ago, is at least 200,000 years old—more than twice as old as any *Homo sapiens* specimen ever found in that part of the world. Moreover, the skull has features resembling those of contemporary Asians. The controversial implication: modern humans may not have evolved just in Africa, as most scientists believe, but may have emerged simultaneously in several regions of the globe.

The *Nature* article came out only a week after an even more surprising report in the competing journal *Science* U.S. and Indonesian researchers said they had redated fossil skull fragments found at two sites on the island of Java. Instead of being a million years old, as earlier analysis suggested, the fossils appear to date back nearly 2 million years. They are from the species known as *Homo erectus*—the first primate to look anything like modern humans and the first to use fire and create sophisticated stone tools. Says F. Clark Howell, an anthropologist at the University of California, Berkeley: "This is just overwhelming. No one expected such an age."

If the evidence from Java holds up, it means that protohumans left their African homeland hundreds of thousands of years earlier than anyone had believed, long before the invention of the advanced stone tools that, according to current textbooks, made the exodus possible. It would also mean that *Homo erectus* had plenty of time to evolve into two different species, one African and one Asian. Most researchers are convinced that the African branch of the family evolved into modern humans. But what about the Asian branch? Did it die out? Or did it also give rise to *Homo sapiens*, as the new Chinese evidence suggests?

Answering such questions requires convincing evidence—which is hard to come by in the contentious world of paleoanthropology. It is difficult to determine directly the age of fossils older than about 200,000 years. Fortunately, many specimens are found in sedimentary rock, laid down in layers through the ages. By developing ways of dating the rock layers, scientists have been able to approximate the age of fossils contained in them. But these methods are far from foolproof. The 200,000-year-old Chinese skull, in particular, is getting only a cautious reception from most scientists, in part because the dating technique used is still experimental.

Confidence is much stronger in the ages put on the Indonesian *Homo erectus* fossils. The leaders of the team that did the analysis, Carl Swisher and Garniss Curtis of the Institute of Human Origins in Berkeley, are acknowledged masters of the art of geochronology, the dating of things from the past. Says Alan Walker of Johns Hopkins University, an expert on early humans: "The IHO is doing world-class stuff." There is always the chance that the bones Swisher and Curtis studied were shifted out of their original position by geologic forces or erosion, ending up in sediments much older than the fossils themselves. But that's probably not the case, since the specimens came from two different sites. "It is highly unlikely," Swisher points out, "that you'd get the same kind of errors in both places." The inescapable conclusion, Swisher maintains, is that *Homo erectus* left Africa nearly a million years earlier than previously thought.

Experts are now scrambling to decide how this discovery changes the already complicated saga of humanity's origins. The longer scientists study the fossil record, the more convinced they become that evolution did not make a simple transition from ape to human. There were probably many false starts and dead ends. At certain times in some parts of the world, two different hominid species may have competed for survival. And the struggle could have taken a different turn at almost any point along the way. Modern *Homo sapiens* was clearly not the inevitable design for an intelligent being. The species seems to have been just one of several rival product lines—the only one successful today in the evolutionary marketplace.

The story of that survivor, who came to dominate the earth, begins in Africa. While many unanswered questions remain about when and where modern humans first appeared, their ancestors almost surely emerged from Africa's lush forests nearly 4 million years ago. The warm climate was right, animal life was abundant, and that's where the oldest hominid fossils have been uncovered.

The crucial piece of evidence came in 1974 with the discovery of the long-sought "missing link" between apes and humans. An expedition to Ethiopia led by Donald Johanson, now president of IHO, painstakingly pieced together a remarkable ancient primate skeleton. Although about 60% of the bones, including much of the skull, were missing, the scientists could tell that the animal stood 3 ft. 6 in. tall. That seemed too short for a hominid, but the animal had an all important human characteristic: unlike any species of primate known to have come before, this creature walked fully upright. How did researchers know? The knee joint was built in such a way that the animal could fully straighten its legs. That would have freed it from the inefficient, bowlegged stride that keeps today's chimps and gorillas from extended periods of two-legged walking. Presuming that this diminutive hominid was a female, Johanson named her Lucy. (While he was examining the first fossils in his tent, the Beatles' *Lucy in the Sky with Diamonds* was playing on his tape recorder.)

Since scientific names don't come from pop songs, Lucy was given the tongue-challenging classification *Australopithecus afarensis*. Many more remains of the species have turned up, including beautifully preserved footprints found in the mid-1970s in Tanzania by a team led by the famed archaeologist Mary Leakey. Set in solidified volcanic ash, the footprints confirmed that Lucy and her kin walked like humans. Some of the *A. afarensis* specimens date back about 3.9 million years B.P. (before the present), making them the oldest known hominid fossils.

The final clue that Lucy was the missing link came when Johanson's team assembled fossil fragments, like a prehistoric jigsaw puzzle, into a fairly complete *A. afarensis* skull. It turned out to be much more apelike than human, with a forward-thrust jaw and chimp-size braincase. These short creatures (males were under five feet tall) were probably no smarter than the average ape. Their upright stance and bipedal locomotion, however, may have given them an advantage by freeing their hands, making them more efficient food gatherers.

That's one theory at least. What matters under the laws of natural selection is that Lucy and her cousins thrived and passed on their genes on to the next evolutionary generation. Between 3 million and 2 million years B.P., a healthy handful of descendants sprang from the *A. afarensis* line, upright primates that were similar to Lucy in overall body design but different in the details of bone structure. *Australo pithecus africanus, Paranthropus robustus, Paranthropus boisei*—all flourished in Africa. But in the evolutionary elimination tournament, the two *Paranthropus* species eventually lost out. Only *A. africanus*, most scientists believe, survived to give rise to the next character in the human drama.

This was a species called *Homo habilis*, or "handy man." Appearing about 2.5 million years B.P., the new hominid probably didn't look terribly different from its predecessors, but it had a somewhat larger brain. And, perhaps as a result of some mental connection other hominids were unable to make, *H. habilis* figured out for the first time how to make tools.

Earlier protohumans had used tools too—bits of horn or bone for digging, sticks for fishing termites out of their mounds (something modern chimps still do). But *H. habilis* deliberately hammered on rocks to crack and flake them into useful shapes. The tools were probably not used for hunting, as anthropologists once thought; *H. habilis*, on average, was less than 5 ft. tall and weighed under 100 lbs., and it could hardly competed with the lions and leopards that stalked the African landscape. The hominids were almost certainly scavengers instead, supplementing a mostly vegetarian diet with meat left over from predators' kills. Even other scavengers—hyenas, jackals and the like—were stronger and tougher than early humans. But *H. habilis* presumably had the intelligence to anticipate the habits of predators and scavengers, and probably used tools to butcher leftovers quickly and get back to safety.

Their adaptations to the rigors of prehistoric African life enabled members of the *H. habilis* clan to survive as a species for 500,000 years or more, and at least one group of them apparently evolved, around 2 million years B.P., into a taller, stronger, smarter variety of human. From the neck down, *Homo erectus*, on average about 5 ft. 6 in. tall, was probably almost indistinguishable from a modern human. Above the neck—well, these were still primitive humans. The skulls have flattened foreheads and prominent brow ridges like those of a gorilla or chimpanzee, and the jawbone shows no hint of anything resembling a chin. Braincases got bigger and bigger over the years, but at first an adult *H. erectus* probably had a brain no larger than that of a modern four-year-old. Anyone who has spent time with a four-year-old, though, knows that such a brain can perform impressive feats of reasoning and creativity.

*H. erectus* was an extraordinarily successful and mobile group, so well traveled, in fact, that fossils from the species were found thousands of miles away from its original home in Africa. In the 1890s, Eugène Dubois, an adventurous Dutch physician, joined his country's army as an excuse to get to the Dutch East Indies (now Indonesia). Dubois agreed with Charles Darwin's idea that early humans and great apes were closely related. Since the East Indies had orangutans, Dubois thought, they might have fossils of the "missing link."

While Dubois didn't find anything like Lucy, he discovered some intriguingly primitive fossils, a skullcap and a leg bone, in eroded sediments along the Solo River in Java. They looked partly human, partly simian, and Dubois decided that they belonged to an ancient race of ape-men. He called his creature *Anthropopithecus erectus*; its popular name was Java man. Over the next several decades, comparable bones were found in China (Peking man) and finally, starting in the 1950s, in Africa.

Gradually, anthropologists realized that all these fossils were from creatures so similar that they could be assigned a single species: *Homo erectus*. Although the African bones were the last to be discovered, some were believed to be much more ancient than those found anywhere else. The most primitive Asian fossils were considered to be a million years old at most, but the African ones went back at least 1.8 million years. The relative ages, plus the fact that *H. erectus'* ancestors were found on that continent and then left sometime later.

When and why did this footloose species take off from Africa? Undoubtedly, reasoned anthropologists, *H. erectus* made a breakthrough that let it thrive in a much broader range of conditions than it was accustomed to. And there was direct evidence that could plausibly have done the trick. Excavations of sites dating back 1.4 million years B.P., 4,000 centuries after *H. erectus* first appeared, uncovered multifaceted hand axes and cleavers much more finely fashioned than the simple stone tools used before. These high-tech implements are called Acheulean tools, after the town of St. Acheul, in France, where they were first discovered. With better tools, goes the theory, *H, erectus* would have had an easier time gathering food. And within a few hundred thousand years, the species moved beyond Africa's borders, spreading first into the Middle East and then into Europe and all the way to the Pacific.

The theory was neat and tidy—as long as everyone overlooked the holes. One problem: if advanced tools were *H. erectus*' ticket out of Africa, why are they not found everywhere the travelers went? Alan Thorne, of the Australian National University in Canberra, suggests that the Asian *H. erectus* built advanced tools from something less durable than stone. "Tools made from bamboo," he observes, "are in many ways superior to stone tools, and more versatile." And bamboo, unlike stone, leaves no trace after a million years.

The most direct evidence of the time *H. erectus* arrived in Asia is obviously the ages of the fossils found there. But accurate dates are elusive, especially in Java. In contrast to East Africa's Rift Valley, where the underground record of geological history has been lifted up and laid bare by faulting and erosion, most Javan deposits are buried under rice paddies. Since the subterranean layers of rock are not so easy to study, scientists have traditionally dated Javan hominids by determining the age of fossilized extinct mammals that crop up nearby. The two fossils cited in the new *Science* paper were originally dated that way. The "Mojokerto child," a juvenile skullcap found in 1936, was estimated to be about 1 million years old. And a crusted face and partial cranium from Sangiran were judged a bit younger.

Theses ages might never have been seriously questioned were it not for a scientific maverick: the IHO's Curtis, one of the authors of the *Science* article. In 1970 he applied a radioactive-dating technique to bits of volcanic pumice from the fossil-bearing sediments at Mojokerto. Curtis' conclusion: the Mojokerto child was not a million years old but closer to 2 million. Nobody took much notice, however, because the technique is prone to errors in the kind of pumice found in Java. Curtis' dates would remain uncertain for more than two decades, until he and Swisher could re-evaluate the pumice with a new, far more accurate method.

The new dates ended up validating Curtis' previous work. The Mojokerto child and the Sangiran fossils were about 1.8 million and 1.7 million years old, respectively, comparable in age to the oldest *Homo erectus* from Africa. Here, then, was a likely solution to one of the greatest mysteries of human evolution. Says Swisher: "We've always wondered why it would take so long for hominids to get out of Africa." The evident answer: it didn't take them much time at all, at least by prehistoric standards—probably no more than 100,000 years, instead of nearly a million.

If that's true, the notion that *H. erectus* needed specialized tools to venture from Africa is completely superseded. But Swisher doesn't find the conclusion all that surprising. "Elephants left Africa several times during their history," he points out. "Lots of animals expand their ranges. The main factor may have been an environmental change that made the expansion easier. No other animal needed stone tools to get out of Africa."

Scientists already have evidence that even the earliest hominids, the australopithecines, could survive in a variety of habitats and climates. Yale paleontologist Elisabeth Vrba believes that their evolutionary success—and the subsequent thriving of the genus *Homo* as well—was tied to climate changes taking place. About 2.5 million to 2.7 million years ago, an ice age sent global temperatures plummeting as much as 20°F, prompting the conversion of moist African woodland into much dryer, open savanna.

By studying fossils, Vrba found that the populations of large mammals in these environments underwent a huge change. Many forest antelopes were replaced by giant buffalo and other grazers. Vrba believes that early hominid evolution can be interpreted the same way. As grasslands continued to expand and tree cover to shrink, forest-dwelling chimpanzees yielded to bipedal creatures better adapted to living in the open. *H. erectus*, finally, was equipped to spread throughout the Old World.

If early humans' adaptability let them move into new environments, Walker of Johns Hopkins believes, it was an increasingly carnivorous diet that drove them to do so. "Once you become a carnivore," he says, "the world is different. Carnivores need immense home ranges." *H. erectus* probably ate both meat and plants, as humans do today. But, says Walker, "there was a qualitative difference between these creatures and other primates. I think they actively hunted. I've always said that they should have gotten out of Africa as soon as possible." Could *H. erectus* have traveled all the way to Asia in just tens of thousands of years? Observes Walker: "If you spread 20 miles every 20 years, it wouldn't take long to go that far."

The big question now: How does the apparent quick exit from Africa affect one of the most heated debates in the field of human evolution? On one side are anthropologists who hold to the "out of Africa" theory—the idea that *Homo sapiens* first arose only in Africa. Their opponents champion the "multiregional hypothesis"—the notion that modern humans evolved in several parts of the world.

Swisher and his colleagues believe that their discovery bolsters the out-of-Africa side. If African and Asian *H. erectus* were separate for almost a million years, the reasoning goes, they could have evolved into two separate species. But it would be virtually impossible for those isolated groups to evolve into one species, *H. sapiens.* Swisher thinks the Asian *H. erectus* died off and *H. sapiens* came from Africa separately.

Not necessarily, says Australia's Thorne, a leading multiregionalist, who offers another interpretation. Whenever *H. erectus* left Africa, the result would have been the same: populations did not evolve in isolation but in concert, trading genetic material by interbreeding with neighboring groups. "Today," says Thorne, "human genes flow between Johannesburg and Beijing and between Paris and Melbourne. Apart from interruptions from ice ages, they have probably been doing this through the entire span of *Homo sapiens'* evolution."

Counters Christopher Stringer of Britain's Natural History Museum: "If we look at the fossil record for the last half-million years, Africa is the only region that has continuity of evolution from primitive to modern humans." The oldest confirmed fossils from modern humans, Stringer points out, are from Africa and the Middle East, up to 120,000 years B.P., and the first modern Europeans and Asians don't show up before 40,000 years B.P.

But what about the new report of the 200,000-year-old human skull in China? Stringer thinks that claim won't stand up to close scrutiny. If it does, he and his colleagues will have a lot of explaining to do.

This, after all, is the arena of human evolution, where no theory dies without a fight and no bit of new evidence is ever interpreted the same way by opposing camps. The next big discovery could tilt the scales toward the multiregional hypothesis, or confirm the out-of-Africa theory, or possibly lend weight to a third idea, discounted by most—but not all—scientists: that *H. erectus* emerged somewhere outside Africa and returned to colonize the continent that spawned its ancestors.

The next fossil find could even point to an unknown branch of the human family tree, perhaps another dead end or maybe another intermediate ancestor. The only certainty in this data-poor, imagination-rich, endlessly fascinating field is that there are plenty of surprises left to come.

**Questions**:

1. How old is the skull of the modern-looking human from China and why is this age significant?

2. *Homo habilis* was the first hominid to do what?

3. If the "multiregional hypothesis" is correct, how did evolving populations in different parts of the world evolve into the same species?

Answers on page 301

# Part Seven

# Extinction—Causes and Consequences

*Many people think of species as arbitrary subdivisions in the evolutionary continuum of life. This leads to considerable trouble when trying to preserve species, such as with the Endangered Species Act. How can we preserve something that is not clearly defined? For example, some endangered "species" have been shown to have living populations elsewhere. Indeed, most of the political wrangling over endangered species involve subspecies, and not true species. This article argues that species are not arbitrary units, but have an objective measurable reality. For example, individuals of different species cannot interbreed, while members of the same species can. This objective reality is important in our struggle to save species because it means that each one is unique. It is a unique branch on a tree that is billions of years old and is the product of a chain of events that will never be exactly duplicated. Unlike many other environmental problems, extinction is thus irreversible.*

# What Is a Species?

## by Stephen Jay Gould

I had visited every state but Idaho. A few months ago, I finally got my opportunity to complete the roster of 50 by driving east from Spokane, Washington, into western Idaho. As I crossed the state line, I made the same feeble attempt at humor that so many of us try in similar situations: "Gee, it doesn't look a bit different from easternmost Washington."

We make such comments because we feel the discomfort of discord between our mental needs and the world's reality. Much of nature (including terrestrial real estate) is continuous, but both our mental and political structures require divisions and categories. We need to break large and continuous items into manageable units.

Many people feel the same way about species as I do about Idaho—but this feeling is wrong. Many people suppose that species must be arbitrary divisions of an evolutionary continuum in the same way that state boundaries are conventional divisions of unbroken land. Moreover, this is not merely an abstract issue of scientific theory but a pressing concern of political reality. The Endangered Species Act, for example, sets policy (with substantial teeth) for the preservation of species. But if species are only arbitrary divisions in nature's continuity, then what are we trying to preserve and how shall we define it? I write this article to argue that such a reading of evolutionary theory is wrong and that species are almost always objective entities in nature.

Let us start with something uncontroversial: the bugs in your backyard. If you go out to make a complete collection of all the kinds of insects living in this small discrete space, you will collect easily definable "packages," not intergrading continua. You might find a kind of bee, three kinds of ants, a butterfly or two, several beetles, and a cicada. You have simply validated the commonsense notion known to all: in any small space during any given moment, the animals we see belong to separate and definable groups—and we call these groups species.

In the eighteenth century this commonsense observation was translated, improperly as we now know, into the creationist taxonomy of Linnaeus. The great Swedish naturalist regarded species as God's created entities, and he gathered them together into genera, genera into orders, and orders into classes, to form the taxonomic hierarchy that we all learned in high school (several more categories, families and phyla, for example, have been added since Linnaeus's time). The creationist version reached its apogee in the writings of America's greatest nineteenth-century naturalist (and last truly scientific creationist), Louis Agassiz. Agassiz argued that species are incarnations of separate ideas in God's mind, and that higher categories (genera, orders, and so forth) are therefore maps of the interrelationships among divine thoughts. Therefore, taxonomy is the most important of all sciences because it gives us direct insight into the structure of God's mind.

Darwin changed this reverie forever by proving that species are related by the physical connection of genealogical descent. But this immensely satisfying resolution for the great puzzle of nature's order engendered a subsidiary problem that Darwin never fully resolved: If all life is interconnected as a genealogical continuum, then what reality can species have? Are they not just arbitrary divisions of evolving lineages? And if so, how can the bugs in my backyard be ordered in separate units? In fact, the two greatest evolutionists of the nineteenth century, Lamarck and Darwin, both questioned the reality of species on the basis of their evolutionary convictions. Lamarck wrote, "In vain do naturalists consume their time in describing new species"; while Darwin lamented: "we shall have to treat species as . . . merely artificial combinations made for convenience. This may not be a cheering prospect; but we shall at least be freed from the vain search for the undiscovered and undiscoverable essence of the term *species*" (from the *Origin of Species*).

But when we examine the technical writings of both Lamarck and Darwin, our sense of paradox is heightened. Darwin produced four long volumes on the taxonomy of barnacles, using conventional species for his divisions. Lamarck spent seven years (1815–1822) publishing his generation's standard, multivolume compendium on the diversity of animal life—*Histoire naturelle des animaux sans vertèbres,* or *Natural History of Invertebrate Animals*—all divided into species, many of which he named for the first time himself. How can these two great evolutionists have denied a concept in theory and then used it so centrally and extensively in practice? To ask the question more generally: If the species is still a useful and necessary concept, how can we define and justify it as evolutionists?

The solution to this question requires a preamble and two steps. For the preamble, let us acknowledge that the conceptual problem arises when we extend the "bugs in my backyard" example into time and space. A momentary slice of any continuum looks tolerably discrete; a slice of salami or a cross section of a tree trunk freezes a complexly changing structure into an apparently stable entity. Modern horses are discrete and separate from all other existing species, but how can we call the horse (*Equus caballus*) a real and definable entity if we can trace an unbroken genealogical series back through time to a dog-size creature with several toes on each foot? Where did this "dawn horse," or "eohippus," stop and the next stage begin; at what moment did the penultimate stage become *Equus caballus*? I now come to the two steps of an answer.

First, if each evolutionary line were like a long salami, then species would not be real and definable in time and space. But in almost all cases large-scale evolution is a story of branching, not of transformation in a single line—bushes, not ladders, in my usual formulation. A branch on a bush is an objective division. One species rarely turns into another by total transformation over its entire geographic range. Rather, a small population becomes geographically isolated from the rest of the species—and this fragment changes to become a new species while the bulk of the parental population does not alter. "Dawn horse" is a misnomer because rhinoceroses evolved from the same parental lineage. The lineage split at an objective branching point into two lines that became (after further events of splitting) the great modern groups of horses (eight species, including asses and zebras) and rhinos (a sadly depleted group of formerly successful species).

Failure to recognize that evolution is a bush and not a ladder leads to one of the most common vernacular misconceptions about human biology. People often challenge me: "If humans evolved from apes, why are apes still around?" To anyone who understands the principle of bushes, there simply is no problem: the human lineage emerged as a branch, while the rest of the trunk continued as apes (and branched several more times to yield modern chimps, gorillas, and so on). But if you think that evolution is a ladder or a salami, then an emergence of humans from apes should mean the elimination of apes by transformation.

Second, you might grasp the principle of bushes and branching but still say: Yes, the ultimate products of a branch become objectively separate, but early on, while the branch is forming, no clear division can be made, and the precursors of the two species that will emerge must blend indefinably. And if evolution is gradual and continuous, and if most of a species' duration is spent in this state of incipient formation, the species will not be objectively definable during most of their geologic lifetimes.

Fair enough as an argument, but the premise is wrong. New species do (and must) have this period of initial ambiguity. But species emerge relatively quickly, compared with their period of later stability, and then live for long periods—often millions of years—with minimal change. Now, suppose that on average (and this is probably a fair estimate), species spend one percent of their geologic lifetimes in this initial state of imperfect separation. Then, on average, about one species in a hundred will encounter problems in definition, while the other 99 will be discrete and objectively separate—cross sections of branches showing no confluence with others. Thus, the principle of bushes, and the speed of branching, resolve the supposed paradox: continuous evolution can and does yield a world in which the vast majority of species are separate from all others and clearly definable at any moment in time. Species are nature's objective packages.

I have given a historical definition of species—as unique and separate branches on nature's bush.

We also need a functional definition, if only because historical evidence (in the form of a complete fossil record) is usually unavailable. The standard criterion, in use at least since the days of the great French naturalist Georges de Buffon (a contemporary of Linnaeus), invokes the capacity for interbreeding. Members of a species can breed with others in the same species but not with individuals belonging to different species.

This functional criterion is a consequence of the historical definition: distinct separateness of a branch emerges only with the attainment of sufficient evolutionary distance to preclude interbreeding, for otherwise the branch is not an irrevocably separate entity and can amalgamate with the parental population. Exceptions exist, but the reproductive criterion generally works well and gives rise to the standard one-liner for a textbook definition of a species: "a population of actually or potentially reproducing organisms sharing a common gene pool."

Much of the ordinary activity of evolutionary biologists is devoted to learning whether or not the groups they study are separate species by this criterion of "reproductive isolation." Such separateness can be based on a variety of factors, collectively termed "isolating mechanisms": for example, genetic programs so different that an embryo cannot form even if egg and sperm unite; behaviors that lead members of one species to shun individuals from other populations; even something so mundane as breeding at different times of the year, or in different parts of the habitat—say, for example, on apple trees rather than on plum trees—so that contact can never take place. (We exclude simple geographic separation—living on different continents for example—because an isolating mechanism must work when actively challenged by a potential for interbreeding through spatial contact. I do not belong to a separate species from my brethren in Brazil just because I have never been there. Similarly, reproductive isolation must be assessed by ordinary behavior in a state of nature. Some truly separate species can be induced to interbreed in zoos and laboratories. The fact that zoos can make tiglons—tiger-lion hybrids—does not challenge the separate status of the two populations as species in nature.)

Modern humans (species *Homo sapiens*) fit these criteria admirably. We are now spread all over the world in great numbers, but we began as a little twig in Africa (the historical criterion). We may look quite different from one another in a few superficially striking aspects of size, skin color, and hair form, but there is astonishingly little overall genetic difference among our so-called races. Above all (the functional criterion), we can all interbreed with one another (and do so with avidity, always, and all over the world), but not with any members of another species (movies about flies notwithstanding). We are often reminded, quite correctly, that we are very similar in overall genetic program to our nearest cousin, the chimpanzee—but no one would mistake a single individual of either species, and we do not hybridize (again, various science fictions notwithstanding).

I do not say that these criteria are free from exceptions; nature is nothing if not a domain of exceptions, where an example against any clean generality can always be found. Some distinct populations of plants, for example, can and frequently do interbreed with others that ought to be separate species by all other standards. (This is why the classification of certain groups—the rhododendrons, for example—is such a mess.) But the criteria work in the vast majority of cases, including humans. Species are not arbitrary units, constructed for human convenience, in dividing continua. Species are the real and objective items of nature's morphology. They are "out there" in the world as historically distinct and functionally separate populations "with their own historical role and tendency" (as the other textbook one-liner proclaims).

Species are unique in the Linnaean hierarchy as the only category with such objectivity. All higher units—genera, families, phyla, et cetera—are human conventions in the following important respect. The evolutionary tree itself is objective; the branches (species) emerge, grow, and form clusters by subsequent branching. The clusters are clearly discernible. But the status we award to these so-called higher taxa (clusters of branches with a single root of common evolutionary ancestry) is partly a matter of human decision. Clusters A and B are groups of species with a common parent. Each branch in each cluster is an objective species. But what are the clusters themselves? Are they two genera or two families? Our decision on this question is partly a matter of human preference constrained by the rules of logic and the facts of nature. (For example, we cannot take one species from cluster A and one from cluster B and put them together as a single genus—for this would violate the rule that all members of a higher taxon must share a common ancestor without excluding other species that are more closely related to the common ancestor. We cannot put domestic cats and dogs in one family while classifying lions and wolves in another.)

The taxonomic hierarchy recognizes only one unit below species—the subspecies. Like higher taxa, subspecies are also partly objective but partly based on human decision. Subspecies are defined as distinctive subpopulations that live in a definite geographic subsection of the entire range of species. I cannot, for example, pluck out all tall members of a species, or all red individuals, wherever they occur over the full geographic range, and establish them as subspecies. A subspecies must be a distinct geographic subpopulation—not yet evolved far enough to become a separate species in its own right but different enough from other subpopulations (in terms of anatomy, genetic structure, physiology, or behavior) that a taxonomist chooses to memorialize the distinction with a name. Yet subspecies cannot be irrevocably unique natural populations (like full species) for two reasons: First, the decision to name them rests with human taxonomists, and isn't solely dictated by nature. Second, they are, by definition, still capable of interbreeding with other subpopulations of the species and are, therefore, impermanent and subject to reamalgamation.

This difference between species and subspecies becomes important in practice because our Endangered Species Act currently mandates the protection of subspecies as well. I do not dispute the act's intention or its teeth, for many subspecies do manifest distinctly evolved properties of great value and wonder (even if these properties do not render them reproductively isolated from other populations of the species). We would not, after all, condone the genocide of all Caucasian human beings because members of the other races would still exist; human races, if formally recognized at all, are subspecies based on our original geographic separations. But since subspecies do not have the same objective status as species (and since not all distinct local populations bear separate names), argument over what does and does not merit protection is inevitable. Most of the major ecological wrangles of recent years—rows over the Mount Graham red squirrel or the Northern spotted owl—involve subspecies, not species.

These taxonomic issues were once abstract, however important. They are now immediate and vital—and all educated people must understand them in the midst of our current crisis in biodiversity and extinction. I therefore close with two observations.

By grasping the objective status of species as real units in nature (and by understanding why they are not arbitrary divisions for human convenience), we may better comprehend the moral rationale for their preservation. You can expunge an arbitrary idea by rearranging your conceptual world. But when a species dies, an item of natural uniqueness is gone forever. Each species is a remarkably complex product of evolution—a branch on a tree that is billions of years old. All the king's horses and men faced an easy problem compared with what we would encounter if we tried to reconstitute a lost species. Reassembling Humpty-Dumpty is just an exceedingly complex jigsaw puzzle, for the pieces lie at the base of the wall. There are no pieces left when the last dodo dies.

But all species eventually die in the fullness of geologic time, so why should we worry? In the words of Tennyson (who died exactly 100 years ago, so the fact is no secret):

> From scarped cliff and quarried stone
> She cries, "A thousand types are gone:
> I care for nothing. All shall go."
> (From *In Memoriam.*)

The argument is true, but the time scale is wrong for our ethical concerns. We live our lives within geologic instants, and we should make our moral decisions at this proper scale—not at the micromoment of thoughtless exploitation for personal profit and public harm; but not at Earth's time scale of billions of years either (a grand irrelevancy for our species' potential tenure of thousands or, at most, a few million years).

We do not let children succumb to easily curable infections just because we know that all people must die eventually. Neither should we condone our current massive wipeout of species because all eventually become extinct. The mass extinctions of our geologic past may have cleared space and created new evolutionary opportunity—but it takes up to 10 million years to reestablish an interesting new world, and what can such an interval mean to us? Mass extinctions may have geologically distant benefits, but life in the midst of such an event is maximally unpleasant—and that, friends, is where we now reside, I fear.

Species are living, breathing items of nature. We lose a bit of our collective soul when we drive species (and their entire lineages with them), prematurely and in large numbers, to oblivion. Tennyson, paraphrasing Goethe, hoped that we could transcend such errors when he wrote, in the same poem:

> I held it truth, with him who sings
> To one clear harp in divers tones
> That men may rise on stepping-stones
> Of their dead selves to higher things.

❑

**Questions**:

1. How long does it take for life to recover and become reestablished after a mass extinction?

2. How many units does the taxonomic hierarchy recognize below species? Name it or them.

3. Do most new species evolve when an old species becomes another over its entire geographic range? Why is the name "dawn horse" a misnomer?

Answers on page 301

# 65

*The mass extinction event at the end of the Cretaceous gets much more attention than any of the other mass extinctions in the history of life. However, it wasn't the largest mass extinction that ever occurred. That honor goes to the end-Permian extinction 250 million years ago. During this mass extinction between 80 and 95% of all marine species went extinct. While the Cretaceous-Tertiary extinction has been tied to a single event, a bolide impact, the Permo-Triassic extinction appears to be much more complicated. First, the time span of the extinctions is still in doubt. Once thought to have occurred over the span of 8 to 10 years, the extinctions are now thought to have taken a million years or less. Second, the exact cause of the extinctions is very uncertain and may involve the interaction of many different factors. Whatever the cause, the Permo-Triassic extinction was one of the most profound events in the history of life.*

## The Greatest Extinction Gets Greater

The mass extinction 250 million years ago that changed the face of life on Earth is looking even more devastating as a pack of killers is invoked to explain the mayhem
by Richard A. Kerr

Life is made of sturdy stuff. A good thing, too, since living creatures have had a rough time of it these past 545 million years. In that time at least five mass extinctions have wracked the biosphere, wiping out major portions of existing species and allowing the survivors to reshape the biological world. While the death of the dinosaurs and the loss of upward of 50% of other species 65 million years ago at the boundary of the Cretaceous and Tertiary periods is the best known extinction, it is by no means the greatest.

In fact, the Cretaceous-Tertiary extinction was little more than a sideshow, says paleontologist Douglas Erwin of the National Museum of Natural History, compared to the devastation that occurred 250 million years ago between the Permian and Triassic periods. At the Permo-Triassic boundary, life in the sea was nearly snuffed out: At least 80%, and perhaps as much as 95%, of marine species disappeared. Life has never been the same. "The organisms you see in a tide pool today are a consequence of what happened 250 million years ago," says Erwin. "If the Permo-Triassic extinction hadn't happened the way it did, you would find a whole different bunch of beasts there."

Despite its size and importance, for many years studies of the Permo-Triassic (P-T) mass extinction lagged far behind those of the Cretaceous-Tertiary boundary. Within recent years, however, that's begun to change as more researchers have begun looking at the P-T boundary. And as the picture of the P-T extinction gets filled in, researchers are recognizing an event even more calamitous than they suspected. Recent results show more victims (insects on land in addition to devastation in the seas) and a briefer duration (shortened by half or more). Rather than indicting a single killer for this calamity, researchers are now thinking that the P-T massacre may have resulted from a fatal combination of deleterious environmental changes that add up to what Erwin calls the "world went to hell" hypothesis.

The current wave of interest in the P-T boundary began in the 1980s, triggered by the realization that the geologic record there had more to tell than had been realized. Earlier studies of the Permian period had focused on sea life as preserved in sedimentary rocks at sites from Pakistan to Greenland. These fossils showed that Permian ocean life was sparse, even sedate. Animals tended to be immobile, standing on stalks or lying on the sea floor and quietly filtering detritus for a meal or waiting for prey to come by. Snails and bivalves were a minor part of the fauna; active predators were scarce.

By the end of the 40-million-year-long Permian, says Erwin, those "well-developed ecosystems were torn apart by the extinction." In their place came a new set of communities that were more diverse. Gone were the easily preyed upon meadows of stalked animals, replaced by armored snails and deep-burrowing bivalves. Predatory free swimmers such as cephalopods and reptiles became abundant. The gigantic extinction as the few hardy survivors eventually filled more niches than the ones left by the victims.

It wasn't until this year that researchers realized that history's most diverse land animals—the insects—had also suffered a major extinction between the Permian and the Triassic. This extinction is doubly notable because it is the only one in the 390-million-year history of these hardy creatures. Writing earlier this year in *Science* (16 July, p.310), Conrad Labandeira of the National Museum of Natural History and John Sepkoski of the University of Chicago reported that buried in some of the less accessible literature—especially the older German and more recent Russian papers—is clear evidence that eight of the 27 orders of insects in the Permian did not survive into the Triassic. Another three orders struggled into the Triassic only to perish there.

Like the marine survivors, the insects that came through the P-T extinction tended to have new, more effective ways of coping with the stresses of life. Survivors were more likely to have an intermediate resting stage in their life cycle, notes Labandeira, or some other means of weathering hard times, such as protective scales or a habit of burrowing. The extraordinary surge of insect diversity that steadily carried the insects to the dominance they enjoy today was "largely attributable to the end-Permian mass extinction," which redirected insect evolution, say Labandeira and Sepkoski.

Naturally, the recognition of a major extinction among another Permian group raises the question of what caused it. But there, paleontologists have been at something of a loss. "This extinction boundary has simply elicited the reaction: 'We know there's an extinction and it's mysterious,'" says paleontologist David Bottjer of the University of Southern California. Paleontologists tended to concentrate on what life was like within a geologic period, notes Bottjer, rather than what happened at the boundaries between them. But approaches developed during the well-publicized work on the Cretaceous-Tertiary mass extinction have provided a helpful model.

That extinction was also a mystery until paleontologists, geologists, and geochemists, prompted by the discovery of traces of a huge cometary impact that might have caused it, began analyzing in minute detail the sediments laid down during the extinction. Now, at the P-T boundary, "what we're seeing is the beneficial spinoff of the [Cretaceous-Tertiary] debate," says paleontologist David Raup of the University of Chicago. "They're getting down on the ground and seeing what it's really like. It's all for the good."

One of the first fruits of that closer examination is the realization that the P-T extinction took place in a much shorter period than was thought, a finding that increases its intensity. A few years ago, the published length was 8 to 10 million years—gradual by any measure. At the time, Erwin thought 5 million years was a better estimate, but "as the data have been getting better, the extinction has appeared to get more rapid," he says.

Exactly how rapid the P-T was is still unclear, because paleontologists are struggling with imperfections in the P-T record—missing sections of the geologic record and poor fossil markers for keeping track of time—that confound efforts to calculate its apparent duration. Despite the difficulties, though, some paleontologists are claiming that at least some of the marine extinctions came in a burst right at the P-T boundary. "It was something like thousands of years, not millions, a big wipe-out," says Anthony Hallam of the University of Birmingham. Hallam's colleague in P-T work, paleontologist/sedimentologist Paul Wignall of the University of Leeds, agrees. "In the field, it seems dramatically sudden. There may have been a decline before that, but I believe there was a sharp event at the very end of the Permian."

Erwin, like many of his colleagues, is open to the possibility of an abrupt pulse of marine extinctions at the P-T boundary, but he also sees a more protracted period of extinction lasting a million years or more—but still far shorter than even his 5-million-year estimate from a few years ago. Evidence from terrestrial vertebrate fossils in South Africa, including those of reptiles and amphibians, also supports the picture of a protracted extinction with a pulse at the boundary lasting about a million years or less, says paleontologist Peter Ward of the University of Washington in Seattle.

With the P-T boundary looking more and more like the site of a prolonged extinction of varying intensity, the experts are now wondering what could cause a massive event of that kind. The one thing all concerned agree on is that—despite claims to the contrary in the mid-1980s—there is no evidence of an asteroid or comet impact having anything to do with the P-T extinction. In place of a single destructive event, researchers are now pondering the possibility that a combination of stresses might be needed to explain the massive extinctions on land and in the sea.

One long-standing proposal is that the extinction was caused when the sea level fell, perhaps as much as 150 meters, during the few million years before the P-T boundary, laying bare most continental rock. While that was a slow decline, it could have produced dramatic climate changes. At the time, all the continents were fused in a single supercontinent, Pangaea, that was 40% covered by water. Because of water's great capacity for absorbing heat, the supercontinent would have experienced few sharp climatic swings. What's more. the high sea level also created an enormous area within the inland seas and on the continental shelf around the edges of Pangaea in which Permian shallow-water communities like reefs flourished.

Life was good in the Permian, the theory goes, until the sea went out. When sea level dropped, the swings from one season to another would have become much more extreme. Increased seasonality of this kind could have been just the sort of change that drove the insect extinctions near the P-T boundary, Labandeira says. And the dwindling area of shallow water could have driven to extinction many marine species, such as the corals that began to decline well before the boundary.

If the recession of sea level wasn't devastating enough to trigger a mass extinction by itself, researchers have lately added another possible cause: volcanic eruption. And not just any volcanic eruption, but the one that produced the Siberian Traps, a frozen puddle of at least 2 million cubic kilometers of lava that flowed across northern Siberia in a million years or less. (For comparison, Mount St. Helens' 1980 eruption drew on a mere 1 cubic kilometer of magma.) Within the limits of accuracy of isotopic dating, it appears that this enormous eruption, the largest of the past 545 million years, coincided with the P-T mass extinction. Another large eruption, which formed the Deccan Traps of India, immediately preceded the Cretaceous-Tertiary boundary (*Science,* 14 June 1991, p.1496).

"It would be irresponsible to ignore the possibility" that the eruption of the Siberian Traps was involved in the extinction, says Gerry Czamanske of the U.S. Geological Survey in Menlo Park, who has worked on the Siberian Traps, although he notes that its precise involvement remains speculative. Perhaps it spewed enough debris to block out sunlight and trigger an ice age, enough sulfuric acid to acidify the oceans, or enough carbon dioxide to fuel greenhouse warming.

To round out the witch's brew of killing agents, Hallam and Wignall are diagnosing their abrupt marine extinctions at the P-T boundary as a case of suffocation. Hallam and Wignall see signs in the sediments that after the sea level slowly dropped, the sea rushed back onto the continental shelves faster than it did in any other marine transgression of the past 545 million years. What caused the sea to rise tens of meters in only some thousands of years "is a problem, a puzzle," notes Hallam, "but the evidence for a global rise at this time is overwhelming."

Whatever caused the rapid return of the sea, chemical traces in the sediments laid down by the rising waters suggest that they were low in oxygen or lacking it altogether, say Hallam and Wignall. Stagnation in the deep sea can build up such anoxic waters, and when a rising sea lifts them onto the continental shelf, they would snuff out the traces of life left after any regression.

Hallam and Wignall's scenario of death by suffocation got a boost this summer when Y. Isozaki of the Tokyo Institute of Technology reported at a Pangaea meeting in Calgary that he had confirmed the likely source of their anoxic waters. Isozaki has found a 50-million-year slice of the deep-sea sedimentary record containing the P-T boundary that was plastered onto Japan. Its chemical composition shows oxygen levels in the deep-sea falling over millions of years until anoxia sets in. At that point, most Permian micro-plankton called radiolarians disappear, and after the worst of the anoxia Triassic ones appear. That seems to tie this exceptional episode of intense anoxia to the P-T boundary in the deep sea and, through Hallam and Wignall's rapid transgression, to the pulse of extinction in shallow waters.

Erwin takes all this in and concludes that the P-T "extinction is not a single event" with a single cause. He sees an unusual convergence of events—possibly including gradual regression of the sea, volcanic eruption, and rapid transgression of oxygen-depleted water—that makes for a uniquely devastating extinction. But all the possible causes are not yet tightly tied to the known effects, he says. Those little beasts in the tide pool have more secrets to divulge.

**Questions**:

1. Besides marine organisms, what other major group of animals was severely affected by the Permo-Triassic extinctions?

2. How might falling sea level cause a mass extinction?

3. What did Hallam and Wignall propose as the cause of the Permo-Triassic extinctions?

Answers on page 301

*For most of this century scientists have viewed the replacement of dinosaurs by mammals as a case of superior mammals outcompeting the inferior dinosaurs. In the last ten years this scenario has changed to one of mere chance. The dinosaurs were just unlucky, wiped out by a meteor that evolution could not prepare them for. If this is true, how did the dinosaurs rise to dominance in the first place? Was it through some superior feature of their biology or just another case of chance? Studies of the earliest dinosaurs suggest that it may have been a lucky break. These studies have looked at Triassic dinosaurs from Argentina, including the earliest known dinosaur* Eoraptor, *which lived at a time very close to the origin of dinosaurs. If the dinosaurs were somehow superior to the other vertebrates of the time, then the dinosaurs should have become more common through time while the other vertebrates became rarer over the same time period. This is not the pattern that is seen in Argentina. It may be that the dinosaurs came to dominance suddenly due to the extinction of other vertebrates. Ultimately, the dinosaurs may owe their reign to the same kind of accident that has given the mammals theirs.*

# The Accidental Reign

## Did a lucky break allow the dinosaurs to take over the Earth?

by Richard Monastersky

Argentina has twice blessed Paul Sereno.

In 1988, the young University of Chicago paleontologist traveled to the beautiful Valley of the Moon in the foothills of the Andes and returned with the best-preserved fossil ever found of *Herrerasaurus,* the oldest and most primitive dinosaur then known. Sereno revisited the same sculpted badlands in 1991 and came home with a new type of dinosaur even more basic in its anatomy than *Herrerasaurus.* Naming the creature *Eoraptor,* Sereno unveiled the specimen earlier this month at a press conference and in the Jan. 7 NATURE.

Because it is the most primitive dinosaur ever found, *Eoraptor* provides an unprecedented look backward into the earliest evolution of these beasts from an ancestral form that spawned all later dinosaurs. "When we look at the skeleton of *Eoraptor,* we have perhaps our best chance to test out theories about dinosaur origins," Sereno says.

Along with *Herrerasaurus,* the new find is also offering insight into how dinosaurs rose from humble beginnings to reign over the continents for a staggering 150 million years. The emerging picture suggests that rather than wresting control of the Earth from the dominant beasts of the time, dinosaurs may have reached the top simply through blind luck.

The *Eoraptor* and *Herrerasaurus* skeletons found by Sereno and his colleagues come from the Ischigualasto Provincial Park in northwestern Argentina. Set against a backdrop of imposing red cliffs, the Valley of the Moon in Ischigualasto contains a lunar landscape of rare rocks laid down during the late Triassic period, roughly 230 million years ago, when dinosaurs first appeared on Earth.

As dinosaurs go, *Eoraptor* was tiny, reaching only a meter long. The dog-size creature walked upright and had a mouthful of curved, serrated teeth—the signature of a carnivorous life-style.

According to Sereno, *Eoraptor's* skull was particularly primitive, displaying almost none of the specialized adaptations that characterize the three main dinosaur subgroups: the meat-eating theropods, which include *Tyrannosaurus;* the gargantuan sauropods, which include brontosaurus; and the herbivorous ornithischians, which include *Triceratops.*

"The skull for all intents and purposes is ancestral. For the most part, it is nearly identical to what we would have expected the ancestral dinosaur to look like," says Sereno.

Even *Herrerasaurus,* a contemporary of the 225-million-year-old *Eoraptor,* seems specialized in comparison. While the new species had a simple jaw, *Herrerasaurus* and later theropods had a sliding jaw joint that helped in subduing live prey, according to Sereno and Fernando E. Novas of the Argentine Museum of Natural Sciences in Buenos Aires. They described the new *Herrerasaurus* specimen in the Nov. 13, 1992 SCIENCE.

Other parts of the *Eoraptor* anatomy also come close to that of the ancestral dinosaur, although the new fossil does show some specialization. *Eoraptor's* hand had three fingers well designed for gripping and slashing open prey—a characteristic that links it with theropods, according to Sereno.

During the dinosaurs' heyday, long after *Eoraptor* had vanished from Earth, the ornithischians, sauropods, and theropods each evolved specialized bodies suited to their modes of life. Because the various groups had such dissimilar anatomies, early dinosaur investigators presumed that the bird-hipped ornithischians had evolved from a different stock of reptiles than had the sauropods and theropods, which together form a group called saurischian dinosaurs, characterized by their lizard-like pelvises.

In 1974, paleontologists Robert T. Bakker and Peter M. Galton challenged the established view by suggesting that all dinosaurs evolved from a single common ancestor. The two researchers showed that some saurischians and ornithischians had remarkably similar joints, indicating that the earliest members of these two lineages had sprung from one line of animals that bore features common to both groups.

To visualize that ancestral dinosaur, many paleontologists have looked to a small reptile called *Lagosuchus,* which lived in the Ischigualasto region in the mid-Triassic period, 10 million years before *Eoraptor.* While not a dinosaur, *Lagosuchus* had evolved some features similar to those of the later beasts. It had a hinged ankle joint that helped it walk erect with its legs underneath its body, as did the dinosaurs. Other animals at the time walked with their legs out to the sides of their bodies, in a more sprawling, crocodile-like posture.

While many paleontologists consider *Lagosuchus* a dinosaur precursor, it can't provide a true picture of the ancestral dinosaur because it lacked many critical innovations that characterize all dinosaurs. "We could really only draw a common stem [ancestral] form of dinosaurs on paper. There wasn't any fossil that even approximated it," says Hans-Dieter Sues of the Royal Ontario Museum in Toronto. "This new find is really a tremendous discovery because we didn't have any idea what the common stem form of dinosaurs really looked like."

According to Sereno, the primitive nature of the *Eoraptor* skull supports the theory that dinosaurs arose from a common ancestor and only later developed specializations that would split them into ornithischians and saurischians. The new discovery also confirms earlier suspicions about what the ancestral dinosaur looked like.

On the basis of *Herrerasaurus* and other finds, researchers have pictured the earliest dinosaurs as small, bipedal carnivores—a description that matches *Eoraptor.* This dinosaur itself could not have been the common ancestor because it had already evolved specializations that place it within the theropod subgroup. But the *Eoraptor* skeleton is as close to a common ancestor as researchers may ever find, Sereno says.

Even paleontologists prepared for finding small dinosaurs might have been surprised by the tiny size of *Eoraptor,* says Sues. Many of the other reptiles living in Ischigualasto at the time would have far outweighed the early dinosaur, which must have survived by preying on still smaller creatures or by taking the young of large animals, Sereno says. The name *Eoraptor,* or "dawn stealer," reflects that presumed lifestyle.

From months of prospecting in the region, Sereno knows that dinosaur skeletons are among the rarest fossils in Ischigualasto's rocks dating from the time of *Eoraptor.* This indicates that *Eoraptor* and the other dinosaurs alive at the time played a relatively small role in the ecology of the region. It was a world dominated by crocodile-like animals, lizards, and ancient relatives of the mammals.

Travel forward 10 million years, however, and the rocks from Ischigualasto record a different picture, one of a world in which many of the once dominant reptiles have disappeared and dinosaurs rank as the most abundant fossils.

Paleontologists have long wondered how dinosaurs rose to the top of the ecosystem—a position they would hold from the late Triassic, roughly 215 million years ago, until the end of the Cretaceous, 65 million years ago, when they disappeared and mammals inherited dominion over the world. The traditional explanation, one in keeping with the classical Darwinian theme of natural selection, holds that dinosaurs established their supremacy through competition with their contemporaries. By dint of some advantage—perhaps their agile posture or their faster metabolism—dinosaurs proved more successful than the existing creatures, which were eventually driven to extinction.

To test the competition theory, Sereno has tabulated the numbers of dinosaurs and their contemporaries living in Ischigualasto through time. If the dinosaurs had driven other reptiles out through competition, the numbers should show ornithischian dinosaurs becoming more common and other herbivorous reptiles less common. Similarly, theropod dinosaurs should have replaced the carnivorous reptiles of the time.

"In fact, we don't see that pattern in this one site. That pattern simply is not manifested," says Sereno.

The recent finds in Ischigualasto support an alternative view, that dinosaurs took control more by luck than anything else, Sereno says. This idea was first proposed a decade ago by paleontologist Michael J. Benton of the University of Bristol in England.

The *Eoraptor* and *Herrerasaurus* skeletons show that dinosaurs had developed many specialized anatomical features early, long before the reptiles became common. By the time of Eoraptor, during the mid-Carnian stage of the Triassic period, dinosaurs had already split into the major groups of carnivorous theropods and herbivorous ornithischians.

"Dinosaurs are rare, and yet they are differentiating. They look like they were waiting in the wings and it was only later that they became the most common," Sereno says.

The dinosaurs may owe their success to something accidental, a major extinction that killed off most of the other animals but allowed the dinosaurs to live on unaffected, says Sereno. That extinction could have stemmed from any number of causes—from a meteorite impact to a severe climate change, he says.

Some scientists have reported finding evidence of mass extinctions and impacts at the end of the Triassic (SN: 2/8/92, p.91). Indeed, two geologists reported last year that the large Manicuoagan crater in eastern Canada formed soon after the time *Eoraptor* lived. But it remains unclear whether these impacts and mass extinctions occurred at the same time that dinosaurs became common, says Sereno.

If dinosaurs did get a lucky break in the beginning, that would bring their story full circle, for it was an accident that brought about their downfall. The dinosaurs died out at the end of the Cretaceous, a time when many scientists believe a meteorite or comet hit Earth. That crash apparently wiped out the last of the dinosaurs, although most species had already started to dwindle before the great impact, perhaps because of severe climate changes in much of their habitat, some paleontologists suggest.

Whatever the cause of the dinosaurs' extinction, it rid the world of the largest land animals, creating the opportunity for mammals to take over. Mammals had evolved in the late Triassic, at roughly the same time as the dinosaurs, but the mammals remained tiny and inconsequential until the dinosaurs disappeared, Sereno says.

Sereno cautions that he has tested the competition hypothesis only in Ischigualasto, which represents just one small part of the supercontinent that existed in the late Triassic period. Researchers must look closely at sites of similar age around the world to see if the same pattern shows up.

Other paleontologists say the cause of the dinosaurs' rise is a difficult problem and probably has a more complex solution, one that may involve competition, changing environmental conditions, and other factors. Kevin Padian of the University of California, Berkeley, comments that "the late Triassic is a very busy time. There are lots of other animals around. The dinosaurs are entering a very crowded world."

Both he and Sues say it is difficult to weigh the validity of the various theories raised to explain the dinosaurs' success. "I don't think any of the existing ideas are really testable at this point," Sues remarks.

Benton does not regard the problem as intractable and points to several advances in refining the dating of rocks at the end of the Triassic. "The greater refinement of stratigraphy and new discoveries like Sereno's are giving us a closer and tighter view of what was going on," he says.

Sereno acknowledges the difficulty in tracing the dinosaurs' rise, but he believes the population data for Ischigualasto argue against the competition theory. He plans to publish his findings later this year.

"The neat thing is that the picture seems to be emerging that the dinosaurs were taking their turn," Sereno says. "They got a lucky break and they stayed in for 150 million years, and then something happened and mammals have been in for 65." ❑

**Questions**:

1. What reptile is most similar to the dinosaurs' presumed ancestor, and what is one feature it had in common with the dinosaurs?

2. Why can't *Eoraptor* be the ancestor to all dinosaurs?

3. According to Paul Sereno, why should scientists be cautious of his findings?

Answers on page 301

# 67

*The Jurassic Period can be called the Dinosaur Mid-life Crisis. This is because the end of the Jurassic (about 140 million years ago) saw the extinction of most of the dominant dinosaur families, although remaining groups went on to diversify in the Cretaceous Period. These extinctions illustrate the importance of ecological roles in determining which groups die out. On land, organisms at the top of the ecological pyramid, such as large meat eaters, were drastically reduced. But organisms lower on the pyramid, such as small mammals and plants, were apparently barely affected. In contrast, marine organisms show the reverse ecological pattern, with such large meat eaters as the giant short-necked pliosaurs surviving the extinctions. Baptanodonts, which fed lower on the food pyramid, became extinct. The fossils also show that new organisms evolved rather rapidly to fill in the ecological roles of those that had died out. Such historical information demonstrates how fossil data can provide insights into ecological dynamics that may prove useful today.*

## Jurassic Sea Monsters

Extinction comes to species on land and in the sea, but not in the same way. As the tales of some remarkable creatures from many millions of years ago show, who goes first is a matter of ecology.

by Robert Bakker

One hundred sixty million years ago, an elegant sea monster lay down on its right side and died on the warm muddy ocean bottom near the present-day town of Lookout, Wyoming. As the creature's 14-foot-long body was convulsed by one last set of involuntary muscle contractions, its powerful sharklike tail twitched and stirred the bottom mud. A faint trail of bubbles escaped from the corner of its mouth and rose 300 feet to the surface, where tropical sunshine was playing on the waves. Then a bottom current gave the beast a decent burial under a blanket of pale green sand, part of the accumulating sediment layers that would become known as the Redwater Shale Member of the Sundance Formation.

Usually the events leading up to any one individual death in geologic time are obscured, the details lost, the exact time of day not recorded by any sign preserved in the rock record. But from what I know about that ancient Wyoming seabed, I have a mental picture of this one animal in its last hours in the Sundance Sea.

The sea monster was *Baptanodon*, the fastest, most advanced species among the ichthyosaurs, the "fish-lizards," seagoing reptiles with large porpoise-shaped bodies and long, narrow snouts. Baptanodons hunted at dusk. Their eyeballs, as big across as dinner plates, could gather up even the faintest light coming into the upper layer of the water. In my mind, I see my fleshed-out Sundance fossil swimming along in a pod of five or ten individuals, cruising silently 50 feet below the surface, scanning the moonlit water above for prey. The leader of the pod catches the telltale speckled light of armored squid, moving in an immense shoal a thousand strong. The squid rise near the surface every evening at this time to feed on small crustaceans and fish larvae.

Armored squid have big eyes, too—for catching their prey and for detecting their predators. But the baptanodons start their attack from the squid's blind quarter, behind and below the shoal. Baptanodons also have the deadly advantage of slashing speed. Their bodies have the 40-knot shape preferred by evolution for all its fastest-swimming creations. Mako sharks and albacore, the speediest fishes today, have the same proportions: a teardrop torso, six times longer than it is thick; short, triangular forefins for steering; reduced hind fins; a narrow tail base; and a deep, graceful, crescent-shaped tail fin.

A burst of muscular tail twitches sends the attacking baptanodons up and forward into the shoal's rear echelons. Squid break formation in every direction. Clouds of camouflaging ink squirt out and swirl in the Sundance water column.

Despite the obfuscatory curtain of ink, the baptanodons' large eyes catch microsecond glimpses of individual targets. Snouts, ultrathin so water resistance is minimal, swing instantly right and left. Jaws snap open and shut. Squid are impaled, still alive, along the close-packed rows of baptanodon teeth. Though the predators make just one five-second pass, that's enough for each of them to snag a half-dozen or more squid. The baptanodons swallow their meal as they breach the water's surface, grabbing quick breaths of air at the rear corners of their jaws. (Baptanodons are true reptiles, after all, descendants of land-living, lizard-like ancestors, and they must breathe air.)

As the fish-lizards churn up the water near the surface, armored mollusks, with large, snaillike shells and dozens of writhing tentacles, move clumsily out of the way. They needn't bother, really—baptanodon jaws and teeth are too weak to crush such well protected calamari. However, there are other sea monsters hunting tonight who relish big, hard-shelled servings of seafood.

An ugly triangular head, two feet long with eyes facing upward, darts from below. A mollusk's coiled shell is shattered in a dozen places by penetrating strikes of long, conical teeth. The jaws release their victim, then grip it again, adding another set of holes to the damage. Again the head strikes, and it now grabs and shakes the stunned victim. The mollusk body, its tentacles writhing helplessly, drifts out of its protective house. The ugly head darts up once more and sucks in the hapless four-pound body.

As soon as the attack began, the pod of baptanodons swerved hard to the right to avoid the scene, while their eyes searched out the body contours of the newly arrived predator. Now they see that the 2-foot head is carried on the end of a 15-foot neck that widens gradually until it merges into a compact 7-foot body. The ugliness of the snout, with its buck-toothed display of crowns sticking out and forward, belies the smooth precision of body movement. Two pairs of backswept flippers beat in syncopated rhythm, the foreflippers on the upbeat when the hind flippers are paused at the downbeat. Like penguins, these animals appear to fly through the water.

The baptanodons relax. They know this shape, and it's not a threat. It's only a long-necked plesiosaur—a swan-lizard. Although it weighs as much as five tons, the plesiosaur has a small mouth and can't inflict a dangerous bite on anything as large as an adult baptanodon. In the Sundance Sea, ichthyosaurs and long-necked plesiosaurs are noncompeting hunters. The ichthyosaurs harvest huge numbers of small squid. The plesiosaurs go for smaller numbers of big-coiled mollusks and medium-size fish.

Reassured, the baptanodon pod maneuvers for another strike at the squid. The lead baptanodon again begins its high-speed run into the shoal. But this time, just as the animal breaks into the squid's rear guard, it sees a huge, 50-foot dark shape curl its body into a tight U. The baptanodon instantly flinches and dives, its brain switching from attack behavior to lifesaving escape tactics.

It has recognized the one reptile in the Sundance ecosystem feared by *Baptanodon* and swan-necked plesiosaur alike: *Pliosaurus*, a short-necked, huge-headed giant kin of the swan-lizards. Pliosaurs are the top predators of the system, the hunters strong enough to kill any other sea creature.

The baptanodon's crash dive almost works. The pliosaur's nine-foot-long head lunges toward the fleeing animal; its jaws, four times stronger than a tyrannosaur's, make a sideways swipe. Though only the front six pliosaur teeth catch the smaller animal, the baptanodon is momentarily hung up on the six-inch fangs. Then it gives a maximum-power twitch of all its body and tail muscles and breaks free.

But blood trails from the wounds. The body wall has been pierced, the lungs skewered. The next day the baptanodon lays itself down on the bottom of the Sundance Sea and dies.

On the evening of August 12, 1992, I walked up a gully in the gently sloping outcrop of the Sundance Formation. I was teaching a dinosaur field course for the Dinamation International Society, a nonprofit group that supplies volunteers for digs (bless their hearts, volunteers make most of the discoveries these days). The six o'clock Wyoming sun was throwing its rays over the dried grass heads, turning everything gold and bronze. And the low-angle light had just hit a ten-foot fossil backbone, half-exposed by wind erosion, lying like a bas-relief in the pale green and gray sandstone. The front flipper bones were there, too, and all the ribs, mapping out the streamlined body form. The back of the skull was just visible.

It was the most beautiful fossil I'd ever seen in the field. A perfect baptanodon. The field party gathered around. They had been a noisy, rambunctious crew, but as they came up to the ichthyosaur, one by one they fell quiet, simply stunned into silence by the extraordinary specimen.

The discovery was paleontological serendipity, the happy result of a half hour's side trip away from the Morrison Formation, the rock layer lying above the Sundance. We were there to help excavate evidence about extinctions at the end of the Jurassic Period—extinctions on land, that is, at what I call the Dinosaur Mid-life Crisis, as preserved in the Morrison rocks. This was a time, about 140 million years ago, when most but not all of the dominant dinosaur families went extinct. They were replaced by new groups evolving early in the next period, the Cretaceous. But it turns out that big-eyed, seagoing Jurassic ichthyosaurs—and their squid prey, too—have their own tale to tell about the fate of Jurassic dinosaurs.

Success and failure—evolution and extinction—are controlled largely by ecological role. Dinosaurs are a clear-cut example. At the Dinosaur Mid-life Crisis, extinctions proceeded from top to bottom, ecologically. The top of the ecological pyramid on land, the biggest predators and the biggest herbivores, suffered terrible extinctions. This was when meat eaters like *Allosaurus* disappeared, along with such giant plant eaters as *Stegosaurus*. But the bottom of the pyramid got away nearly unscathed. Tiny insect-eating mammals and lizards and frogs survived with little change, and so did most groups of land plants.

This top-to-bottom extinction schedule works for all other mass extinction episodes in the terrestrial ecosystem. The most recent episode was the Ice Age die-offs of 11,000 years ago, when such monster meat eaters as saber-toothed cats were exterminated, as were most multiton herbivores, among them mammoths and giant sloths. But the Ice Age event left little mark on moles, voles, frogs, trees, shrubs, and grasses.

This is not what happened to the ichthyosaurs and their fellow swimmers. These extinctions proceeded from bottom to top, the reverse of the dinosaur extinctions. The baptanodonts (meaning not just *Baptanodon* alone but all the similar, related species) were the most advanced ichthyosaurs, featuring the largest eyes, smallest teeth, and fastest body design—adaptations for exploiting small prey near the bottom of the ecological pyramid. And it was the baptanodonts that went extinct at the end of the Jurassic. More-primitive ichthyosaurs, like *Grendelius*, retained bigger teeth, stronger jaws, and longer bodies, fed on larger prey higher in the food chain, and survived.

At the top of the pyramid were the giant, short-necked pliosaurs. If pliosaurs obeyed the extinction rules we see in the land ecosystem, then these huge, speedy carnivores should have had the highest vulnerability to complete extermination. Just the opposite turns out to be true. Giant pliosaurs survived the massive Jurassic-Cretaceous boundary extinctions that terminated the baptanodonts. When another partial extinction struck halfway through the Cretaceous, giant pliosaurs survived this crisis too, with little visible effect. The giant pliosaurs of the Late Cretaceous are in fact only slightly modified versions of the Jurassic ancestor *Pliosaurus*, who had hunted *Baptanodon* in the Sundance Sea nearly 100 million years earlier.

While the top-predator pliosaurs thumbed their noses at major extinction events, the ichthyosaurs displayed much greater ecological fragility. Those that managed to survive into the Early Cretaceous hunted bigger prey than the baptanodonts did, but they were still much farther down the food chain than the giant pliosaurs. When the Mid-Cretaceous extinction event hit, all the remaining ichthyosaurs died out, while the top-predator guild was left unshaken.

So the extinction schedules for ichthyosaurs and pliosaurs were the reverse of those for land ecosystems—the top predators were least vulnerable. If we imagine these sea-creature schedules working on land, then during the Ice Age, foxes and weasels would have died out while saber-toothed cats survived. Or at the end of the Cretaceous, *Tyrannosaurus* and its kin would have survived while birds died out. That obviously didn't happen. That *never* happens.

There is a way to make sense of this topsy-turvy world of sea-monster extinctions: we have to put them in a box—a guild box, to be precise. A guild comprises species living together and performing the same ecological function. A guild box is a simple three-dimensional graph that locates each creature along ecological axes that define key elements of habitat, food choice, and behavior.

All our Sundance creatures fit snugly into a guild box as defined by these axes: axis 1 (left to right), how far from shore the animals hunted; axis 2 (bottom to top), swimming speed; axis 3 (front to back), the size of the average prey. Giant pliosaurs are at the back of the box, occupying the top predator spot (offshore, fast, big prey). Long-necked plesiosaurs, the swan-lizards, grew nearly as big and swam nearly as fast, but they took relatively small prey, so they would be farther forward, at the right-middle-middle (offshore, fast, small prey). And the 40-knot baptanodons would fit at the extreme right-front-top. Far to the left are the long-bodied sea crocodiles, shore-hugging reptilian predators that filled the top-predator spot in the shallower habitats of the Sundance Sea.

Mass extinction opens gaping ecological holes in the guild box, opportunities for some surviving group to expand its adaptive diversity. Tyrannosaur extinction ultimately opened the way for lions and tigers and bears. Oceanic extinctions evidently followed inverted rules, but nonetheless gaps were opened. How did the sea-monster menagerie respond when the food chain was disturbed by ichthyosaur extinction? Did some other reptile group play the role of evolutionary carpetbagger, rushing in to exploit the resources near the bottom of the ecological pyramid?

Yes.

On a shelf at the University of Colorado Museum is the delicately snouted skull of a small Late Cretaceous pliosaur, dug from near Red Bird, Wyoming, and known appropriately as *Dolichorhynchops*, or "long-beaked face." Unlike the giant pliosaur species, the long-faced pliosaurs had lightly built heads. When I first saw that jet black fossil face in Boulder, I had a feeling of osteological déjà vu. I'd seen that exact face design before, in the Jurassic, tens of millions of years earlier: the ultrathin snout, the long rows of very small, tightly packed teeth, the huge eye sockets, the reduced jaw muscles that would permit a weak but very quick bite.

This delicate Cretaceous pliosaur had the face of a baptanodont. The basic adaptive geometry of pliosaurs had been re-shaped into a form fit for chasing small, swift squid. And the body size was a perfect match, too. In fact, *Dolichorhynchops* was small for a pliosaur—only 10 to 20 feet—but its body-size range overlapped perfectly with that of the baptanodonts.

All a pliosaur would need to fill a baptanodont hole in the guild box would be the correct combination of body size and jaw armament. *Dolichorhynchops* had that combination, and that made it a Darwinian carpetbagger, playing the baptanodont role in the Late Cretaceous.

Back in the Late Jurassic, all the pliosaurs had large, widely spaced tooth rows, suitable for subduing large fish and reptiles but not as efficient as the tiny baptanodont teeth for snapping up smaller, quick-turning squid. There's no anatomical reason Jurassic pliosaurs couldn't have evolved small teeth, but they didn't do it. The explanation is probably ecological. Since the small-toothed, fast-swimming guild was already filled by baptanodonts, there was no evolutionary vacancy for pliosaurs here. But the extinction at the Jurassic-Cretaceous boundary opened up to pliosaurs the fast-squid-eater corner of the guild box.

A similar scenario played itself out with the shore-hugging, shallow-water predators (roles played by seals and sea lions today). In the Jurassic seas there were two families of advanced sea crocodiles, 6 to 20 feet long, with elongated, sinuous bodies. Who filled their near-shore top-predator role in the Late Cretaceous? Not the sea crocodiles themselves—they went extinct during the Mid-Cretaceous event. Instead a totally new group enters the sea-monster guild, the long-bodied sea lizards of the mosasaur family. Mosasaurs have a body form astonishingly similar to that of the sea crocodiles, but they aren't related at all. Sea lizards evolved from land lizards early in the Cretaceous and were yet another case of evolutionary opportunism, taking advantage of gaps made in guilds by extinction events.

Thinking about all these creatures—pliosaurs and baptanodonts and sea crocodiles and sea lizards—made me suspicious of the long-necked-plesiosaur story that is usually told in textbooks. For their entire history, these swan-lizards certainly were far below giant pliosaurs in the food chain. The upside-down rules of extinction would seem to predict that their fate would be linked to baptanodont fate, since both groups were specialists in small-to-medium prey, and baptanodonts had high ecological vulnerability. But according to the standard textbook view, plesiosaurs were remarkably resistant to die-off events. The swan-lizards supposedly survived through the Late Jurassic extinction and through the Mid-Cretaceous crisis as well.

I didn't believe it. The case of the *Dolichorhynchops*-style pliosaurs replacing baptanodonts made me wonder whether the Cretaceous swan-lizards might be carpetbaggers, too. How sure were we that the long-necked plesiosaurs of the Cretaceous were direct descendants of their Jurassic counterparts? All my experience studying extinctions has led me to expect rather rigid ecological rules of survivorship. The swan-lizard scenario didn't add up.

In body form Cretaceous and Jurassic long-necked plesiosaurs did look identical. Flippers fore and aft, shoulders and hips, and proportions of neck and torso were built the same way. What I needed was a Darwinian marker, some anatomical feature that would expose Cretaceous swan-lizards if they were the product of opportunistic evolution from a non-swan-lizard ancestor. I needed what nineteenth-century anatomists called a heritage character.

In nineteenth-century parlance "habitus" is the obvious adaptive part of the body design, how the leg shape and tooth shape fit the beast to its particular environment. "Heritage" is the deeper, more fundamental part of anatomy, those features that evolve early in the history of a group and become fixed, not changing when the family tree branches out into many different ecological roles.

Here's a good example: from the outside, an African spotted hyena looks like a tall version of the African hyena dog (also called the hunting dog)—both have massive muzzles, long legs, and compact feet, and both hunt in large packs. Early-nineteenth-century zoologists were duped by hyenas and classified them among the true canids. But the heritage details of the skull anatomy around the ear and jaw joint show that these two hunting mammals evolved their similar habitus independently from different ancestors. The hyena dog is a genuine, bona fide member of the dog family, a close kin of the wolf, jackal, and coyote. But the spotted hyena is related to cats, civets, and mongooses, an entirely different branch of the carnivore family tree.

I applied habitus-heritage analysis to swan-lizards of the Jurassic and Cretaceous. I set skulls side by side. I ignored the teeth (habitus features—quick to change in evolution). I focused on the roof of the mouth (the palate), the first two neck bones, and how the snout attached to the back of the skull (heritage features).

This straightforward, old-fashioned exercise in anatomizing confirmed my suspicions. Evolutionary carpetbaggery had shaped long-necked plesiosaur history. The swan-lizards of the Cretaceous were *not* direct descendants of the swan-lizards of the Jurassic seas. In fact, the

Cretaceous swan-lizards were taxonomic impostors, long-necked opportunists from the pliosaur family!

The evidence goes like this: the Cretaceous swan-lizards had the bones of the roof of the mouth (pterygoid bones) wrapped around the underside of the braincase bones, exactly as in short-necked pliosaurs. And in the Cretaceous swan-lizards a snout bone and a rear skull bone meet, covering up the forehead bone above the eye—again exactly as in pliosaurs. Finally, the Cretaceous swan-lizards and the pliosaurs displayed the same relative proportions in the first two bones of the neck. The Jurassic swan-lizards were completely different.

Conclusion: Cretaceous long-necked plesiosaurs evolved from Jurassic pliosaurs. And so the Jurassic long-necked plesiosaurs had gone extinct without issue.

I find all this sea-monster history deliciously counterintuitive. It's intuitively obvious that an ecosystem should collapse from the top down, with the base of the food pyramid surviving longest because the base is where most of the species and individuals—all the small prey and plankton—are located, and the base is closest to the ultimate source of energy, photosynthesis. Intuitively obvious but historically wrong. Giant pliosaurs are the rule in oceanic history, not the exception. In the post-Cretaceous oceans, giant open-water sharks filled the role of fast predator in tropical waters. And these families of sharks, including species akin to the surviving white shark, tiger shark, and mako, ignored the worldwide partial extinctions that terminated families occupying other guild corners.

Does the Jurassic-Cretaceous sea-monster story help solve the general riddle of extinctions on land and water? Yes, definitely. I've been persuaded that pioneering paleontologist Henry Osborn's theory of 90 years ago is the best to explain terrestrial extinctions. He argued that the development of land bridges allowed massive mixing of big-bodied species across different continents, causing extinction through the introduction of new predators, competitors, and diseases. I think the flip side of his argument works in the water.

It's a venerable idea to see extinctions at sea as being caused by a disturbance of the whole oceanic habitat. There may be a destruction of the rich, shallow-water habitats, as occurred when the Sundance Sea drained off the midcontinent. Or there may be a sudden influx of cold polar water toward the equator. Or there may be severe changes in the upwelling of nutrients in the ocean. Sometimes all three events happen simultaneously. But all such changes should coincide with cycles of land bridges, since widespread land bridges act to impede normal ocean circulation.

Now back to the top-predator paradox. On land, top predators should be hypervulnerable because they are quick travelers, spreading faster than any other guild across land bridges and into all available habitats. During the Ice Age, for example, saber-toothed cats spread quickly from Europe to the southern tip of Patagonia. Fast, complete penetration of new areas guarantees that predators will inflict maximum ecological damage and that they will meet the greatest assortment of new enemies among the competitors and microorganisms native to the lands they've invaded.

But in the oceans, the far-ranging habits of the top predator can work in the exact opposite way—minimizing extinction probabilities. If ocean extinctions are caused by physical changes in the oceans, then guilds tied to localized habitat patches are on unstable ecological ground. A baptanodont preying mainly on small, fast-moving squid is exterminated when the squid numbers are cut by planktonic disturbance.

But the top predator is not locked into any one habitat. Giant Pliosaurs were already worldwide in distribution before the Late Jurassic and Mid-Cretaceous extinctions began; individual pliosaurs probably had immense hunting ranges and could subsist on prey from nearly any ocean guild structure, just as modern white sharks are catholic in their tastes and cosmopolitan in their species range. As long as some prey populations existed here and there, the top predator could survive the crisis. Only the most profound and prolonged sea crisis could eliminate the giant pliosaur guild corner—a crisis like the apocalypse at the end of the Cretaceous.

The incomplete sea-monster extinctions of the Late Jurassic and Mid-Cretaceous are wonderful opportunities to get at the rules of mass die-offs. But we have to ask the right questions. Extinctions on land and in the sea are not random. No, extinctions have precision. Survival and recovery depend on where each species fits into the network of ecological interdependencies. To appreciate the real meaning of speedy ichthyosaurs, long-necked plesiosaurs, and huge, short-necked plesiosaurs, we must first place these marvelous creatures in their proper context—a box, with a slyly shifting pattern of occupancy. ❑

## Questions:

1. Are top predators usually more widespread than organisms lower on the ecological pyramid? Give an Ice Age and Jurassic example.

2. What is a "guild box"?

3. Do the ichthyosaurs show greater ecological fragility than the pliosaurs? When did the ichthyosaur group go extinct?

Answers on page 301

# 68

*The following two articles discuss recent findings regarding the extinction of the dinosaurs 65 million years ago. Most geologists appear to have accepted the theory that an impact played some role in the end Cretaceous extinction. Recent research has indicated that the impact was even bigger than previously thought, while other research has cast some doubt on the likelihood that the impact was the primary cause of the dinosaur's extinction. The previous estimate of the Chicxulub crater in the Yucatan was about 180 kilometers in diameter. Recent studies suggest that it may have been as much as 300 kilometers in diameter. An impact of this size would be a once in a billion years event, emphasizing the element of bad luck in the dinosaur's extinction. Other recent studies, however, suggest that the expected dark and cold following an impact might not have been enough to wipe out the dinosaurs. This is because some dinosaurs appear to have lived in the polar regions where they would have had to contend with darkness and cold on a yearly basis. Also, some organism, such as plankton and turtles, which should be very climate-sensitive, survived the extinction relatively unscathed. When considering any theory for the end Cretaceous extinction, it must be remembered that the dinosaurs weren't the only organisms that went extinct and that many other organisms survived. The fate of all Late Cretaceous species must be accounted for.*

## How Lethal Was the K-T Impact?

### The asteroid that hit Earth 65 million years ago appears bigger than previously thought, but scientists have new doubts about its ability to kill the dinosaurs

by Virginia Morell

More than a decade has passed since Luis W. Alvarez and his colleagues first proposed that an asteroid impact 65 million years ago kicked up enough of a dust cloud to darken and chill the planet for months, killing off the dinosaurs (*Science*, 6 June 1980, p. 1095). Since then, the theory has gathered force as scientists have gathered more evidence pointing to an impact at that time—the boundary between the Cretaceous and Tertiary periods (K-T). Some researchers now believe the asteroid was so big it was a once-in-a-billion-year event.

Ironically, as more scientists satisfy themselves that an impact did occur, other researchers have begun raising tough questions about whether that impact packed enough punch to make the dinosaurs disappear. Alvarez' original hypothesis has been refined, with global wildfires and other proposed disasters being added to the mechanism of dinosaur destruction, but most catastrophists still include some period of winter darkness in their models. Yet several months of dark and cold might not have been enough to kill off the dinosaurs.

The reason: New findings show some dinosaurs thrived in cold climates. Dramatic evidence comes from Alaska's North Slope, where William A. Clemens and L. Gayle Nelms of the University of California, Berkeley, have found signs, which they reported in the June issue of *Geology* (vol. 21, p. 503-506), that several dinosaur species spent the winters in cold, dark climes. When added to the survival of organisms such as plankton and turtles across the K-T boundary, this creates a puzzle: Whatever happened at the boundary seems to have destroyed cold-adapted dinosaurs while leaving climate-sensitive creatures untouched. And how can you explain that with an asteroid impact? "Not with the simplistic idea of turning off the lights for 3 months," says Tom Rich, a paleontologist at the University of Monash in Australia. "Clemens' paper shoots that down." But catastrophe theorists, far from giving up the argument, now contend that the lights may have gone off for a much longer period.

Clemens and Nelms' arguments begin with the climate on the North Slope in the late Cretaceous. As indicated by plant fossils found only in cold areas, the region had a mean annual temperature of 2 to 8 degrees Centigrade, not unlike the climate of present-day Anchorage. The Slope lay well within the Arctic Circle and experienced 3 months of darkness annually. Previously, researchers had suggested that the dinosaurs spent only summers in Alaska, migrating south before winter. Yet Clemens and Nelms argue that their fossils show some dinosaur species were year-round residents.

Their primary evidence comes from a collection of teeth of young hypsilophodonts—lightly built, agile creatures that stood a little less than 2 meters tall. "If they were migratory and hatched that far north," says Nelms, "then within a few months, when they were still quite small, they would have had to travel thousands of miles—something I have a hard time imagining." Nelms and others believe the juveniles' small size argues against a 2100 kilometer journey. The scientists actually suspect that hypsilophodonts may not have migrated at all; the fossil record indicates these creatures did not live in herds, and herding behavior is a signature of most migratory animals.

Further, it seems that hypsilophodonts may have been particularly adapted to cooler, more seasonal environments. Not only are their fossils found in Alaska, but recently five new genera have been unearthed in southern Australia—from an area that in the Early Cretaceous lay within the Antarctic Circle. Clemens and Nelms also suspect that several other species of dinosaurs, including the carnivorous troödon and vegetarian dromaeo-saur, overwintered in cold, dark regions.

The absence of amphibians and reptiles in the Alaskan deposits reinforces Clemens' and Nelms' doubts. These species are common only in Cretaceous deposits from more temperate latitudes, indicating that they were not adapted for cold. But, says Clemens, "it's the dinosaurs (except for birds) that went extinct, while these other creatures survived—which is exactly the opposite of what you would expect" if the cold and dark actually did the dinosaurs in.

Given some past testy exchanges between paleontologists and impact theorists on this subject, some of those theorists seem surprisingly easily persuaded of trouble with the cold, dark killer theory. "The biggest uncertainty we face is the kill mechanism," says H. Jay Melosh, a planetary scientist at the University of Arizona. Adds Owen B. Toon, an atmospheric scientist at NASA Ames Research Center, "Clemens' paper does make it harder to understand the extinction of small dinosaurs from the cold and dark."

Toon and Melosh, however, don't think that cold and dark were the only potentially lethal consequences of the impact. Melosh favors death-by-acid-rain, and Toon suggests global wildfires generated by the asteroid would have caused considerable trouble for dinosaurs large and small. Other catastrophists argue that the period of dark cold may have been 5 to 10 years, rather than just a few months—and even the most highly winterized dinosaur would have had trouble coping with that. At a geochemistry meeting this spring, Kevin Pope of Geo Eco Arc Research in La Canada, California, and his colleagues suggested such an extended winter would result if the impact triggered the formation of clouds of sulfuric acid, which stay in the atmosphere for years.

But any universal mechanism, be it fire or 10 years of ice, faces the same problems, says Clemens. "The real question is, How did the others—how did any animal—manage to survive? [Impact theorists] have got to come up with a hypothesis that puts equal weight on survival. So many of these catastrophists want to kill the dinosaurs, they forget the rest of the biota. Birds, mammals, and amphibians managed to survive, and that tells you that there is something wrong with most of these hypothetical horrors." ❑

## A Bigger Death Knell for the Dinosaurs?

by Richard A. Kerr

Most geologists agree that a sizable comet or a smallish asteroid struck the Yucatan coast at the same geological instant as the last of the dinosaurs disappeared 65 million years ago, along with many other creatures. But while many geologists hold the impact responsible for the extinctions, paleontologists tend to think the impact and the extinctions coincided by chance. For one thing, the diameter of the buried crater—180 kilometers—seemed to imply a catastrophe of the size that planetary scientists calculate should occur every 100 million years or so. What's so startling, ask these doubters, about the association of a relatively commonplace cataclysm with the extinction? But a paper in this issue of *Science* argues that the impact 65 million years ago was truly extraordinary.

On page 1564, a group led by Virgil Sharpton of the Lunar and Planetary Institute in Houston reports a new analysis of the surface traces of the Chicxulub crater, which lies buried beneath almost 2 kilometers of Yucatan sediment. These researchers see the faint outlines of a scar that is nearly twice as large as had been thought—300 kilometers in diameter. That would imply an impact so large as to be "very unusual, perhaps unique," says Sharpton. "Perhaps Earth has not

con't on next page

experienced such an event since complex life appeared" 1 billion years ago, he says. If he and his colleagues are right—and other geologists are intrigued, but far from convinced—the impact could easily have left a unique mark in the history of life.

It might seem that a crater's size is easy to determine, but when the crater is as deeply buried as Chicxulub, evidence has to come from subtle surface effects of the buried rocks, such as variations in the strength of Earth's magnetic field and gravity. In the studies that revealed the crater's existence in 1991, researchers detected a bull's eye pattern of gravity variations amounting to one ten-thousandth of total gravity across a circle some 180 kilometers wide. Sharpton and his colleagues have now removed spurious points from those gravity data, added new points, and reprocessed the refined data. The ring of ever so slightly heightened gravity outlining the 180-kilometer circle remains. But added to it is an even fainter ring at 300 kilometers—the outer limit, Sharpton and his colleagues think, of the crater.

"The outer ring is very subtle" admits Sharpton. "But its subtlety is consistent with what we know about how craters form." Large ones like Chicxulub form in two steps. The impact creates a deep "transient" crater. The walls of that initial crater collapse within minutes of the impact, partially filling the transient bowl and enlarging the crater to its final size. Geologists had thought that the 180-kilometer ring at Chicxulub marked the outer bounds of the collapse. Sharpton, though, thinks that ring is too prominent to be the rim of the collapse and interprets it as the edge of the smaller, transient crater. The subsequent collapse, he says, apparently extends all the way out to 300 kilometers.

"This is very exciting," says planetary scientist William Hartmann of the Planetary Science Institute in Tucson, although he and others stress that the larger size is far from certain. Other researchers such as Richard Pike of the U.S. Geological Survey in Menlo Park, who has studied the new analysis, aren't convinced that the subtle outer ring is real. And an analysis of similar, though less thoroughly processed, gravity data by Mark Pilkington and Alan Hildebrand of the Geological Survey of Canada in Ottawa and C. Ortiz Aleman of City University in Mexico City failed to reveal a 300-kilometer ring. "We see no reason at the moment to suggest" that there are as many rings as Sharpton and company suggest, says Pilkington.

Even if the new gravity ring is ultimately accepted, says Hartmann, a convincing argument that the ring represents the outer limit of the crater will also require independent evidence of shattered rock out to 300 kilometers. A few wells were drilled across Chicxulub to search for oil long before the crater was recognized, but assessments of the rock retrieved from them differ. Pilkington notes that a drill hole just outside the 180-kilometer ring but well inside any ring at 300 kilometers revealed undisturbed rock, according to the initial evaluation by the drillers. Sharpton counters that the drillers were wrong. He has inspected the drill samples himself in Mexico City and saw plenty of impact debris, he says.

Another sort of geologic evidence—a ring of sinkholes concentric with the gravity rings—is also being interpreted as a sign of a larger crater, albeit not as large as the one Sharpton infers. Kevin Pope of Geo Eco Arc Research in La Canada, California, and his colleagues are suggesting that the 170-kilometer sinkhole ring delineates the flat inner floor of the crater. If that's the case, then by analogy with large craters on other planets, the full crater would be at least 240 kilometers.

Sharpton's version of the crater, spanning a full 300 kilometers, would imply an impact eight times as powerful as was thought. But so little is known about the global effects of a large impact that impact specialists can't say how much more devastating an impact of that magnitude would have been. And though a bigger impact could strengthen the case for a causal link to the extinctions, it won't help paleontologists explain the puzzling pattern of extinctions (see main story).

Nevertheless, a larger impact could help explain why, after more than a decade of looking for other impact-extinction pairs, researchers have turned up only two other possible examples, both much less convincing so far than the cataclysm 65 million years ago (*Science*, 8 January, p. 175, and 11 January 1991, p. 161). With a 300-kilometer crater, the event could be unparalleled on Earth. Indeed, only one other impact known to have taken place in the inner solar system over the past few billion years—Mead Crater on Venus—would be as large, says Sharpton. And that would make the dinosaurs the unluckiest beasts of all. ❑

**Questions**:

1. What evidence suggests that the Chicxulub impact crater is 300 kilometers in diameter?

2. What evidence suggests that dinosaurs lived year round within the Arctic Circle?

3. If the cold and darkness from an impact did not kill the dinosaurs, what other impact-related effects could have killed them?

Answers on page 301

# 69

*If you follow the popular press you would think that all scientists have concluded that an impact caused the extinction of the dinosaurs and other organisms at the end of the Cretaceous. However, not all scientists agree with this position. In an effort to solve some of this disagreement, one scientist, Robert Ginsburg, proposed an interesting test. Using one particularly contentious group of fossils, the marine protozoans known as forams, he sent samples which were not labeled as to their stratigraphic position to four different specialists on forams. The scientists receiving these samples identified the species present in each sample and then returned them to Ginsburg who put the results into the proper stratigraphic order. This is known as a blind test because the specialists did not know what order the samples were supposed to be in, preventing them from biasing the results as to when a particular species went extinct. The results of this test appear to have been accepted by most workers as proof that the extinctions at the end of the Cretaceous were in fact abrupt.*

## Testing an Ancient Impact's Punch

Did the impact at the end of the dinosaur age deliver a haymaker to life on Earth? Results newly reported from a "blind test" of the marine fossil record suggest it did
by Richard A. Kerr

HOUSTON—The provocative idea that a huge meteorite blasted Earth 65 million years ago and wiped out the dinosaurs and other creatures faced a formidable struggle when it was proposed 15 years ago this spring. Its proponents were forced to fight on two fronts at once. On one, they did battle with geologists and geochemists who disputed the evidence of the impact; on the other, they engaged paleontologists who doubted that the mass extinction 65 million years ago took place in a geologic instant, as the impact hypothesis requires.

That the first battle has finally ended was clear at last month's conference here on catastrophes in Earth history (dubbed Snowbird III after the Utah location of the first two meetings). In something of a first, not a single researcher at the meeting publicly questioned the reality of a giant impact on the Yucatán coast. Even a peripheral question about the origin of 65-million-year-old deposits around the Gulf of Mexico seemed settled in favor of an impact.

The second dispute continues, but Snowbird III saw a major shift in its battle lines. The fossil record of microscopic marine protozoans called forams, which should provide the most reliable measure of the pace of extinction, has for the first time yielded a widely, though not universally, accepted verdict. "It sure looks catastrophic to me," says paleontologist Peter Ward of the University of Washington, who once viewed the extinctions as gradual and has since seen evidence for both gradual and abrupt disappearances, depending on the species.

There are holdouts, but the innovative strategy that yielded this initial verdict on the pace of the extinctions may have the potential to resolve the issue once and for all. The results, first presented at the Snowbird meeting, are from a blind test, in which investigators examined samples and identified the species in them without having any idea of the samples' ages in relation to the impact. While investigators working on their own haven't been able to agree on whether or not forams died out gradually, the blind test showed all of the forams persisting until the impact, when at least half suddenly disappeared.

In a field often rife with subjective judgments, that novel strategy generated as much excitement as the results. "Paleontology has finally entered the 20th century," says Ward. "It was a true scientific test, a watershed event for my field." Adds University of Chicago paleontologist David Jablonski: "It's marvelous it was done; we should do more of this."

The new results from marine microorganisms add to the mounting evidence of an abrupt extinction from other fossils. When the impact hypothesis was first proposed, paleontologists tended to view the mass extinction that ended the Cretaceous Period and the age of the dinosaurs as a gradual affair, taking place over hundreds of thousands if not millions of years—a pattern more likely to have resulted from sea level fall or global cooling than an impact. But in the

1982 proceedings of the first Snowbird conference, two marine micropaleontologists, Philip Signor and Jere Lipps of the University of California, Davis, cautioned their colleagues not to take the fossil record at face value. They pointed out that how abrupt a mass extinction appears in the record can depend on how closely paleontologists examine it. The rarer the fossil—dinosaurs are the worst case—the less likely paleontologists are to find the last remains of that species before it vanished. As a result, rarer species can appear to die out before they actually do.

In the following years, some paleontologists tried to overcome the Signor-Lipps effect by sampling up and down their favorite fossil records every few centimeters or even millimeters, rather than at the usual intervals of a few meters. In these new higher-resolution studies, some extinctions that had seemed to be gradual, such as that of plants in North America and coil-shelled ammonites from the Bay of Biscay, now looked relatively quick (*Science*, 11 January 1991, p.161).

But the microscopic fossils in the ocean, which because of their abundance should provide the strongest evidence about the pace of the extinctions, yielded an ambiguous verdict. Gerta Keller of Princeton University argued in a 1989 paper that 29% of the Cretaceous foram species she identified at El Kef in Tunisia became extinct over the 300,000 years leading up to the impact. Therefore, it must have been global cooling or the sea level drop that did them in, she said. Since only 26% of the species become extinct right at the end of the Cretaceous—the K-T boundary—"the effect of the impact was of more limited scope than generally assumed," she wrote. But Jan Smit of the Free University of Amsterdam couldn't find any forams disappearing before the boundary at El Kef, where he saw all but a few species going extinct.

To resolve the dispute, sedimentologist Robert Ginsburg of the University of Miami took up a novel proposal that had been made at the previous Snowbird meeting: a blind test of gradual versus abrupt extinctions. With the assistance of Smit and Keller, he collected new samples at El Kef, split them into coded subsamples, and distributed them to four foram investigators. Unaware of how far below or above the impact each sample had been collected, each analyst identified the species present. The investigators then sent their results back to Ginsburg.

When the results were unveiled at the meeting, both sides claimed victory. Keller pointed out that each of the blind investigators had some fraction of the Cretaceous species—ranging from 2% to 21%—disappearing before the K-T boundary. That "basically confirms the pattern" of gradual extinctions, Keller told the meeting.

Smit saw it differently. "That's typical Signor-Lipps effect," he says. To minimize the influence of rare or misidentified species on the results, Smit combined all four efforts, including only those species that two or more of the blind investigators spotted somewhere in their sample set. In the case of the seven species that, by Keller's analysis, disappeared before the impact, one or another of the blind investigators found all seven in the last sample before the boundary. "Taken together, they found them all," says Smit. "This eliminates any evidence for pre-impact extinctions in the [open-ocean] realm."

Many others at the meeting agreed that the results seem to point to abrupt extinctions. James Pospichal of Florida State University, for example, had already concluded for his own high-resolution work that marine nannofossils, the remains of planktonic algae, had continued to be abundant right up to a disastrous extinction at the time of the impact, but he says he was open minded about the fate of the protozoans. To judge by the blind test results, he says, the forams behaved the same way.

Keller, though, thinks the evidence for abrupt extinctions still involves "major taxonomic problems." For example, if the blind investigators lumped together separate species that look similar, she says, what was actually a series of extinctions could appear to be a single, abrupt extinction. But now her own taxonomy is under fire. Brian Huber of the National Museum of Natural History had examined forams from a deep-sea sediment core of K-T age, drilled from the far South Atlantic, that Keller used in a 1993 *Marine Micropaleontology* paper to support a claim of gradual extinctions. "None of her taxonomy or quantitative studies [of this core] can be reproduced," says Huber. "The gradual side of the debate doesn't hold water because of her inconsistencies" in identifying foram species.

Keller isn't conceding anything, however. She presented her latest analyses of the El Kef forams at the meeting and will be presenting a reply to Huber's comments, which he is now preparing for publication. "The data stand and the data will be published," she told *Science*.

An extension of the test might settle the sticky points of taxonomy—if all the combatants were willing. Ideally, the adversaries would gather around a single microscope and examine each disputed species, conferring until everyone agreed on how it should be identified. As a more practical solution, Ginsburg may circulate the samples among the investigators and tally their votes.

Even if further tests can definitively resolve the gradual-versus-abrupt dispute at El Kef, however, plenty of disputes would remain about the K-T extinctions. Were they really less severe at high latitudes, as Keller and others suggested at the meeting? Did many foram species survive the impact, as Keller argues? And once the nature of the K-T extinctions has been settled, the fossil record has plenty of other mysteries to which investigators might turn a blind eye.

**Questions**:

1. What is the Signor-Lipps effect?
2. Why does Jan Smit conclude that the results of the blind test indicate an abrupt extinction?
3. Besides the Signor-Lipps effect, what potential problem might cause the four foram specialists to get different results?

Answers on page 301

# 70

*The "turnover pulse" hypothesis states that large-scale climatic changes can cause bursts of evolutionary change. This controversial idea is rooted in the older idea of "punctuated equilibrium," which says most new species evolve during short periods of rapid change. The turnover pulse hypothesis has important implications for human evolution because there is evidence of major climatic change in Africa over the last few million years. These changes may correlate with evolutionary bursts in many mammals, ranging from hoofed mammals to apes. Specialized organisms should be especially susceptible to such bursts because they are more sensitive to environmental changes. However, the hypothesis is challenged by some evolutionary biologists. They argue that, while environmental shifts certainly cause much evolutionary change, there is no convincing evidence that a clear-cut series of periodic pulses occurred.*

## Taking the Pulse of Evolution

Do we owe our existence to short periods of change in the world's climate?

by Colin Tudge

Suppose God had blown his whistle about 10 million years ago, and said, "Let there be no more change in climate: let there be no more volcanoes; let tectonic movements cease, and the continents remain exactly where they are." Would animals have ceased to evolve? Specifically, would the chimp-like apes who began to evolve into humans between 7 and 5 million years ago still be swinging in their trees?

Charles Darwin's answer would have been no. While he acknowledged that evolution speeded up when conditions changed, he also supposed that animals and plants must continue to evolve whatever the conditions. They would always be competing with each other: leopards becoming ever more cunning, the better to capture antelopes; antelopes ever swifter, to avoid the leopards. But some modern biologists feel that evolution would indeed grind to a halt in the absence of environmental change—especially climatic change. This controversial idea is rooted in the notion of "punctuated equilibrium", championed by Niles Eldredge and Stephen Gould in the 1970s. Its strongest proponent today is Elisabeth Vrba of Yale University.

Vrba postulates that big evolutionary changes—migrations, extinctions and the creation of new species—are triggered by climate changes. To the extent that climate changes tend to happen in bursts and at regular intervals, major evolutionary changes also occur in bursts, or "pulses". Vrba calls this idea the "turnover pulse hypothesis". She has spent much of her professional life in Africa studying the fossil record. One aspect that has intrigued her is the apparently dramatic changes in evolution and climate which occurred some 2.5 million years ago. At that time, so increasing evidence shows, there were huge turnovers in faunas worldwide, and, most important for us, the earliest ancestors of true humans appeared. What's more, says Vrba, this burst of evolutionary activity is only one of several that occurred in Africa and elsewhere.

If Vrba is right, humanity may owe its existence to a relatively short period of change in the world's climate. But can her hypothesis be proved? While some researchers believe that the turnover pulse hypothesis should be written, at least provisionally, into the textbooks, others profoundly disagree. Nobody disputes the notion that sudden environmental changes could trigger bursts of evolutionary upheaval. The contentious question is whether such bursts could at times be global. Vrba says yes; her critics, no.

"The turnover pulse hypothesis is effectively dead," says Andrew Hill, a biologist at Yale, who believes that, in arguing for global bursts of evolutionary change, Vrba is going out on a limb. "Yes, environmental shift leads to evolutionary change. But a series of 'pulses' in faunas caused by episodic shifts of climate?—No."

Such differences of opinion are perhaps only to be expected, given the vagaries of the fossil record and the difficulty of reconstructing trends in the world's climate over the past few million years. In the last decade or so, researchers have developed some brilliant techniques for detecting past climate changes, most based on studies of fossilised pollen and protozoa. But the endeavour as a whole has suffered some irritating drawbacks.

Nick Shackleton of the University of Cambridge, for example, studies sediment cores from the Pacific that contain fossils of the benthic foraminifera, or "forams"—single-celled protozoa which live on the seabed. These fossils contain the two common isotopes of oxygen, $^{16}O$ and $^{18}O$, in different ratios, depending (so Shackleton and others infer) on how much ice there is in the world at the time. An increase in $^{18}O$ indicates more ice and therefore a cooler Earth. The deeper the core the older the fossils, so by measuring the ratio of oxygen isotopes through a foram core, researchers can trace the Earth's iciness back through time—and hence piece together a picture of past global temperature. Shackleton says that his findings support the idea that the Earth suddenly cooled about 2.5 million years ago, consistent with Vrba's hypothesis.

Other researchers have reached a similar conclusion from pollen studies. Cores dug from the bottoms of lakes (or places that used to be the bottoms of lakes) still contain the pollen of local plants, in various degrees of fossilisation. Most of the sampled species still exist, or have close modern relatives, enabling researchers to reconstruct their ecology. This knowledge, combined with other information about dates (in general, the pollen layers must correlate with datable strata), reveals what the climate did in the past, and when.

One proponent of this approach is Henry Hooghiemstra of the University of Amsterdam, who has studied the sediments of the high plain of Bogotá in the Eastern Cordillera of Colombia. Among other things, concludes Hooghiemstra, there was "a long period of strong climatic fluctuations that started about 2.7 million years ago". In East Africa, too, there is evidence of cooling in the highlands 2.5 million years ago, and possibly 3.4 million years ago, according to pollen studies by Raymonde Bonnefille of the CNRS (France's national agency for scientific research at Marseilles. Cooling 2.7 to 2.4 million years ago is also suggested by the appearance of Saharan dust at around that time, and by the build-up of loess—sediment formed from wind-blown dust in China. Dust indicates an increase in aridity, which in turn correlates with cooling. And Michael Archer of the University of New South Wales suggests that extensive grassland, which is usually linked to a cooling and drying climate, first appeared in Australia some time after 3 million years ago.

**The Panama Connection**

"The evidence for a global climate shift between 2.7 and 2.4 million years ago is quite strong, and is becoming stronger," says Vrba, who links this shift with major changes in animal life in the fossil record. But there are complications. Some researchers, including Bill Ruddiman of the University of Virginia, do not find a sharp fall in temperature 2.5 million years ago in their foram data. Instead, they perceive a steady decline in temperature lasting 1 million years and starting 3 million years ago. Furthermore, the foram research tells you only what happened at the bottom of the sea, not on land.

Nor is there any firm evidence, says Hill, to suggest that the climate changes detected on individual continents are part of a global change in the environment. The case of South America illustrates just how hard it is to disentangle local and global influences on past climatic upheavals. South America did not merely grow cooler between 3 and 2 million years ago. Until 3 million years ago it was an island, a piece of Gondwanaland, drifting through the Southern Ocean, like Australia. But around 3 million years ago it joined up with North America, via what is now the Isthmus of Panama. This joining resulted in part from climatic change, since the isthmus appeared as ice accumulated at the poles, and the sea level fell. But it must itself have had serious climatic consequences, since it cut off any ocean currents that might previously have flowed between what are now the Atlantic and the Pacific. And in East Africa, where our ancestors seem first to have arisen, the cooling was accompanied by a huge tectonic shift that lifted the eastern part of the continent to a height of hundreds of metres—itself enough to produce significant cooling.

"I don't doubt that physical changes lead to major evolutionary events," says Hill. "I am even happy with the idea that evolution might grind to a stop in the absence of external change. But I do not think that the physical changes are necessarily climatic; when they are climatic I do not believe that they are necessarily global." However, climatologists like Shackleton and Ruddiman insist that there was a major shift in the pattern of the global climate system near 2.5 million years ago.

So everyone agrees that there was widespread cooling between 3 and 2 million years ago. The uncertainty is over whether the climate changes detected on individual continents are part of a global change, how sudden that change might have been, and whether it really precipitated evolutionary events on different continents. And in theory even this dispute can be resolved as more data accumulate. What about the crucial part of the turnover pulse hypothesis? Does the fossil record show that animals changed dramatically, en masse, 2.5 million years ago?

Hill is again skeptical: "Some groups change, like shrews or rodents, and some, like African pigs, do not." With other groups such as hominids, he says, the data are simply too sparse to make any judgment. "In short, you don't see a 'pulse' of evolutionary change, you see a spread-out response. It's difficult to see how this differs from the kinds of stepping up of pace that Darwin said would happen."

In Vrba's view, neither pigs nor hominids can make the case one way or the other. Her hypothesis does not require that all animal groups would respond to climatic change at exactly the same moment. It is not surprising, she says, that animals like shrews and rodents might respond more rapidly than pigs to climate change. Being small, they would be especially sensitive to temperature; and having specialised feeding needs, they might react particularly strongly to a sudden change in vegetation. But the same may not be true of pigs. Most living pigs are omnivores. Only a few—like the wart hog of Africa—have become specialist grazers. And even if the pigs had responded rapidly to climate change, argues Vrba, detecting the resulting turnover in species would be difficult, because there are so few species in the fossil record—a mere 13. "In such small samples it is very difficult to see any bursts of change that would stand up to statistical analysis," says Vrba.

But if this should seem like a cop-out, Vrba has looked at the fossil records of 20 different groups of closely related large animals, from giraffes to spring hares, and from elephants to African monkeys. The clearest data come from the fossil remains of the Bovidae, a group of hoofed animals with an even number of toes, which includes cattle, sheep, goats and antelopes.

There are hundreds of known species of bovids going back more than 20 million years, and their fossil records are extremely rich. One particularly dramatic period of innovation occurred between 7 and 5 million years ago. Cattle migrated into Africa from Eurasia, leading to the first appearance in Africa of the ancestors of modern buffalo, the Bovini "tribe" of cattle. Grazing antelopes such as reedbuck, water-buck and rhebok (the Reduncini) also appeared for the first time. Other grazers—hartebeest, blesbok and gnu (the Alcelaphini)—evolved for the first time; and so too did the sable, roan and the oryxes (the Hippotragini), the kudus and nyalas (the Tragelaphini), and the impala (the single species of the Aepycer-otini tribe). But this period of diversification seems to have given way to a wave of extinctions and replacements between 3 and 2.5 million years ago. A huge loss of genera occurred within the various bovid tribes, and today's modern genera first evolved.

An earlier turnover of species, between 7 and 5 million years ago, is intriguing. For one thing, compa-rable turnovers seem to have taken place at the same time in North America and the Siwalik mountains in Pakistan. Moreover, both these regions, according to fossil evidence, seem to have experienced previous upheavals in fauna around 11 million years ago and again 14 million years ago. On the basis of this pattern, Vrba speculates that there may be a three million year cycle in global climate.

Nobody knows what might cause such long climatic cycles—or indeed, whether they truly occur. But one possible origin is in the periodic movements of the Earth's tectonic plates. Certainly, the cooling has nothing to do with the kind of shifts in the Earth's orbit, or in its angle of tilt—the so-called Milankovitch shifts—that cause ice ages. Ice ages come and go in a relatively rapid cycle (around 100,000 years), superimposed on an underlying temperature decline. It is only in the Earth's recent, and relatively cool, past that Milankovitch shifts have triggered ice ages.

**A Dance through Time**

Vrba's vision of evolutionary pulses leads to various specific predictions. Vrba predicted that specialist creatures, like shrews, which feed on insects, slugs and worms which are very sensitive to climate, should respond to climatic change more quickly than general-ists, such as pigs. "And that," says Vrba, "is precisely what you see within the African antelopes." Contrast, she says, the explosion of new species that formed between 2.7 and 2.4 million years ago within the Alcelaphini tribe—specialist grazers—with the impala, which feeds on anything and has remained virtually unchanged for 3 million years. It matters to the Alcelaphini when the vegetation alters. They have to change (speciate) or become extinct. "Specialist species replace each other as the climate changes, like partners in a gavotte—a dance through time. The impala simply alters its diet."

Traditional Darwinism would predict a steady modification of the impala over 3 million years, even without climatic change, because it still needs to outrun leopards. But, in fact, neither impalas nor leopards have changed very much. They are both too versatile to be worried by climatic change, and competition between them and with their own kind does not—as Darwin sup-posed—provide sufficient selective pressure to cause them to alter. "In the absence of sufficient extraneous stimulus," says Vrba, "they have reached equilibrium."

It also follows from the turnover pulse hypothesis that animals should change in ways which correspond to a particular climatic change. In cold conditions, for instance, natural selection favours large body size in warm-blooded animals, so that heat can be conserved more easily. Many of the animals that first appeared 2.5 million years ago were the biggest of their kind, including giant cattle and antelopes in Africa, and giant deer (like the Irish elk) in Europe. Even *Homo habilis*, traditionally regarded as the first bona fide member of the genus *Homo* was considerably bigger than *Homo*'s likely ancestor, *Australopithecus afarensis*.

Insights into the environment of our earliest probable ancestor come from Hank Wesselman of the American River College in Sacramento, California, who has studied fossil rodents from the same East African deposits, near the River Omo, that have yielded remains of *A. afarensis*. He found that about half of the rodents dating from around 2.9 million years ago—the time of *A. afarensis*—are of species associated with moist forest, suggesting, as does pollen evidence, that *A. afarensis* arose first in forest. Later fossils, from around 3 million years ago, suggest drier, more open country.

Yet, if the response to climatic change takes years to unfold, as seems probable, can we really call it a "pulse"? The point, says Vrba, is not how long the response takes but whether, while it lasts, the rate of change through extinction, migration and speciation, is faster than at other times, and whether such discrete periods of rapid change follow shifts in physical environment. "A million years of nothing followed by 100,000 years of change, and then another million years of stability is a pulse." Vrba maintains that the fossil record might reveal such pulses once it has been more thoroughly studied, and that modern statistical methods are sensitive enough to distinguish them from random clustering.

In a nutshell, then, Vrba's theory is this. Evolutionary pulses should not be seen as sudden, across-the-board changes. Neither should they be seen as general increases in evolutionary pace, as Darwin predicted. They should be seen as discrete periods of evolutionary change that accompany periodic changes in the physical environment, usually in the climate. Within each pulse, as the climatic change intensifies, animals are likely to change in a definite order: specialists first, generalists less readily, perhaps not at all. The hypothesis further predicts that new species should only form when initiated by physical changes. So far, Vrba maintains, the evidence is reinforcing that hypothesis.

If she is right, we have indeed to concede that if God had decided to keep the world stable five million years ago, then our arboreal ancestors would not have been forced to change their ways. I would not be writing this and you would not be reading it. We would still be involved in border disputes with chimpanzees. ❑

**Questions**:

1. What kind of climate favors large body size in warm-blooded animals? When did large animals such as giant cattle first appear?

2. What is a "foram"? What does increased O-18 tell us?

3. When did South America cease to be an island? Why did it affect climate when it no longer was?

Answers on page 301

*Most of the major extinction events during Earth history affected many different organisms, both on land and in the sea. The last major extinction event, which occurred at the end of the Pleistocene ice ages, affected primarily large mammals. Throughout most of the world these extinctions appear to correlate with the spread of humans across the globe. Some archeological sites show clear evidence of intense hunting of these large mammals just prior to their extinction. In other cases human alteration of the landscape and introduction of exotic animals may have contributed to the extinctions. Climate changes combined with human activities may have enhanced extinction rates. All of these factors are at work in our present world and many scientists are concerned that our present activities may lead to a mass extinction event to rival or surpass those of the geologic past. Investigating how humans may have contributed to the extinction of the large Pleistocene mammals and other animals may help us to understand what effect our current activities may have on the biosphere.*

# Recent Animal Extinctions: Recipes for Disaster

## Many late-prehistoric extinctions share ingredients: climate and vegetation change, human hunting, and the arrival of exotic animals

by David A. Burney

Perhaps we are all growing slightly weary of being reminded that we are in the midst of a crisis in species survival, a decline in diversity that may ultimately exceed the greatest catastrophes in the earth's long history of mass extinctions. But whether or not the point has been overstated, the arena of public life is crowded with debates over management decisions concerning species in danger of extinction. It is useful to ask whether these debates can be informed by studying extinctions of the past.

Among the small group of paleoecologists, archaeologists, paleontologists and other scientists around the world who have focused their research on multidisciplinary investigations of the extinction record, there are some who have also been looking for ways to bridge the gap between the past that they study and the current public concern over biodiversity. The present period is likely to be viewed in the near future as a time when humanity made, or failed to make, some tough choices between environmental and economic priorities that will decide the fate of much of the planet's biota. There are good practical reasons to know all we can about past extinctions.

Unfortunately, as skeptics would be quick to point out, few extinctions have been well documented. Many of the most interesting, such as the events that claimed many large-animal species or a large portion of regional flora and fauna, took place in prehistoric times. They left only vague stratigraphic clues that may be hard to read, if they can be detected at all. The challenge is to uncover evidence that can be interpreted with sufficient detail and in the proper chronological sequence.

Like detectives on a murder case with only a few stale and murky clues, my colleagues and I have had a frustrating, but always fascinating time. The scarce and sometimes ambiguous details of the lives and deaths of an array of extinct creatures seem to be falling into place now, thanks to dogged persistence and recent scientific advances. The patterns that we see emerging are not only relevant to the present crisis but also, to a chilling extent, remarkably similar. It appears possible that the spread of early human populations interacted with changes in climate and habitat to contribute to a series of extinctions that shaped the biological communities that surround us. Unfortunately, the same factors can be seen at work today in some of the richest of those communities.

Only 12,000 years ago, the landscape from Florida to California provided a spectacle to rival the Serengeti Plains of Africa in large-animal diversity. Extinct Pleistocene animals of North America are the stuff of schoolchild fantasies: woolly mammoths, mastodons, saber-tooth cats, giant ground sloths and the like. In addition to these seemingly exotic animals, the North American fauna also included many forms that have survived elsewhere, such as cheetahs, lions, zebras, yaks, tapirs, capybaras and spectacled bears.

***Reprinted with permission from AMERICAN SCIENTIST, November/December 1993, Journal of Sigma Xi, The Scientific Research Society.***

This wildlife spectacle ended with apparent suddenness. Since about 30,000 years ago—as human populations started to expand into previously uninhabited regions, but well before modern technology—most continents and large islands throughout the world lost half or more of their large-animal species. On some oceanic islands, virtually all the larger species, and many smaller ones as well, disappeared, often rapidly.

It stands to reason that the past extinctions most applicable to the present situation are these relatively recent ones. There is much interest in the extinction of the dinosaurs, but the great age of this event and the types of sites and dating methods available dictate that causes and timing are difficult to study, except on a long geological time scale. Fortunately we can date fossils of the past 30,000 years or so reliably enough to know critical details about timing. Ecological events that have taken place in this time frame provide the most direct and concrete information available on which to base predictions concerning future environments and human impacts.

**Eurasia: The Art Lives On**

Some of the most elegant evidence that early human beings and extinct animals crossed paths can be seen in the caves of southern France and northern Spain. Cave paintings, engravings and sculptures created 10,000 to 30,000 years ago portray a fauna far more diverse than that surviving today in Europe's parks and reserves. Some of these animals, such as the horse, the cow and the reindeer, have survived as domesticated animals. But the wooly mammoth (*Mammuthus primigenius*), wooly rhinoceros (*Coelodonta antiquitatis*), cave bear (*Ursus spelaeus*), and giant deer (*Megaloceros gigantea*, also called the Irish elk), among others, survive only as fossils and in the artwork of Cro-Magnon Man.

Since these and some other extinct Eurasian animals were common during the last Ice Age and gradually disappeared about 10,000 to 15,000 years ago (about the time the climate became warmer), early investigators very reasonably suggested that climate change was the probable cause of their extinction. Work in recent decades, however, has cast doubt on such a simple explanation. Paleoclimatological data from deep-sea sediment cores show that while these animals thrived, the climate made more than 20 shifts from glacial to interglacial conditions over the last 2.5 million years. It is quite likely that open-country grazers, as the mammoth and rhinoceros certainly must have been, may have been forced to move far to the north when temperate-zone forests began to recolonize Europe after each thaw. Recent archaeological evidence suggests that it was only within the last 30,000 to 40,000 years that human beings mastered the considerable technical skills needed to colonize the far-northern tundra regions.

Thus, suggests anthropologist Richard Klein of Stanford University, human beings may have triggered these extinctions through a combination of hunting and habitat modification in northern areas that may have served as refuges during earlier periods of warm climate. In a sense, both the climate and humanity were responsible, because they happened to combine effects in a way that was deadly for these species.

**Australia: Disaster Down Under**

While the northern continents were still under the cold grip of the last Ice Age, Australia and New Guinea were in the midst of a different kind of struggle. During times in the Pleistocene when sea level was more than 100 meters lower than it is now, Australia and New Guinea were joined by a broad land bridge, and the channel separating New Guinea from the Malaysian Archipelago was much narrower. Perhaps some time between 30,000 and 60,000 years ago, the ancestors of the Australian aborigines crossed this water gap and spread southward on the island continent. Assessing the age of fossils with radiocarbon dating techniques is difficult in this time period, but it is clear that, by the end of the last Ice Age, Australia and New Guinea had been swept by a wave of large-animal extinctions to rival that of any other continent in the late Pleistocene.

Australia lost 19 genera, totaling more than 50 species. The marsupial fauna associated with the region today is diverse, but it only includes one genus of large animals, the *Macropus* kangaroos. Thirteen other genera of giant marsupials became extinct, including the rhinoceros-like marsupial *Diprotodon*, which weighed perhaps two tons or more. *Thylacoleo* (sometimes called the "giant killer possum," since it may have been a fierce carnivore) was represented by two species the size of a leopard. One giant kangaroo, *Palorchestes*, stood 3-1/2 meters tall. In addition to the extinct giant marsupials, there was a huge lizard (even larger than the Komodo dragon), giant horned tortoises, and an ostrichlike bird similar to the surviving emu.

The Australian case remains puzzling; some of these species are thought to have survived more than 10,000 years after the arrival of human beings. Climatic change is often invoked in these extinctions, since there is not much archaeological evidence for large-scale hunting. It has been difficult, however, to visualize how continent-wide climate changes could have been severe enough to do the job. It is tempting to look for an animal that might have been introduced by the prehistoric aborigines that could have proliferated and disrupted native populations. The dingo, Australia's wild dog, might seem a likely candidate, but it appears from the archaeological record that the dingo was introduced long after some of these animals became extinct. It may have played a role in later extinctions, but much work remains to be done in Australia, as there is no entirely convincing theory for the cause of the massive die-offs down under.

**The *Blitzkreig* Hypothesis**

The debate over the causes of the late-Pleistocene extinctions has centered in the Americas, perhaps primarily because theories proposed by Paul Martin and his colleagues at the University of Arizona have sparked a worldwide interest in systematically testing competing ideas with data from multidisciplinary research. The losses in North America are truly staggering to contemplate, dwarfing even more modern environmental catastrophes: 33 genera of large animals, including seven families and one order (elephants and their relatives).

Radiocarbon dates from charcoal associated with mammoth remains in archaeological sites, from protein-rich bones found at the Rancho La Brea tar pits of southern California, and from ground-sloth dung from western caves, all show the same trend: that many of these animals disappeared more or less simultaneously within a few centuries of 11,000 years ago. At the same time or just before, the first solid evidence of human activity appears at many sites in the form of specialized hunting tools, the distinctive Clovis projectile points associated with Paleoindian hunters. Several controversial sites around the continent have been reported to show earlier evidence of Native Americans, but the first widespread, fully accepted human evidence in North America seems to coincide with this extinction event.

More than two decades ago, this prompted Martin to propose one of the more plausible versions of "Pleistocene overkill" to explain these extinctions. Overkill theories hinge on the notion that it was primarily the hunting activities of the earliest Native Americans that precipitated the faunal crash. These early inhabitants presumably crossed the Bering Land Bridge from Siberia to Alaska when sea level was lower, near the end of the last Ice Age. One of the most serious scientific criticisms of Martin's theory is that these first Americans were probably few in number and were technically incapable of mounting such mass slaughter in so short a time. But in the late 1960s the Russian scientist M.I. Budyko published a series of equations showing that, theoretically at least, such a human-caused faunal crash was possible. He calculated that a small founding population of human beings could multiply fast enough to eventually outstrip the capacity of very large mammals (with their slow birthrate) to replenish their numbers under a steady hunting regime. Then in the mid-1970s James Mosimann of the National Institutes of Health teamed up with Paul Martin to develop a computer model incorporating several key ideas that together came to be known as the *Blitzkrieg* Hypothesis.

In this model, a large human population would not be required to do the job, if the extinctions spread over the Americas as a "front." The early American big-game hunters, they theorized, may have multiplied rapidly (for people) under conditions of unlimited food supply, low incidence of human disease, and an efficient social organization. Beginning in Alaska, where the extinctions and first occurrence of projectile points appear slightly earlier than on the rest of the continent, human hunters encountered a fauna of large animals unfamiliar with the danger they posed. As these animals declined and human beings multiplied, the hunters could have moved like a wave southward and eastward, laying waste to the naive animals as they went. Some support for the idea comes from two kinds of evidence. First, the southern and eastern parts of the continent seem to show progressively later dates of the arrival of the first Native Americans and the latest occurrences of some of the extinct animals. Second, careful analysis of archaeological sites suggests only a brief period of overlap between the first hunters and the last mammoths.

For a while, it looked as if South America also conformed to the pattern of *Blitzkrieg*. Now this looks less certain, as some sites on that continent seem to be dating to times that are earlier rather than later than dates from the southwestern United States and other areas closer to the Bering Strait. This may not be a definitive test, however, as some Native American groups may have arrived earlier, but may have been less interested in big-game hunting or too few in number to have an effect. Alternatively, the small number of radiocarbon dates that do not agree with the *Blitzkrieg* pattern could simply be wrong.

Other overkill models allow for small human populations in the Americas as much as 20,000 years before their population levels became a threat to the large-mammal prey populations. It is still a matter of debate the extent to which hunting, versus other human activities and natural climatic change, may have played a role. One thing seems fairly certain: North America's ecology was drastically changed, long before European contact, by the loss of so many important faunal elements. The giant beasts that must have modified their environment on the scale of elephants were an especially critical loss. Giant ground sloths, for instance, probably satisfied their dietary need for tree leaves by breaking down large limbs and whole trees. This particular kind of disturbance has been largely absent from the Holocene forests of recent millennia—for the first time in millions of years of American environmental history.

South America has probably suffered from this type of change even more. No other continent lost as many animals in the late Pleistocene. In one article, Dan Janzen of the University of Pennsylvania and Paul Martin speculate reasonably that the structure of many forests in Central and South America may have been fundamentally changed by the loss of large animals that played a major role in the dispersion of large seeds of forest trees and in opening up the forest by breaking down limbs and small trees.

The list of animals that became extinct during this period in South America is staggering: It includes 46 or more genera, more extinctions than are known to have taken place during all of the previous nearly 3 million years on the continent. Among these are four whole families of sloth relatives, one genus of giant rodents, four of large carnivores, four of mastodons, three of horses, and 11 of cloven-hoofed animals including peccaries, camels and deer.

**Losing the Sweepstakes**

Perhaps the supreme irony of the global pattern of late-Pleistocene extinctions is that isolated oceanic islands, generally regarded by island biogeographers as particularly extinction-prone ecosystems, were the last places to be affected. On the surface, this fact seems to run contrary to nearly everything that biologists think they know about evolution. The faunas of oceanic islands (islands not connected to a continent even when sea level is at its lowest during ice ages) are generally regarded as "unbalanced." That is, they are descended from a few founding populations that are not representative of the full array of genetic material available on the continents. Through a process that the late G.G. Simpson described as a "sweepstakes," the ancestors of animals native to oceanic islands reach the remote land mass by swimming, rafting on floating debris, or some other highly uncertain means. Most native oceanic-island mammal faunas are therefore dominated by animals that can swim well, such as elephants, hippopotami and deer, or that are small enough to cling to debris, such as rodents and insectivores. In addition, especially on the islands that are farthest from the continents, birds and reptiles may occupy a disproportionate share (from the continental viewpoint) of ecological niches. Plants face similar limitations and challenges on remote oceanic islands.

Thanks to the unique genetic configurations and evolutionary pressures posed by isolated islands, however, these founding animals soon become quite different from their continental ancestors who "won" the sweepstakes. Large mammals tend to evolve into dwarf forms, small mammals may get larger, birds may become flightless and reptiles may become huge, lumbering herbivores like the island tortoises, or fierce top carnivores such as the giant monitor lizards of Komodo Island.

For millennia after the late-Pleistocene extinctions, these seemingly fragile and ungainly ecosystems apparently thrived throughout the world. This is one of the soundest pieces of evidence against purely climatic explanations for the passing of the Pleistocene faunas. In Europe, for instance, the last members of the elephant family survived climatic warming not on the vast Eurasian land mass, but on small islands in the Mediterranean—an even warmer climate. Radiocarbon dates suggest, for instance, that dwarf elephants persisted on Tilos, a tiny island in the Aegean, until perhaps 4,000 to 7,000 years ago. On Mallorca, in the Balearic Islands, the endemic goat-antelope *Myotragus* survived until 5,000 years ago or less and turns up in some Bronze Age archaeological sites.

But these Mediterranean island megafaunas did eventually become extinct. Thirteen genera of large and small mammals disappeared, along with giant tortoises. Many investigators believe the cause was probably a combination of human activities (hunting and landscape modification) and the introduction of rats, dogs, goats and other animals that came with the people. But a major problem in substantiating this hypothesis has been establishing exactly when people arrived. Only a few archaeological sites have turned up so far that show an overlap between human beings and extinct fauna. I was fortunate a few years ago to visit one of these, Corbeddu Cave on the island of Sardinia. Here, Paul Sondaar of Rijksuniversiteit Utrecht in the Netherlands and his Dutch and Italian colleagues have excavated Late Stone Age artifacts and fragments of human bone in a layer containing many bones, some possibly modified by human beings, of the extinct deer *Megaceros cazioti*. In younger strata above this level, the deer disappears, but the extinct rabbit relative *Prolagus sardus* persists until a later time, then disappears too. Dates from the level containing the earliest human bones are as recent as about 9,000 years.

An even more spectacular example of human interaction with an extinct Mediterranean island megafauna comes from the recent work on Cyprus of A.H. Simmons of the University of Chicago and D.S. Reese of the Field Museum in Chicago. The site of Akrotiri-Aetokremnos provides evidence that, a little over 10,000 years ago, hunger-gatherers camped there and hunted the endemic pig-size pygmy hippopotamus. More than 240,000 bones representing at least 200 individuals have been found in the archaeological site, some cut and burned. A few pygmy elephants are also represented, along with shellfish and bird bones. In the slightly younger upper strata, the proportion of extinct megafauna in the middle assemblage decreases, implying that these hunters may have been rather efficient at rounding up the unsuspecting hippopotami and depleting this resource. So far, however, sites like this are extremely rare, and detailed chronologies remain to be worked out.

Even less is known about the extinctions on islands of the West Indies. Twenty-one genera of endemic mammals have become extinct since the late Pleistocene, although few dates have been obtained. Among the extinct denizens of Caribbean islands were giant rodents, dwarf ground sloths and large insectivores. Huge birds of prey were probably the top carnivores in some of these peculiar ecosystems. Only a few archaeological sites have been found with bones of extinct animals, but radiocarbon dates from Antigua and the Greater Antilles do show that some of the extinct fauna were present as recently as 4,000 to 7,000 years ago, which is also the time period when the first Native Americans are believed to have reached the islands.

G.K. Pregill at the University of California at San Diego and Storrs Olson of the Smithsonian Institution have suggested that some of the extinctions may have been caused by the combined effects of rising sea level (decreasing the size of some of these islands to less than half) and climatic change. Until more of the extinct animals and the timing of human arrival on these islands are firmly dated, this idea cannot be fully evaluated. Mammalogist Ross MacPhee of the American Museum of Natural History and his colleagues have begun a search for sites in the West Indies containing clues to the fate of the extinct animals. They are now working on some promising sites in Cuba. One interesting clue has turned up from our collaboration in Puerto Rico. MacPhee and I collected a 7,000-year sediment core from Laguna Tortuguero, a freshwater lake on the north coast of the island. Lida Pigott Burney of Fordham University analyzed the core for stratigraphic charcoal particles, a record of past burning in the environment. The charcoal record shows that, about the time Native Americans are thought to have begun colonizing the Great Antilles, charcoal values make a sudden increase of several orders of magnitude above background (presumably prehuman) values. These fires seem to have predominated in the area from about 5,200 years ago, gradually decreasing to lower levels about three millennia ago. Perhaps the earliest settlers had a large-scale impact on the islands with a slash-and-burn approach to agriculture, in addition to their other presumed impacts.

Beginning 6,000 years ago or more, prehistoric mariners spread across the South Pacific from west to east, colonizing islands as they went. David Steadman of the New York State Museum and other avian paleontologists have discovered that many islands in the South Pacific had a much richer bird fauna before human arrival, often two times or more as diverse as the modern native fauna. Steadman and his colleagues, archaeologist Pat Kirch of the University of California at Berkeley and the palynologist John Flenley of Massey University in New Zealand, believe that, on these relatively small and extremely isolated islands where there are few if any large native animals, human beings may have had their most profound impact not by hunting, but by modifying the environment and introducing rats and domestic animals.

**Some of the Last Places on Earth**

Among the last major land masses to be reached by prehistoric human beings were the Hawaiian Islands, New Zealand and Madagascar. True to the patterns outlined above, these were also among the last places to record major prehistoric extinction events. Because of the freshness of the subfossil record of these relatively recent events, they have received considerable attention from scientists interested in catastrophic extinctions.

Storrs Olson and Helen James of the Smithsonian Institution continue (literally, up to the present moment) to add new birds to the list of extinctions from Hawaii. One thing is certain now: Their work shows that more native bird species have become extinct in the Hawaiian Islands since the arrival of the Polynesians some 1,500 years ago than now survive. Among them are peculiar flightless relatives of mainland forms, including *Thambetochen* and *Ptaiochen*, giant waterfowl, their "turtle-jawed" relative *Chelychelynechen* on Kauai, and *Apteribus*, an ibis resembling the kiwi of New Zealand in shape and presumed habits.

The list of extinct Hawaiian birds dwarfs the considerable list of bird extinctions that have taken place since European arrival on the place since European arrival on the islands. Olson and James believe that some of the extinctions may have been precipitated by hunting, and others by introduced rats, pigs and other exotic animals. But a host of small forest birds also became extinct, prompting a reasonable speculation that deforestation by prehistoric agriculturalists could have played a major role. I have recently begun a project with Olson and James to use fossil pollen analysis to learn more about past environments and the dietary habits of some of the extinct birds. Perhaps eventually, we will be able to determine more accurately the extent to which various human and natural factors may have played a role in this relatively recent catastrophe.

Lacking terrestrial mammals other than bats, New Zealand's primary large animals were giant birds, known as moas to the Polynesians who arrived about 1,000 years ago. These relatives of ostriches diversified considerably, various authors recognizing between 13 and more than 30 species.

Although the North Island has few archaeological sites with moa bones, the South Island is quite a different matter. Atholl Anderson of the Australian National University and a host of other investigators have excavated over 100 human sites with moa bones, sometimes in sizeable heaps representing many individuals and species. Radiocarbon dates on these camps of moa hunters are as recent as 300 to 400 years ago, suggesting that it might have taken the native Maoris about 600 years to hunt the moas to extinction. Pollen data from lakes and swamps in New Zealand also show that the Maori probably caused major vegetation changes, perhaps primarily by setting fires.

Madagascar, the world's largest oceanic island, has been separated from the continents for 165 million years or more. It is perhaps not surprising that its late prehistoric fauna was dominated by an array of giant birds and reptiles, pygmy hippopotami, and unique rodents, insectivores and mongoose relatives. The most unusual part of the lineup, however, was the lemurs, including several giant species of these primates as large as chimps and small gorillas.

The first Malagasy people, arriving between perhaps 1,500 and 2,000 years ago, must have played some role in the extinction of this entire endemic large-animal fauna, including at least 12 genera and more than two dozen species of large and small animals. But as our research group, composed of archaeologists Bob Dewar of the University of Connecticut and Henry Wright of the University of Michigan, primatologist Elwyn Simons at the Duke University Primate Center, paleontologists Ross MacPhee of the American Museum of Natural History and Helen James of the Smithsonian Institution, and other experts in the United States and Madagascar have been learning for over a decade, figuring out exactly how the early settlers of Madagascar contributed to one of the last major prehistoric extinction events on the plant has not been easy. Despite concerted searching, only a handful of bones of extinct animals have turned up in archaeological sites around the island—nothing approaching the hauls from moa-hunting parties taking place in New Zealand about the same time. The favorite old theory about the extinctions in Madagascar was that man caused the extinctions less directly, primarily by burning the forests. In recent years, such a simplistic explanation has become less compelling, as fossil pollen and charcoal in my sediment cores from throughout the island have shown that wildfires and vegetation changes were a normal part of many of the highly seasonal environments for 35,000 years or more before the arrival of humankind. The new evidence also shows that many major climatic changes took place prior to human arrival.

Fruitful collaborations with my students and colleagues have documented that major environmental changes took place in prehuman Madagascar. But many kinds of fairly drastic changes occurred more or less simultaneously in the period between about 2,000 and 1,000 years ago, a period when, radiocarbon dating shows, many of the now-extinct creatures went from apparently abundant to essentially absent in the subfossil record. Dates on human-modified hippopotamus bones, the first occurrence in sediment cores of pollen of introduced *Cannabis* and a sudden rise in background levels (which were already fairly high in some sites) of stratigraphic charcoal, all support the notion that people probably arrived in Madagascar about two thousand years ago. Paleolimnological collaborations with Norman Reyes at Fordham University, George Kling of the University of Michigan and Katsumi Matsumoto of Brown University have uncovered evidence in our sediment cores for changes in microscopic diatom floras and lake levels. These stratigraphic records indicate that late-Holocene climates in Madagascar have become increasingly arid, and especially rapid changes taking place about 2,000 years ago.

In an effort to disentangle the various potential causes and their effects, we have increasingly concentrated our multidisciplinary efforts on a very special kind of subfossil site—one where fossil pollen and charcoal, paleolimnological and sedimentological changes, bones of extinct animals, and archaeological materials turn up in the same or nearby (and cross-datable) stratigraphic sequences. Such sites are generally regarded as rare, but we have had some initial successes. A hypersaline pond in the arid southwest known as Andolonomby, for instance, has shown that the region's climate began to turn drier about 3,000 years ago, with maximum aridity perhaps 1,000 to 2,000 years ago. As the forests were declining, the extinct Malagasy hippopotami, giant tortoises and other large animals become more scarce in the paleontological record for the site. Human-modified hippopotamus bones recovered from the vicinity date to about 2,000 years ago, and no younger hippopotamus bones have been detected here. At the same time, charcoal values peak at about 2,000 years ago, and the pollen of woody plants continues to decline through this hyperarid phase.

A similar story, showing abundant evidence for the extinct fauna until about 2,000 years ago, then a decline over the next millennium, seems to be emerging from recent work with Helen James and Fred Grady of the Smithsonian and our Malagasy colleagues Jean-Gervais Rafamantanantsoa and Ramily Ramilisonina. We concentrated on another site with many types of evidence in close proximity, the spectacular cavern complex of Anjohibe in the northwestern part of the island. This area, and other northern cave sites excavated by Robert Dewar and Elwyn Simons, show intriguingly late evidence for some of the extinct giant lemurs, including a few radiocarbon dates around the beginning of the present millennium. It is beginning to appear that, whatever happened to the Malagasy fauna, it took over 1,000 years for the catastrophe to fully develop—a longer time than would have been predicted from the strictest interpretation of the *Blitzkrieg* hypothesis.

The case is not closed in Madagascar, just as in many of the other sites I have mentioned. Because many of our sites show evidence for a combination of simultaneous changes, including natural climate change, activities of the first human hunters, changes in fire regime and vegetation structure, and the arrival of exotic species, I have come to refer to the extinction event in Madagascar about 1,000 years ago as a "recipe for disaster." Instead of finding overwhelming evidence for the actions of a single cause in the extinctions, these four factors, and perhaps others, seem to have been functioning simultaneously on the island. Perhaps none of these agents alone would have wiped out all the extinct animals, but each could have been responsible for some of the extinctions. What I suspect is more likely, though, is that the primary lethal effect was the combination of these factors, acting in a synergistic manner on a fauna unaccustomed to so many disruptions at once.

**Cautionary Tales**

Late prehistoric extinctions, in my opinion, constitute a series of important cautionary tales for conservation biologists. First, the record shows that the native faunas in most parts of the world today are only damaged fragments of what existed in the recent past. Most ecosystems were first disrupted by people not in recent centuries or decades but thousands of years ago. Most of the "communities" we seek to preserve are not highly integrated entities that have reached some long-standing equilibrium. Instead they are primarily of two types: continental biotas with the largest herbivores and the large and medium-sized carnivores "skimmed off," and island biotas lacking many original components of all sizes and heavily invaded by cosmopolitan exotics. This in no way diminishes the value of saving what is left. Instead, the record of recent millennia should serve to remind us of how much has been lost, rather abruptly and without the benefit in most cases of industrial-era lethal capabilities.

Second, environmentalists should be quick to respond to critics who say that "extinction is natural" as justification for allowing a species to go extinct today. Most extinction in recent millennia is fundamentally different from earlier extinction, because it is extinction without replacement, at least on human time scales. Evolution continues to work only if, as the "unfit" become extinct, new species arise to fill their place. On the other hand, the fossil record shows that earlier mass extinctions, such as that of the dinosaurs, gave rise over the longer geologic time scale to new evolutionary opportunities, such as the great early-Cenozoic radiations of mammals in the apparent vacuum left by the passing of the dinosaurs. This was an evolutionary "recovery" that was apparently millions of years in the making, though.

Third, it is humbling to realize that we still understand very little about extinction, one of biology's most fundamental concepts. When wildlife scientists and managers go through the sad but seemingly unavoidable business of calculating how little habitat or how few individuals are necessary to maintain an endangered population in the face of competing land-use or funding priorities, they should keep in mind the lesson of the mammoths and mastodons: For millions of years, these mighty beasts were abundant and widespread practically throughout the world. Yet, in a relatively short time, a poorly understood combination of circumstances wiped them out. I doubt if anyone around in those days would have believed such a thing possible.

Fourth, studying past extinctions may be leading us to some important new discoveries that relate directly to future conservation efforts. One clear example is a new hypothesis for explaining catastrophic late-Pleistocene continental extinctions put forward by Norman Owen-Smith, which he calls the "Keystone Herbivore Hypothesis." Simply put, he suggests that the apparent suddenness with which virtually whole faunas may disappear without a clear cause is not really surprising, since the fossil record shows that the first animals to go extinct are often the very large, slow-breeding herbivores. These animals play a role in shaping their own environment, creating physical diversity in the habitat and biotic turnover that allows many smaller animals to find niches. If he is correct that the loss of elephants and giant ground sloths in the Americas resulted in a "cascade" of extinctions of the middle-sized grazers and browsers and their predators, what can we expect in Africa in the wake of the decline of the rhino and the elephant?

Which leads to my final point. I have scarcely mentioned Africa, the one place on earth that seems to have virtually escaped the late-prehistoric extinction crisis. Over 80 percent of the late Pleistocene megafauna of Africa is still with us, if somewhat tenuously. However, all the ingredients in the "recipe for disaster" outlined above are now present in Africa in great measure. Human populations are exploding, and lethal technologies (guns, chainsaws and bulldozers) are available as never before. Increasingly confined to parks, many animals species have become naive to the dangers posed by humans with guns instead of cameras. Exotic animals, especially livestock, are having their own population explosion. Fire, although a natural element, has increased in frequency with the help of human beings in many pastoral areas, to the point of degrading much of the soil and vegetation. Finally, climates are certainly changing. Climate change, which for so long has been the major alternative explanation for extinctions otherwise ascribed to humans, may now be rearing its head in an ironic new form: People themselves are changing the climate. The impacts of deforestation, fossil-fuel burning, acid rain, ozone depletion and a host of toxic substances on climates and ecosystems are almost certain to become major factors in species survival. Just as in prehistoric times, the combined effects of all these ingredients may add up to a recipe far more disastrous than each of them taken separately.

**Bibliography**

Burney, D.A. 1987. Late Quaternary stratigraphic charcoal records from Madagascar. *Quaternary Research* 28:274-280.

Burney, D.A. 1987. Pre-settlement vegetation changes at Lake Tritrivakely, Madagascar. *Palaeoecology of Africa* 18:357-381.

Burney, D.A. 1993. Late Holocene environmental changes in arid southwestern Madagascar. *Quaternary Research* 40:98-106.

Burney, D.A., L.P. Burney and R.D.E. MacPhee. In press. Holocene charcoal stratigraphy from Laguna Tortuguero, Puerto Rico, and the timing of human arrival on the island. *Journal of Archaeological Science.*

Diamond J. 1989. Overview of recent extinctions. In *Conservation for the Twenty-first Century*, ed. D. Western and M. Pearl. New York: Oxford University Press, pp.37-41.

Eldredge, N. 1991. *The Miner's Canary*. New York: Prentice Hall.

James, H.F., T. W. Stafford, Jr., D.W. Steadman, S.L. Olson, P.S. Martin, A.J.T. Jull and P.C. McCoy. 1987. Radiocarbon dates on bones of extinct birds from Hawaii. *Proceedings of the National Academy of Sciences* 84:2350-2354.

Janzen, D.H. and P.S. Martin. 1982. Neotropical anachronisms: The fruits the gomphotheres ate. *Science* 215:19-27.

Kirch, P.V., J.R. Flenley, D.W. Steadman, F. Lamont and S. Dawson. 1992. Ancient environmental degradation. *National Geographic Research and Exploration* 8:166-179.

MacPhee, R.D.E., D.C. Ford and D.A. MacFarlane. 1989. Pre-Wisconsinan mammals from Jamaica and models of late Quaternary extinction in the Greater Antilles. *Quaternary Research* 31:94-106.

Martin, P.S. 1990. 40,000 years of extinctions on the "planet of doom." *Palaeogeography, Palaeoclimatology, Palaeoecology* 82:187-201.

Martin, P.S., and R.G. Klein, eds. 1984. *Quaternary Extinctions: A Prehistoric Revolution*. Tucson: University of Arizona Press.

Mosimann, J., and P.S. Martin. 1975. Simulating overkill by Paleoindians. *American Scientist* 63:304-313.

Olson, S.L. 1989. Extinction on islands: Man as a catastrophe. In *Conservation for the Twenty-first Century*, ed. D.Western and M. Pearl. New York: Oxford University Press, pp.50-53.

Owen-Smith, N. 1987. Pleistocene extinctions: the pivotal role of megaherbivores. *Paleobiology* 12:351-362.

Steadman, D.W. 1989. Extinction of birds in eastern Polynesia: A review of the record, and comparisions with other Pacific Island groups. *Journal of Archaeological Science* 16:177-205.

Sutcliffe, A.J. 1985. *On the Track of Ice Age Mammals*. Cambridge, Mass.: Harvard University Press.

## Questions:

1. What is the *Blitzkrieg* Hypothesis for extinction in North America?
2. What does the survival of island faunas well past the end of the Pleistocene tell us about possible causes of the extinctions?
3. What were the last places to experience extinction events and how long ago did they occur?

Answers on page 301

# Answer Section

## Part One: The Origin of the Earth and Its Internal Processes

### 1. A Rocky Start

*Answers*:

1. The different types of grains in a meteorite have different freezing temperatures indicating that they came from different regions of the solar system.

2. It would take about one million years.

3. The moon formed as a result of the collision of a Mars-size object with the Earth.

### 2. The Fast Young Earth

*Answers*:

1. Because the sun was dimmer 4 billion years ago all surface water should have been frozen but there are rocks four billion years old that were deposited by water.

2. Water reflects less sunlight back into space, absorbing more heat, so the larger the surface area of the ocean, the warmer the earth. The faster rotation of the earth would have confined clouds to the equatorial and subtropical regions leaving more clear sky for the sun to penetrate.

3. The tidal drag of the moon has slowed the earth's spin.

### 3. Cooling the Vision of Earth's Hot Core

*Answers*:

1. The iron is squeezed between two diamond crystals and is heated by a laser.

2. There is a large drop in temperature from 4,000 kelvins to 2,700 kelvins. This transfer of heat causes plumes of hot rock to move up through the mantle causing the movement of crustal plates.

3. These findings suggest that there is little chemical reaction between the core and the mantle.

### 4. Bits of the Lower Mantle Found in Brazilian Diamonds

*Answers*:

1. The two lower mantle minerals are ferropericlase and calcium silicate.

2. The eruptions that bring diamonds to the surface may develop below the lower mantle-upper mantle boundary, meaning that all of the source material that formed the diamond originated in the lower mantle.

3. The abundance of iron and silicon should be greater for lower-mantle-derived minerals. If mixing occurs, then some mineral inclusions should have lower mantle abundances and some should have upper mantle abundances.

**5. Scrambled Earth**

*Answers*:

1. The mantle comprises 83 percent of the Earth's volume.

2. The mineral peridotite may not be what makes up the lower mantle because its density is 3 to 5 percent lower than the density of the lower mantle as measured by seismologists.

3. The Earth would cool faster if it was mixed from top to bottom. Stratification would slow the rate at which internal heat could escape by 5 to 10 times.

**6. The Flap Over Magnetic Reversals**

*Answers*:

1. The Earth's magnetic field originates in the iron-rich outer core.

2. The mantle moves very slowly—at about the pace of a growing fingernail. This means that variations in the structure of the mantle will be preserved for millions of years.

3. Kenneth Hoffman found that the reversals did not follow a line but were concentrated in two locations: one off the southeast coast of South America and the other off the west coast of Australia.

**7. Breaking Up is Hard to Understand**

*Answers*:

1. Continental rifting is occuring in the Middle East where Africa and the Arabian Peninsula are being separated creating the Red Sea.

2. The molten rock is more dense then the continental crust so that most of it cannot rise to the surface and gets trapped beneath the crust.

3. As a hot rift splits apart it rises due to the addition of so much igneous rock. It may even rise above sea level, just as Iceland is today.

**8. The Ocean Within**

*Answers*:

1. A hot spot.

2. Helium would quickly float into the upper atmosphere, so it must have been released recently from a large volume of water.

3. As the Earth cools less water will be released through volcanoes while water is still being drawn in with the subducting slabs. Ultimately all of the ocean's water may be drained away, leaving the Earth's surface dry.

**9. Looking Deeply Into the Earth's Crust in Europe**

*Answers*:

1. The deepest well at present is 12 kilometers deep and was drilled on the Kola Peninsula in the former Soviet Union.

2. Graphite easily conducts electricity, and geophysical surveys had indicated that the conductivity of the crust in this area was similar to other areas containing graphite.

3. This suggests that the upper crust carries most of the stress of the entire 100-kilometer-thick plate, and that the upper crust may play an important role in plate movements.

## Part Two: Earthquakes and Volcanoes

**10. Finding Fault**

*Answers*:

1. The magnitude of the Landers earthquake was 7.1 on the Richter scale.

2. They caused a triangular block of land adjacent to the San Andreas Fault to move northward, thus releasing the geologic "brake pad" on the San Andreas Fault.

3. The new fault would have to cut through mountains where the crust is particularly thick.

**11. Midcontinent Heat May Explain Great Quakes**

*Answers*:

1. Cooler mantle is stiffer and flows less easily. This prevents tectonic forces from compressing the crust above such regions.

2. The amount of heat coming out of the crust is larger than in surrounding areas and seismic waves move slower which is characteristic of hotter crust.

3. It is 30 times more powerful.

**12. Predicting Earthquake Effects—Learning from Northridge and Loma Prieta.**

*Answers:*

1. Both the Loma Prieta and the Northridge earthquakes occurred on a reverse fault.

2. The Northridge earthquake was unusual because the peak horizontal ground acceleration was 1.7 times larger than predicted.

3. The Loma Prieta earthquake was unusual because approximately 70% of the property damage occurred over 100 km away from the epicente.

**13. Waves of Destruction**

*Answers*:

1. During tsunami-generating earthquakes slippage along faults may be slow with movement occurring over a minute or two.

2. Tsunami waves are not noticeable on the open ocean.

3. Tsunami waves may last for days, even lasting for over a week.

**14. Abandoning Richter**

*Answers*:

1. It measures the earthquake source.

2. No.

3. Cause, effect.

**15. Distant Effects of Vocanism–How Big and How Often?**

*Answers*:

1. The largest historic eruption was Indonesia's Tambora in 1815.

2. There were decreases in the number of volcanic eruptions reported during World War I and World War II, probably because less people were looking for eruptions.

3. The total amount of $SO_2$ released and the total volume of erupted material are both important in determining the eruption's affect on climate.

**16. First Direct Measure of Volcano's Blast**

*Answers*:

1. The velocity was determined by the amount of deformation to a lead plate in the meter.

2. The collapse of the lava dome caused the shock wave.

3. The measurements can be used to check the accuracy of theoretical models of volcanic behavior.

## Part Three: External Earth Processes and Extraterrestrial Geology

**17. Deadly Eruption Yields Prediction Clues**

*Answers*:

1. The long-period tremors are created by pressurized gases flowing through underground fractures.

2. During the second phase, sulfur dioxide emissions fall and the energy of long-period tremors increases.

3. The emission of gases did not change before the eruption of Alaska's Mt. Redoubt.

**18. Dante Conquers the Crater, Then Stumbles**

*Answers*:

1. Dante spent eight days in the crater.

2. The crater was not emitting hydrogen sulfide or sulfur dioxide suggesting that no fresh magma is present at shallow depths.

3. Robots such as Dante will be very useful for future planetary exploration.

**19. New Views of Old Carbonate Sediments**

*Answers*:

1. The presence of widespread beds of pure precipitated carbonate.

2. There was either an increase in atmospheric oxygen or a change in the calcium-to-bicarbonate ratio of seawater.

3. Evolution of new organisms or changes in seawater composition.

**20. Water, Water Everywhere**

*Answers*:

1. They become less polar and they may break apart.

2. As an acid, water speeds up reactions.

3. Hot water would add hydrogen to the oil. This would loosen the oil allowing it to flow more easily.

**21. From Brown to Black**

*Answers*:

1. Heat and pressure cause plants to turn to coal.

2. During the transformation into coal, lignin molecules lose their oxygen atoms.

3. The lignin molecule may occur as a helix.

**22. Geology, Geologists and Geologic Hazards**

*Answers*:

1. California. 1968.

2. Very long, geologic timespans. Yes.

3. They are not well covered. 30 million/358 million = 8.4%.

**23. Landslides and Avalanches**

*Answers*:

1. The ratio of forces that tend to hold material in place to those forces that tend to drive it downslope.

2. Schistosity and loess soils. Slope less than the angle of repose.

3. Creep. No.

**24. This Beach Boy Sings A Song Developers Don't Want to Hear**

*Answers*:

1. "Hard" stabilization (such as groins), replenishment of beach sand, relocation and retreat.

2. It is expensive and very temporary. No. Relocation and retreat.

3. The Kissimmee River of Florida was turned into a canal that suffocated the river basin ecosystems. The jetty that keeps Charleston Harbor open is starving downshore beaches of sand. This also happened in Westhampton, Long Island, causing houses downshore to be covered in surf.

**25. Geology of Mars**

*Answers*:

1. This division may be due to a very large impact which occurred at the end of accretion.

2. Volcanism may have caused massive melting of ground ice which would have been released to the surface creating some of the channel features.

3. Mars' valley networks have tributaries and increase in size downstream just like Earth's river valleys, but they are much shorter.

**26. New Evidence of Ancient Sea on Venus**

*Answers*:

1. The ratio of deuterium-to-hydrogen is 150 times higher than elsewhere in the solar system, suggesting that there was once a much greater abundance of hydrogen, which would easily bond with oxygen to form water.

2. Because hydrogen molecules are so light, half the mass of deuterium, they could easily escape the atmosphere of Venus.

3. The sun was 30 percent dimmer and the atmosphere of Venus contained less gasses that would trap heat.

## 27. Shoemaker-Levy Dazzles, Bewilders

*Answers*:

1. The largest fragments may have been 2 to 3 kilometers in diameter.

2. Paul Weissman thinks that the comet fragments were each composed of many small fragments.

3. Eugene Shoemaker thinks that the fragments were coherent but that they started to break up just before impact due to Jupiter's gravity.

# Part Four: Resources and Pollution

## 28. Population: The View from Cairo

*Answers*:

1. 24 years, 1025 years.

2. 21.

3. 55, 3.5.

## 29. A Decade of Discontinuity

*Answers*:

1. Capacity of the oceans to yield seafood, of grasslands to produce mutton, of the hydrological cycle to produce fresh water, of crops to use fertilizer, of the atmosphere to absorb CFs, carbon dioxide, and other greenhouse gasses, of people to breathe polluted air, and of forests to withstand acid rain.

2. From 14 to 126 million tons, a 9-fold increase. It has declined 10%.

3. Natural gas. Wind power. 91 million.

## 30. The Future of Nuclear Power

*Answers*:

1. He doubts it will play a significant role.

2. It breeds fuel as it runs. U-238 and plutonium. It can be easily used in weapons manufacture.

3. Ratio of electricity produced to what a plant is designed to produce if it runs 100% of the time. 60.5, which says that it has not been reliable compared to other countries.

## 31. Renewable Energy: Economic and Environmental Issues

*Answers*:

1. 4.7%, 25%.
2. 8 cents, 10 cents.
3. 23%.

## 32. 1872 Mining Law: Meet 1993 Reform

*Answers*:

1. 10,000. 424,000.
2. A handful. Companies file bankruptcy to avoid future liability costs.
3. 33%. Private, state, tribal. Yes.

## 33. Subterranean Blues

*Answers*:

1. To protect them from vandals and to avoid having federal officials prohibit access to their discoveries.
2. Lechuagilla, New Mexico. Trampling rock formations, introducing foreign materials, including food.
3. Environmental Impact Statement. Yes.

## 34. New Life for the 'Glades

*Answers*:

1. $2 billion.
2. 1988.
3. Interior Secretary Bruce Babbitt.

## 35. Troubled Waters

*Answers*:

1. 50 to 70 percent. 1990 Farm Bill.
2. Yes. Counting the number of different aquatic species and the abundance of each. Yes.
3. They are filter-feeders that have little tolerance for water-borne pollutants.

## 36. 20 Years of the Clean Water Act

*Answers*:

1. Cropland.

2. Do not measure toxics and ecological indicators.

3. Silent Spring by Rachel Carson, in 1962.

## 37. Off-the-Shelf Bugs Hungrily Gobble Our Nastiest Pollutants

*Answers*:

1. Cellcap. Potato beetles and corn borers.

2. Warm water is more hospitable to the bacteria, promoting biochemical activity.

3. Air is filtered through material like compost containing bacteria. The bacteria convert the VOCs to carbon dioxide and water.

## 38. Can Superfund Get on Track

*Answers*:

1. Bioremediation.

2. 1,192, 221.

3. New Bedford, Mass.

## 39. Waste Not, Want Not

*Answers*:

1. 30%. 25%.

2. Reducing waste flow, saving energy and resources, creating jobs. Yes, twice as many.

3. It replaces the usual flat fee for garbage disposal with a charge for each bag or can of refuse. By charging people for how much waste they produce, they provide an incentive to reduce waste. 60%.

## 40. Asbestos

*Answers*:

1. Sheet minerals and amphiboles.

2. If the fibrous mineral is inhaled, it lodges in the lung. The mineral cannot be broken down by the body and causes irritation over 10 to 40 years, causing cancerous cells to develop at the site of irritation.

3. The proven toxic asbestos minerals are the amphiboles; in particular, the mineral crocidolite. These

minerals comprise about 5 percent of the total asbestos in use.

### 41. Radon and Other Hazardous Gasses

*Answers*:

1. Radon-222 is commonly released from mylonites, granites, gneisses, schists, slates, sandstones, and glacial deposits.

2. Radon-222 is a decay product of uranium-238.

3. Radon-222 is taken into the lungs where it decays into polonium-218, which quickly begins to decay.

### 42. Ozone

*Answers*:

1. Chlorine. 1973.

2. Yes. Dobson.

3. No. It will cause more human suffering and economic mayhem than a more managed phaseout.

### 43. The Price of Everything

*Answers*:

1. 1990 Clean Air Act.

2. $140 billion.

3. Banditry, drug trade, frontier societies, Somalia.

## Part Five: Global Climate Change—Past, Present, and Future

### 44. Asteroid Impacts, Microbes, and the Cooling of the Atmosphere

*Answers*:

1. Continents promote cooling by removing carbon dioxide from the atmosphere during the formation of carbonate rocks.

2. First, impacts may have caused topographic variations leading to the separation of continents and occan basins. Second, impacts caused the release of carbon dioxide from impacted rocks. Third, heavy bombardment prevented the continual existence of life.

3. Chemoautotrophs remove carbon dioxide from the atmosphere promoting cooling.

**45. The War Between Plants and Animals**

*Answers*:

1. Evolving a digestive system to handle terrestrial plants was difficult.

2. 20 million years ago. The current cycle of ice ages began.

3. Plants help produce soil, which produces carbonic acid and holds the acid next to rocks.

**46. Location, Location, Location**

*Answers*:

1. Most of the continents formed a single large land mass which was situated over the South Pole. Cold waters off the coasts would have decreased summer temperatures preventing winter ice buildup from melting.

2. During the Cretaceous much of the continental interiors were flooded by shallow seas which brought humid air into the continents and contributed to the buildup of clouds. The cloud cover would have prevented solar energy from reaching the Earth's surface making it cooler than expected.

3. Computer models used to predict future global warming do not accurately depict ocean currents or the outline of the continents which could affect their results.

**47. The Leaf Annals**

*Answers*:

1. Plants would decrease the number of unnecessary stomata to prevent loss of moisture.

2. Their stomatal density has decreased by 40 percent.

3. Cerling examines the ratio of carbon isotopes in ancient soils to determine past carbon dioxide levels.

**48. Did Tibet Cool the World?**

*Answers*:

1. Carbon dioxide allows the sun's radiation to enter the Earth's atmosphere but traps radiation trying to leave the Earth's atmosphere. By trapping this radiation it keeps the Earth's surface warm.

2. Precipitation contains not only water but carbon dioxide, so that increased rainfall also means increased removal of carbon dioxide from the atmosphere. The carbon dioxide causes weathering of the mountains and the conversion of the carbon dioxide to bicarbonate that is then transported to the oceans.

3. Silicates have a higher ratio of strontium-87 to strontium-86 than do limestones. During periods of increased weathering this strontium ratio in ocean sediments would increase due to the weathering of silicate rocks such as granite.

## 49. Staggering Through the Ice Ages

*Answers*:

1. When an ice sheet grows thick it insulates the bottom layers from the cold air. These bottom layers may then melt due to heat rising from the interior of the Earth.

2. Global temperatures became warm (balmy) after the release of the icebergs.

3. The circulation of ocean currents in the North Atlantic carries warm air from the tropics towards the poles allowing global temperatures to stay relatively warm. When this current is stopped warm air would not reach the polar regions causing them to cool.

## 50. Good News and Bad News

*Answers*:

1. The shifts in wind and ocean currents cause drought in some areas, which leads to the death and decay of vegetation, which in turn releases $CO_2$ into the atmosphere.

2. First, the release of iron from the eruption may have fertilized the ocean causing the growth of microscopic plants which would draw $CO_2$ out of the atmosphere during photosynthesis. Second, sulfuric acid droplets would have blocked some sunlight slowing the growth of soil bacteria which normally would break down plant matter, releasing $CO_2$ into the atmosphere.

3. First, the repair of a natural gas pipeline in the former Soviet Union may have prevented the escape of large quantities of methane into the atmosphere. Second, the same effect from Mount Pinatubo that slowed the growth of soil bacteria may have slowed the activities of methane-producing bacteria in wetlands.

## 51. A Vision of the End

*Answers*:

1. Increased solar heat causes increased evaporation of the oceans. This water vapor then rains out of the atmosphere taking $CO_2$ with it. This carbon is then precipitated out as carbonate rocks.

2. Grasses can survive because they are very efficient in their use of $CO_2$.

3. Volcanoes will add $CO_2$ to the atmosphere, but without any water left the $CO_2$ will stay in the atmosphere and continue to build up.

# Part Six: The History of Life—Origin, Evolution, and Ecology

**52. How Did Life Begin?**

*Answers*:

1. The sulfur-eating microorganisms around modern vents are genetically the most primitive organisms on Earth.

2. Interplanetary dust whose collective mass is much larger than that of meteorites hitting the Earth.

3. Clays have regularly repeating structures that may have acted as a template to organize any organic molecules into regular patterns.

**53. Static Evolution**

*Answers*:

1. Of the 300 fossil species, 90 have modern counterparts.

2. Cyanobacteria are able to withstand X-ray, ultraviolet, and gamma-ray exposure, which normally damages the DNA of other organisms.

3. Cyanobacteria and the chloroplasts of plants have a similar genetic signature, suggesting that the chloroplasts of plants were derived from cyanobacteria.

**54. Evolution's Big Bang Gets Even More Explosive**

*Answers*:

1. The radioactive elements were uranium and lead, which were found in the mineral zircon.

2. About 3 billion years passed between the first single-celled organism and the Cambrian explosion.

3. Oxygen is required in many of the metabolic reactions of multicellular animals.

**55. The First Monsters**

*Answers*:

1. Before the beginning of the Cambrian some of the most complex organisms were jellyfish, corals, and wormlike animals.

2. *Anomalocaris* was similar to the arthropods because it had jointed appendages and a segmented body.

3. The mouth of *Anomalocaris* was a circular ring of teeth. Usually more than one ring of teeth were found, with one stacked on top of the other.

**56. It's Alive and It's a Graptolite**

*Answers*:

1. Living colonial animals, similar to corals. Individuals making up a pterobranch colony.

2. Between 570 to 300 million years ago.

3. The long thin (nema) spines of graptolites could not, they said, have been formed in the same way that pterobranchs form their skeleton, through successive layering.

**57. Evolution of the Early Vertebrates**

*Answers*:

1. Both lampreys and gnathostomes are fairly active, whereas the hagfishes are lethargic.

2. One of the first developments of vertebrates was the sense of smell which is indicated by the presence of nasal openings in fossil skulls.

3. Conodonts are the only group of vertebrates known to have existed in the Cambrian based on fossil evidence.

**58. Insects Ascendant**

*Answers*:

1. Tunnels, ramps and chambers preserved in sandstone are features characteristic of termite nests.

2. Mouth parts. Insects have a very wide diversity of mouth parts that are specialized for eating specific parts of plants.

3. Insects have a very low rate of extinction, which allows them to reach high levels of diversity even though they evolve slowly.

**59. Masters of an Ancient Sky**

*Answers*:

1. The earliest known pterosaurs lived 225 million years ago.

2. The presence of fibers within the wing membrane on some fossil pterosaurs suggests that they may not have needed to support the wing through attachment to the leg or foot.

3. Pterosaurs were better designed for soaring than birds. Some of the last pterosaurs to go extinct were giant soarers.

**60. The Pulse of T Rex**

*Answers*:

1. John H. Ostrom of Yale University first proposed the idea in the 1960s.

2. Growth rings indicate that an organism stopped growing at certain times of the year.

3. Mammals have nasal bones called maxilloturbinals which help mammals retain moisture.

**61. Dino DNA: The Hunt and the Hype**

*Answers*:

1. Dinosaur DNA could be retrieved from fossil bones or from blood taken from insects preserved in amber.

2. The technique is so sensitive that it could easily pick up contamination from modern DNA.

3. The oldest known DNA is from a 120-million-year-old weevil which was preserved in amber.

**62. Saber-Toothed Tales**

*Answers*:

1. The throat bones of Smilodon most closely resemble those of modern cats that roar.

2. Smilodon had short legs, a bulky body, and a puny tail. It probably hunted juvenile mastodons or mammoths.

3. Fractured lower limbs, dislocated hips, and crushed chests suggest this method of obtaining prey.

**63. How Man Began**

*Answers*:

1. The skull was dated at 200,000 years before the present which is twice as old as any other *Homo sapiens* fossil from that area of the world.

2. *Homo habilis* was the first hominid to make tools.

3. Interbreeding between populations may have allowed *Homo sapiens* to evolve simultaneously in different parts of the world.

## Part Seven: Extinction—Causes and Consequences

**64. What is a Species?**

*Answers*:

1. Up to 10 million years.

2. One. Subspecies.

3. No. Rhinoceroses evolved from the same lineage.

**65. The Greatest Extinction Gets Greater**

*Answers*:

1. Insects suffered their only mass extinction during the Permo-Triassic.

2. A sea level fall could have caused mass extinction by increasing seasonality on land and by exposing a large shelf area which was previously occupied by marine organisms.

3. Hallam and Wignall proposed that a rapid rise in sea level brought anoxic waters onto the shelf causing marine organisms to suffocate.

**66. The Accidental Reign**

*Answers*:

1. The reptile most similar to the dinosaur's ancestor is *Lagosuchus,* which had a hinged ankle joint similar to that of dinosaurs.

2. *Eoraptor* had already developed features common to only one lineage of dinosaurs, the theropods. If it had features common to only one group of dinosaurs these features must have evolved after the two groups separated.

3. Sereno's sample size is very small. His study was conducted at only one site. Further studies should be conducted at other sites of similar age.

**67. Jurassic Sea Monsters**

*Answers*:

1. Yes. Saber-toothed cats, pliosaurs.

2. A 3-dimensional graph that locates each creature along ecological axes that define key elements of habitat, food choice, and behavior.

3. Yes. Mid-Cretaceous.

**68. How Lethal Was the K-T Impact? and A Bigger Death Knell for the Dinosaurs?**

*Answers*:

1. Gravity data shows a bull's eye pattern that extends for 300 kilometers in diameter. The outer rim of this pattern may be the actual edge of the crater.

2. The presence of young hypsilophodonts which may have been too small to migrate 2000 kilometers south. Also, hypsilophodonts do not appear to have traveled in herds, which most migratory animals do.

3. Other effects of an impact could have been acidic rain and global wildfires.

## 69. Testing an Ancient Impact's Punch

*Answers*:

1. The Signor-Lipps effect says that the rarer a species is the less likely it will be to find the last occurrence of that species.

2. Jan Smit concludes that the extinctions were abrupt because, although each of the four foram specialists found some species going extinct before the Cretaceous-Tertiary boundary, when the four results are combined each species existed right up until the boundary.

3. A potential problem with the results would occur if one of the four scientists classify some forams as different species than the other scientists do.

## 70. Taking the Pulse of Evolution

*Answers*:

1. Cool. 2.5 million years ago.

2. A single-celled protozoan that lives on the seabed. More ice and a cooler Earth.

3. 3 million years ago. It cut off ocean currents between the Atlantic and Pacific Oceans.

## 71. Recent Animal Extinctions: Recipe for Disaster

*Answers*:

1. The *Blitzkrieg* hypothesis says that a large human population is not necessary to cause the extinction of the large mammals if the humans moved as a front across the continent hunting the mammals as they went.

2. The survival of the island faunas is strong evidence that climatic change was not the major cause of the extinctions.

3. The last places to suffer extinctions were the Hawaiian Islands, New Zealand, and Madagascar. These extinctions occurred between 500 and 2,000 years ago.